The Conservation of Cultural Landscapes

The Conservation of Cultural Landscapes

Edited by

Mauro Agnoletti

Faculty of Agriculture
University of Florence, Italy

www.cabi.org

GF
90
.C66
2006

CABI is a trading name of CAB International

CABI Head Office
Nosworthy Way
Wallingford
Oxfordshire OX10 8DE
UK

CABI North American Office
875 Massachusetts Avenue
7th Floor
Cambridge, MA 02139
USA

Tel: +44 (0)1491 832111
Fax: +44 (0)1491 833508
E-mail: cabi@cabi.org
Website: www.cabi.org

Tel: +1 617 395 4056
Fax: +1 617 354 6875
E-mail: cabi-nao@cabi.org

A catalogue record for this book is available from the British Library, London, UK.
A catalogue record for this book is available from the Library of Congress, Washington, DC.

ISBN 10: 1 84593 074 6
ISBN 13: 978 1 84593 074 5

Typeset by Phoenix Photosetting, Chatham, UK
Printed and bound in the UK by Biddles Ltd, King's Lynn

Contents

The colour plate section is inserted after p. 52

The Editor

———————————

Mauro Agnoletti is associate professor at the Faculty of Agriculture, University of Florence (Italy) where he teaches the courses of 'landscape analysis and management' and 'environmental and forest history'. He also teaches at the Faculty of Architecture, and has taught in the USA and in several European countries. He is the coordinator of the research group 'Forest and woodland history' of the International Union of Forest Research Organization (IUFRO) and vice-president of the European Society for Environmental History (ESEH). He is the coordinator of the committee on landscape of the Italian National Strategic Plan for Rural Development (2007–2013) and the representative for history, culture and landscape of the Italian Ministry of Agriculture and Forestry Policies at the Ministerial Conference for the Protection of Forest in Europe. He has directed national and international research projects and organized several international meetings on landscape history and management.
www.forestlandscape.unifi.it

Contributors

M. Agnoletti, *Department of Environmental Forestry Science and Technology, University of Florence, Via San Bonaventura 13, I-50145, Florence, Italy.*
E-mail: mauro.agnoletti@unifi.it

M. Ala, *Dipartimento di Colture Arboree, Università di Palermo, Viale delle Scienze, 90128 Palermo, Italy.*

S. Anderson, *President, Forest History Society, 701 William Vickers Avenue, Durham, North Carolina 27701-3162, USA. E-mail: stevena@duke.edu*

P. Angelstam, *Faculty of Forest Sciences, School for Forest Engineers, Swedish University of Agricultural Sciences, SE-739 21 Skinnskatteberg, Sweden.*
E-mail: per.angelstam@smsk.slu.se

G. Barbera, *Dipartimento di Colture Arboree, Università di Palermo, Viale delle Scienze, 90128 Palermo, Italy. E-mail: barbera@unipa.it*

I. Bergman, *Associate Professor, Silvermuseet, 930 90 Arjeplog, Sweden.*
E-mail: Ingela.Bergman@arjeplog.se

G.B. Blank, *Associate Professor, North Carolina State University, Department of Forestry, Raleigh, North Carolina 27695-8002, USA. E-mail: gblank@unity.ncsu.edu*

R.H.W. Bradshaw, *Environmental History Research Group, Geocenter – Øster Volgrade 10, DK-1350 Copenhagen K, Denmark. E-mail: rhwb@geus.dk*

G. Chirici, *geoLAB – Laboratory of Geomatica, Dipartimento di Scienze dell'Ambiente Forestale, University of Florence, Via San Bonaventura 13, I-50145, Florence, Italy.*
E-mail: gherardo.chirici@unifi.it

P. Corona, *Dipartimento di Scienze dell'Ambiente Forestale e delle Sue Risorse, University of Tuscia, Via San Camillo de Lellis, 01100 Viterbo, Italy.*
Email: piermaria.corona@unitus.it

X. Cussó, *Departament d'Economia i d'Història Econòmica, Universitat Autònoma de Barcelona, Edifici B, Campus de la UAB, 08193 Bellaterra (Cerdanyola del Vallès), Spain. E-mail: Xavier.Cusso@uab.es*

R. Diamant, *Marsh–Billings–Rockefeller National Historical Park, Woodstock, Vermont 05091, USA. E-mail: Rolf_diamant@nps.gov*

R. Garrabou, *Departament d'Economia i d'Història Econòmica, Universitat Autònoma de Barcelona, Edifici B, Campus de la UAB, 08193 Bellaterra (Cerdanyola del Vallès), Spain. E-mail: Ramon.Garrabou@uab.es*

G.E. Hannon, *Environmental History Research Group, Geocenter – Øster Volgrade 10, DK-1350 Copenhagen K, Denmark.*

E. Johann, *University of Natural Resources and Applied Life Sciences Vienna, Wlassakstrasse 56, 1130 Vienna, Austria. E-mail: elis.johann@utanet.at*

M. Köhl, *Department of Wood Science, World Forestry, University of Hamburg, Leuschnerstrasse 91, 21031 Hamburg, Germany. E-mail: weltforstwirtschaft@holz.uni-hamburg.de*

T. La Mantia, *Dipartimento di Colture Arboree, Università di Palermo, Viale delle Scienze, 90128 Palermo, Italy. tommasolamantia@unipa.it*

D.S. La Mela Veca, *Dipartimento di Colture Arboree, Università di Palermo, Viale delle Scienze, 90128 Palermo, Italy.*

N. Langston, *Associate Professor, Department of Forest Ecology and Management, University of Wisconsin-Madison, Madison, WI 53706, USA. E-mail: nelangst@wisc.edu*

G. Latz, *Vice Provost for International Affairs, Professor of Geography and International Studies, Office of International Affairs, Portland State University, PO Box 751, Portland, Oregon 97207-0751, USA. E-mail: latzg@pdx.edu*

V. Marinai, *Department of Environmental Forestry Science and Technology, University of Florence, Via San Bonaventura 13, I-50145, Florence, Italy.*

C. Marts, *Marsh–Billings–Rockefeller National Historical Park, Woodstock, Vermont 05091, USA. christina_marts@nps.gov*

J.-P. Métailié, *GEODE – UMR 5602 CNRS, Maison de la Recherche, Université Toulouse–le Mirail, 31058 Toulouse, France. E-mail: Metailie@univ-tlse2.fr*

N. Mitchell, *Marsh–Billings–Rockefeller National Historical Park, Woodstock, Vermont 05091, USA. nora_mitchell@nps.gov*

C. Montiel Molina, *Complutense, University of Madrid, Faculty of Geography and History, and Department of Regional Geographic Analysis and Physical Geography, Ciudad Universitaria, s/n, E-28040 Madrid, Spain. E-mail: crismont@ghis.ucm.es*

L. Östlund, *Associate Professor, Department of Vegetation Ecology, Swedish University of Agricultural Sciences, 901 83 Umeå, Sweden. E-mail: Lars.Ostlund@svek.slu.se*

S. Paoletti, *Department of Environmental Forestry Science and Technology, University of Florence, Via San Bonaventura 13, I-50145, Florence, Italy.*

I.D. Rotherham, *Tourism, Leisure and Environmental Change Research Unit, Sheffield Hallam University, City Campus – Pond Street, Sheffield S1 1WB, UK. E-mail: i.d.rotherham@shu.ac.uk*

W. Schenk, *Institute for Geography, Historical Geography, Meckenheimer Allee 166, D-53115 Bonn, Germany. E-mail: winfried.schenk@giub.uni-bonn.de*

E. Tello, *Departament d'Història i Institucions Econòmiques, Facultat de Ciences Econòmiques i Empresarials, Universitat de Barcelona, Diagonal 690, 08034 Barcelona, Spain. E-mail: tello@ub.edu*

S. Weizenegger, *Institute for Economic Geography, Ludwigstrasse 28, D-80530 Munich, Germany. sabine.weizenegger@gmx.net*

'... now in these things, a large part of what we call natural, is not; it is even quite artificial: that is to say, the tilled fields, the trees and other domesticated plants that are placed in order, the rivers kept within bounds and directed toward a certain course, and such, lack both the state and the appearance that they would have in nature. In this way the appearance of any land inhabited for a few generations by civilized people, not to say in cities and other places where men live together, is artificial, and much different from what it would be in nature.'

Giacomo Leopardi (1798–1837), *In Praise of Birds*

Introduction: Framing the Issue – a Trans-disciplinary Reflection on Cultural Landscapes

M. Agnoletti

Cultural landscapes are today a resource whose preservation represents a most modern theme, relevant to a great number of sectors such as planning, cultural heritage preservation, rural development, nature conservation and forestry, to cite just a few. The role of the landscape and therefore its perception has changed through time; it is no longer just a 'cultural' aspect, intended as an elitist phenomenon, isolated from the socio-economic aspect, but emerges as an essential element in the interpretation of a modern approach to sustainable development, far from paradigmatic views, but close to the needs of a large part of society in the whole world. It is not useful to report all the many definitions used to describe cultural landscapes, a matter for which more than a single chapter would be needed, but there are words that live in our minds more than others, also affecting the way to view research matters. In this respect, the definition given by Carl Sauer in 1926 has the advantage of summarizing much of the core concept: 'The cultural landscape is fashioned from a natural landscape by a culture group. Culture is the agent, the natural area the medium, the cultural landscapes the result'. It is interesting to note that in those same years the Italian philosopher Benedetto Croce promoted the first law protecting landscape in Italy,

mostly based on the concept of the preservation of aesthetic values, an interesting but very different approach, close to what Italy has represented since the end of the 18th century when the travellers of the 'Grand tour' journeyed across Europe, a trip undertaken by many European men of letters and philosophers to widen their cultural background. A more modern concept considers cultural landscapes to be the expression of historical integration between social, economical and environmental factors, influencing all aspects of development. According to the European Landscape Convention, landscape constitutes a resource favourable to economic activity, contributing to human well being and consolidation of cultural identity. At world level there is an evident trend towards degradation and the creation of less valuable landscapes, up to the point that cultural landscapes are often more endangered than nature. Their conservation imposes choices that are not easily made, along with the revision of some past orientations in the fields of agriculture, forestry and nature conservation.

The Conservation of Cultural Heritage

Until recently, international documents regarding sustainable development said little about cultural landscapes. The Stockholm declaration of 1972 and the Bruntland Report in 1987 did not refer to landscape, while Agenda 21 (1992) had some reference to this matter, but without clearly addressing it. In 2003, the FAO GIAHS project (Globally Important Ingenious Agricultural Heritage Systems) clearly addressed the relationships between agricultural heritage systems and their landscape. However, the main specific tool available at world level for the conservation of cultural landscapes is surely the World Heritage Convention (WHC) of UNESCO (1972). At a European level the European Landscape Convention (ELC) is the most comprehensive proposal applying to the entire territory, while the Pan-European Biological and Landscape Diversity Strategy set up for the period 1996–2016 offers a more specific approach, as also described in the chapter by Schenk.

It is very significant that cultural landscapes (CL) have only recently been introduced into the World Heritage Convention (1992). Before that time the convention was mainly protecting natural heritage and cultural heritage, the latter concerning mostly monuments or architectural assets, with emphasis placed on the aesthetic. According to the WHC, cultural landscapes represent the 'combined work of nature and of man. They are illustrative of the evolution of human society and settlements over time, under the influence of the physical constraints and/or opportunities presented by their natural environment and of successive social, economic and cultural forces, both external and internal'. The three categories selected to classify CL and include them in the World Heritage List (WHL) of protected properties are: (i) 'a clearly defined landscape', mostly referring to parks and gardens; (ii) 'organically evolved landscapes', divided into two subcategories (a) 'relict and fossil landscapes' and (b) 'continuing landscape'; and (iii) 'associative landscapes'. Item (ii) more specifically refers to the features resulting from the action of forestry and agriculture shaping the land, while item (iii) is more linked to the intangible value created by man–nature relations.

There are six crucial cultural criteria used for the inclusion of properties in the WHL, which must also meet the test of 'authenticity' to satisfy the criterion of being of universal value, as well as possessing 'uniqueness', 'significance' and 'integrity'. All these criteria suggest the crucial role of landscape assessment and the importance of the methodologies used for their assesment. As stated in the recent UNESCO world paper n. 6, evaluation has already become increasingly difficult, mostly because of the absence of comparative studies, an 'outstanding need' for places like Europe characterized by an 'extraordinary' variety of farmed landscapes (Fowler, 2003). However, attention to methodologies is still needed. During some lectures at the ICCROM in Rome in 2002, one of the advisory bodies of WHC, it was evident how the background of students coming from many different countries of the world was basically well suited to analyse proposed properties, mostly using classic historical, archaeological or ethno-anthropological research. On the other hand, modern techniques based on multi-temporal analysis, remote sensing and assessments of land use changes, together with all the techniques today available in the field of forest and woodland history or historical ecology, necessary for understanding landscape changes and assessing authenticity, significance, and integrity, were much less known and mastered. In this respect this book offers a contribution proposing not only specific methods as explained in the first part by Östlund and Bergman, Tello et al. and Chirici et al., but also presenting comparative approaches, as explained for Tuscany. In fact, although different groups of people may have different perceptions of what a valuable landscape is, it is clear that a dynamic evaluation is needed to understand the trajectory of a landscape system and recognize which elements have become a value and which have not, while similar methodologies applied to different sites offer better chances to assess landscape, especially when trying to evaluate specific elements (e.g. chestnut woods, vineyards, terraces, etc.) that may present different values in different contexts. On the other hand, this evaluation also meets the requirements of the ELC whose specific measures require identification, analysis of changes and assessment of landscapes in each country. It is also worth noting the potential offered by the approach of Tello *et al.*, where a traditional cultural landscape is also analysed from the point of view of energy flows, offering a possibly valuable tool for multiple analysis, particularly important for assessing sustainability.

The approach of UNESCO is meant to save specific sites and in fact the criticism coming from the Council of Europe, suggesting an 'elitist' approach, is symptomatic of the different views, perspectives and goals of the European Landscape Convention (ELC) and WHL. Actually, the two processes are not comparable. The WHL cannot be used as a primary instrument for the simple matter of the difficulty of including all valuable cultural landscapes in the list and submitting them all to a management system as requested. In this respect the ELC might be a more powerful tool, addressing the matters of policies, quality objectives, protection, management and planning for European Union (EU) landscapes at national and regional level, facilitating a process that is going to affect governance. As at 7 June 2006, ten states had signed it and 23 more had signed and ratified, accepted or approved. The fact that landscape must be recognized and protected independently from its value does not prevent the selection of specific properties of special significance, also because although there are many opportunities to establish nature protection areas in Europe, there are no similar instruments to protect cultural landscapes. This simple consideration could open a long discussion about the idea of 'nature' and 'culture' in modern societies. However, for now the WHL represents the only chance to protect cultural landscapes of special importance, unless proponents accept a sort of unclear mixture between natural and cultural values, also proposed in some definitions of WH cultural landscapes and the EU Habitats directive, where cultural values

could be saved within the framework of nature conservation, obviously a quite peculiar angle from which to look at the problem. These are some of the problems encountered with the proposal of creating a landscape park in the Apennine mountains in Moscheta (Italy) described in the book, amplified by the fact that restoring a wood pasture from a wood is simply forbidden by law, since woodlands cannot be reduced in their extension. It is also time to reduce the artificial separation between natural and cultural values typical of many conservation approaches; in the world today the natural system is well embedded into the socio-economic system, affecting all its features. However, an effective conservation of cultural landscapes cannot be done without interfering with the processes affecting their dynamics and also with the way sustainability is perceived and applied.

Agriculture and Rural Development

The dynamics of rural landscape are triggered by socio-economic developments affecting the rural world. The techniques used in traditional societies, usually before the technological development of mechanization and chemical fertilizers, created valuable cultural landscapes where the strict relationship between man and the land over a long time period has accumulated values, stratifying them in the physical components of the territory. The different forms of fields and woodlands, the use of tree species for hedgerows and mixed cultivation, the use of fire – an often misunderstood ecological agent, as very well evidenced by Metailiè – have created an extraordinary variety of landscapes. Europe is a good example of this diversity if we look closely at landscapes along a gradient north–south and east–west, as presented by several chapters in this book; but a great degree of diversity can be understood also in the chapters describing North America. The changes in technology, culture and economy at world level are threatening traditional landscapes, including the biodiversity on which they are based, but also the structure of rural society. Complex cultural landscapes typical of densely populated regions and landscapes where the population had to establish specific management practices to adapt to the local environment are rapidly disappearing. Farmers are often compelled to develop innovation and to adopt unsustainable practices, overexploiting resources and also contributing to the genetic erosion and loss of the cultural identity of places, interrupting the transmission of important heritage from one generation to another. This accumulated knowledge and experience in the management and use of local resources is a significant wealth at world level expressing the cultural identity of each ethnic group.

Unfortunately, there has not been a clear recognition of the significance of landscape resources in rural development strategies, as well as the role of rural landscape for society, not only in the international documents concerning sustainability, but also in policies. In this respect, political entities like the EU are very interesting case studies of the effects of a Common Agricultural Policy (CAP) affecting different nations with different histories. According to the OECD (Organization for Economic Co-operation and Development) definition, rural regions in the new EU, open to 25 countries, represent 92% of the territory; 56% of the population lives in rural regions, generating 45% of gross value, providing 53% of the employment, while agriculture and forestry represent 77% of land use. Despite the evident importance of the rural regions for landscape quality and socio-economic aspects, we can easily conclude that agricultural policy (including forestry) in the past decades has favoured the degradation of cultural landscapes. The contribution given to technological development, production, setaside[1] and the measures favouring tree plantations in areas removed from production have contributed to the disappearance of traditional cultivation practices, homogenizing landscape and often introducing new degradation, as described in the chapters about Tuscany and in the

1. Removal of cultivated land from production.

case of Spain described by Montiel Molina. There has been little recognition of the relationships between typical products and local landscapes, or services like agritourism, based on landscape, and few actions for the conservation of the cultural values of traditional landscapes, representing the cultural identity of the European regions. From this point of view the new EU countries in Eastern Europe, described in the chapter by Angelstam, will probably experience the same trends.

The new EU agriculture reform (CAP) offers further possibilities, but also new threats to cultural landscapes, although much will depend on the way regions use this instrument. The new 'single farm payment' independent from production is probably going to again favour the abandonment of traditional cultivations, usually less remunerative for farmers, who will not be interested in saving these types of cultivation without specific measures. The lack of important initiatives regarding the landscape is also tied to the will to defend the interests, however lawful, of economic activities which consider regulations about the landscape limiting or possibly damaging to their activity, not only in the industrial sector, but also in the agricultural one. There is also the idea that farming activities always preserve landscape quality, an attitude historically criticizable, but symptomatic of an opinion shared by many people.

The Axis 1 of CAP EU Rural Development Regulation 2007–2013 – 'improving competitiveness of the agriculture and forestry sector' – holds no indications concerning the development of the quality of agricultural production promoting landscape as an added value. There is in fact an underestimation of the role of landscape within several productive sectors, foremost of which is viticulture, which represents a power point for several countries. As shown by recent research carried out in Tuscany, the market value of the product 'wine' largely consists of immaterial factors, among which the landscape (expression of culture, history and environmental quality) represents the main component. Therefore, the producer bases much of his/her earnings on the exploitation of a resource for whose maintenance one should invest some resources. The wine-makers know it very well, but few of them are investing money in this resource. The 'key actions' described in the CAP address the need to face increasing global competition, and landscape resources could easily help to reach this goal. From this aspect, the effort of several countries to include some wine regions in the World Heritage List of UNESCO is quite significant, as has already occurred for the Tokaji region in Hungary. For the Chianti region of Italy this has not been successful as yet because of disagreement on the possible advantages and disadvantages, not among the winemakers, but among towns for the effects of the management system requested by UNESCO on their urban planning. It is also a matter of accepting the concept of a gradual evolution from a merely productive role to a role of territory preservation, which some farmers themselves still find hard to grasp. That is also one of the reasons why much of the EU budget goes into agriculture. In this respect, the reference to environmental services in Axis 2, recognizing the farmers' role in delivering services such as water and soil protection with no specific indications about landscape, is also neglecting the fact that the abandonmemt of traditional practices may also increase hydro-geological risk.

Axis 2 of EU rural development – 'improving environment and countryside' – might offer some possibilities, especially when it refers to preserving farmed landscapes. However, it is not clear what is meant exactly by protecting both natural resources and landscape in rural areas and the indicators suggested in the guidelines for rural development do not help in this respect. The rural landscape is a cultural creation, therefore there is the need for careful evaluation when promoting more nature; pushing for more renaturalization might work for heavily industrialized areas, but the use of agri-environmental measures to recreate traditional mixed cultivations, wood pastures, tree rows, pollard trees, hedges and landscape mosaics would often

be better than recreating 'pristine forests' even for ecological networks. We do not necessarily need large forest areas to connect habitats, while we should not confuse the role of a 'network' with its physical structure. Not only the chapters on Southern Europe but also the one by Bradshaw and Hannon on Sweden describe traditional cultivations, raising the issue of what is meant by the term 'biodiversity' and the preservation of 'high nature value farming and forestry systems', both in CAP and nature conservation strategies. It should be remembered that the loss of biodiversity is also linked to the reduction of species introduced by farmers in some periods of history, as the Romans did also by importing 'non-native' species from the orient. Biodiversity should also consider 'spaces' created by the different land uses, typical of many traditional landscapes. This diversity is today dramatically reduced by abandonment and consequent advancement of forest vegetation on old fields, or by the extension of mechanized monocultures. Therefore, measures concerning afforestation and also agri-environmental measures need to be carefully evaluated since many would simply use these subsidies because they are there, despite the fact that what is really needed is something else. With regard to organic farming, which is a very positive initiative in many ways, it must be remembered that organic products can be made in Sicily or in Sweden, but their production does not ensure the conservation of the landscape they come from. It is instead time to close the circle 'quality of food – quality of the landscape', favouring a strong correlation between the two.

Axis 3 of EU rural development – 'the quality of life in rural areas and diversification of rural economy' – could actually represent a good opportunity if diversification into non-agricultural activities could include services like the restoration and management of landscape and the promotion of agritourism, which would create new jobs. Tourism has not always been seen as the direct result of farming activities – in other words, 'services' are sometimes placed outside the rural world, but especially when they are tied to the appreciation of landscape resources, they are clearly linked to the activity of farmers. In this respect, the conservation of cultural landscapes might represent an economic opportunity even without having a productive landscape in terms of crops. In other words, the simple maintenance of cultural landscape represents an economic activity in itself, with people employed just for this purpose. Unfortunately, there is not a clear understanding of how landscape affects even the usual tourism forms, such as those linked to museums or historic city centres, because when many visitors are travelling from one place to another it is also to appreciate the landscape. In this respect it is interesting to see how a large portion of this tourism is presented as 'ecotourism', clearly proposing the issue of the appreciation of the natural values of territories, whereas they are mostly cultural. From this perspective it is useful to view what has happened in countries like the USA, where employment in landscape services has seen a spectacular growth between 1972 and 2003, accompanied by a strong decrease of entrepreneurs and employees in the traditional productive activities in agriculture or forestry. It would not be unrealistic to imagine a similar development especially for those regions offering important landscape resources. One example of interpreting the new CAP for landscape conservation could be the development of strategies and actions for preserving and developing landscape resources through the national rural development plans. This has occurred in Italy, with the establishment of a commission for this purpose.

Forest Strategies

The strategies concerning forestry, and the way sustainable forest management (SFM) has been interpreted, are playing an important role in view of the increasing extension of forest areas in the EU. Forest and woodlands have been mostly regarded as a source of timber or of ecological value. There has rarely been an appreciation of their cultural significance, unless when referring to recreation or social values. Most of the traditional knowledge related to local management forms, timber assortments and the relationships between woodland and agriculture typical of agro-forestry systems has been lost. Today this knowledge survives in some niches in Western Europe and in developing countries, as described by Angelstam for Eastern Europe. The tendency to use species better suited for productive functions, also through afforestation, and to re-naturalize forests are rapidly deleting the evidence of past cultural influences. The United Nations Conference on Environment and Development held in Rio in 1992 adopted several 'forest principles', meant to submit the forests of the world to specific management, supporting the development and implementation of criteria and indicators to clearly define elements of SFM and to monitor progress towards it.

Several international meetings have suggested some thematic elements as key components of SFM: extent of forest resources; biological diversity; forest health and vitality; productive functions of forest resources; protective functions of forest resources; socio-economic functions; legal, policy and institutional frameworks. These elements, also acknowledged by the United Nations Forum on Forests (UNFF), reflect the criteria of the nine ongoing regional/international processes on criteria and indicators for SFM and were acknowledged by the International Conference on Criteria and Indicators in Guatemala in February 2003 and by the FAO Committee on Forestry in 2003. In Europe, these indications have been included in the SFM criteria endorsed by the Ministerial Conference for the Protection of Forest in Europe (MCPFE), and are basically reflecting the same view, including one more item concerning the role of forests as CO_2 sinks in criterion No. 1. Similar criteria are used in the main certification standards like PEFC (Pan-European Forest Certification) or FSC (Forest Stewardship Council). The thematic elements, the criteria of MCPFE, and the certification standards represent a sort of hierarchy of values for forests and woodlands. It is easy to note the absence of a real consideration of cultural values in the list of indicators. We must look through the various chapters, particularly on socio-economic functions, to find reference to this issue. This problem is well addressed for FSC, SFI and PEFC by Anderson. The first main criteria, 'extent of forests', which in MCPFE has been modified to 'maintenance and appropriate enhancement of forests and their contribution to carbon cycle', should probably meet the necessities of countries with scarce forest cover to increase their extent and the general feeling that more forest will mitigate climatic change in respect to global warming. However, especially in Europe, the further extension of forest land is probably not the main problem in SFM, considering the increase occurring at least from the second half of the 20th century. In Italy, forests have increased almost threefold since 1900, but it is questionable that this increase had a significant effect on mitigating global or local warming since they represent about 0.21–0.25% of world forests; therefore their possible contribution can only be estimated according to this figure. On the contrary, the cultural value of the forest landscape and its contribution to the world's cultural heritage would be probably higher. The relevance of cultural, historical and landscape values in Europe could well represent one of the main criteria of SFM. Recently the MCPFE, according to an initiative started by the Austrian government during the Vienna conference of 2003, has started a reflection on this issue, reflected by Resolution 3. Two international meetings to discuss the implementation of cultural, historical and landscape values in SFM were organized in 2005 and 2006, involving IUFRO, MCPFE, UNESCO, WHC, UNFF, European Landscape Convention and FAO. After the last meeting in Florence, a group of experts for the enhancement of indicators and guidelines was created (Agnoletti et al., 2006). It has been acknowledged that recognizing the cultural origin of EU forests cannot be reduced to the matter of saving particular sites somewhere on the continent, but acknowledge the cultural origin of EU forests and find a way of managing them according to this perspective.

In this respect, the recent vision and strategies of the EU Forest Action Plan, are worth comment. In the document the role of culture is not really addressed, although the text recalls the MCPFE meeting in Vienna of 2003. Therefore, the international activity previously mentioned has not been taken much into account, as well as almost a century of investigations in forest history and cultural heritage. Concerning the multiple functions of forestry, economic, ecological and social functions are the main themes. Among the environmental functions, landscape is recalled as an element of ecological stability and integrity, but the document seems to refer to natural values rather than cultural landscapes. It must be remembered that more nature might also mean deleting cultural values, expressed by traditional practices, like different forms of coppice woods, pollard trees, chestnut orchards, and wood pastures, which are going to disappear if left to natural processes.

Moreover, as in other directives, using the word 'nature' to describe cultural landscapes creates mental associations putting landscape management into nature conservation. This often leads to the erroneous assumption that by protecting nature we always protect landscapes.

In the chapter about the 'values of forests and forestry', 'social functions' might offer some opportunities to protect landscapes. Especially when referred to make rural areas attractive for living, providing recreational opportunities, but also when mentioning that forests represent a cultural heritage. Concerning the 'key actions', the document proposes the structure of the CAP and its many axes, but it is worth noting that culture and landscape are not mentioned in the 18 key actions listed. An important opportunity, at least for monitoring landscape, could be the implementation of the proposed 'European Forest Monitoring System' if properly designed. The environmental objectives are recalling the goal to halt the loss of biodiversity by 2010 but, according to past experiences, biodiversity will probably be mostly referred to species and natural habitats, not to landscape diversity. Together with the protection of 'integrity' and prevention of 'fragmentation' of forests, this will probably favour the further extension of forest land, and the strong reduction of the residual landscape diversity. These views, however, are also a result of the general approach to environmental conservation.

Nature Conservation

The European Union has achieved a really important objective, establishing a network of protected areas for the management of Europe's natural heritage. The European Community has gradually been implementing a policy on its territory starting from 1973. The priorities were established in the first Action Programme for the Environment. In the following decades, specific financial instruments were created for nature conservation and a long series of directives have been enhanced. Among the most interesting ones are the Habitat Directive and the Bird Directive aimed at protecting wildlife species and their habitats. Member states have identified special areas of conservation and should draw up management plans combining their long-term preservation 'with people's economic and social activities', to create a sustainable development strategy. The directives identify some 200 types of habitats, 200 animals and over 500 plant species as being of community interest and requiring protection. A scientific assessment on a national level of each habitat or species of community interest was made for this purpose and protected sites have been identified and proposed in the form of national lists presented to the European Commission, now forming the Natura 2000 network. Any action not directly connected with the management of the site, but likely to have a significant effect, must be subjected to appropriate assessment in view of the site's conservation objectives.

Although the Habitat Directive contains specific reference to socio-economic development and seems to offer a chance for a positive integration of the protected areas with cultural landscapes, the establishment of these areas is presenting contradictions. The need to identify the areas, according to a fixed list of habitats mainly focused on natural 'species' and not on 'species' and 'spaces' related to land uses, has created situations in which the cultural origin of many areas is sometimes neglected presenting naturalness where it plays a minor role. Furthermore, the rules clearly state that any action that is going to fragment or affect density and composition of these habitats is to be avoided. Therefore, not only the fragmentation typical of many historical landscape mosaics, but also the action of man needed to preserve these areas can be seen as potentially dangerous, or even forbidden according to the way the 'evaluation of incidence', a sort of environmental impact assessment for protected areas, is carried out.

The potential, but in many cases already effective, conflicts that have arisen are the result of a certain view interpreting sustainability at global level. The original problem probably relates to the approach to sustainability that has been largely affected by the 'degradation' paradigm, emphasizing the negative role of man in the environment, as an agent depleting the ideal state of 'naturalness', considered the most desirable for the life of living organisms and the overall quality of the biosphere. Although the degradation of the environment is undoubtedly a reality and a threat affecting the world, several investigations carried out in the field of forest and woodland history and historical ecology, but today generally included in the wider framework of environmental history, indicated a wide number of cases where man has created valuable landscapes, not only from a cultural value standpoint, but also from an ecological point of view, enhancing biodiversity and improving the conditions of the environment.

The approach to biodiversity has often neglected the diversity of spaces, generating a reductive interpretation of nature conservation and also promoting views pertaining only to certain scientific groups. The chapter on Tuscany in Part I shows that

48% of the diversity has been lost in the last 200 years, but no agencies are evaluating the problem, even in the monitoring of environmental conditions, or in the management of protected areas. However, the result of the research presented in this volume has convinced the service of nature conservation in Tuscany to promote new guidelines for the conservation and management of landscape in protected areas, proposing also to establish a monitoring system for this specific problem. There are clearly uncertainties in the way biodiversity should be managed. However, it is not surprising that many forest and woodland areas are described as having mostly natural and semi-natural features and how policies descending from this interpretation will apply an ecological approach to their management. There would be instead the need for promoting projects for landscape restoration considering that the 'combination of human intervention and natural processes has often created places for utility and beauty, of nature and culture', as stated by Diamant, Marts and Mitchell in their chapter. However, there is also the need for a dynamic management of cultural landscapes, including them in the socio-economic process and not only in special conservation areas. In this respect, the chapter about Moscheta and those by Rotherham, Latz, Barbera *et al.*, Métailié and Johann present approaches and methodologies suited for different cases.

Conclusions

Besides the economic aspects, there is also the matter of cultural sensibility, which still has to be stimulated for a full understanding of the problem by administrators, scholars and the public. It is often said that the landscape is a perceptive category, which cannot be objectified, as if its values were exclusively immaterial and could not find their concrete representation in the territory. The limit of such a concept is clearly identifiable in the analogy with the urban sector, where in a not-too-distant past it was not a foregone conclusion that the preservation of the architectural structure of a historical city centre had to come first. The recognition of the values tied to architectural assets was also the result of a cultural maturation, which considers the structure of a building as a value to be preserved, not different, from a conceptual point of view, from that represented by a terracing, a row of maples and vines or an ancient chestnut grove. An important difference lies in the fact that the rural landscape, with its woods, fields and, therefore, living elements, is often characterized by a higher dynamism, whereas both of them can be used for economic ends. The desire to live in a farmhouse or restore a historic building in a city centre is certainly not connected with a higher management economy (if anything it is the contrary), while the maintenance of an old mixed olive orchard, a wood pasture, or a hedge cannot be proposed from a perspective which aims at maximizing profits, but for the whole range of values they represent. Certainly cultural landscapes reflect the evolution of humanity and its interrelationships with nature, resulting not only in outstanding aesthetic beauty, the maintenance of biodiversity and valuable ecosystems, but also providing multiple goods, services and quality of life, which is a fairly good way of interpreting sustainability. As recalled by the FAO, it is perhaps time to pay more attention to cultural landscapes and their contribution to the natural and cultural heritage of the world, establishing the basis for their global recognition, dynamic conservation and management in the face of economic and cultural globalization.

Reference

Fowler, P.J. (2003) *World Heritage Cultural Landscapes.* WH papers 6, UNESCO WHC, Paris.
Agnoletti, M., Parrotta, J. and Johann, E. (2006) *Cultural Heritage and Sustainable Forest Management: the Role of Traditional Knowledge.* MCPFE, Warsaw.

Part I

Analysis

There are many ways of interpreting and studying a landscape. The analysis of single species populating a small plot of land and remote sensing techniques analysing entire countries are both common tools available to the many scholars involved in landscape studies around the world, allowing consideration of different geographical and time scales. Indeed, one of the most difficult things is to combine the approaches, choosing the right combination of methods to put into evidence the complexities that landscape resources always represent.

The approaches presented in Part I offer a continental perspective for Europe, analysing the different roles of human influence in shaping the land in two very different areas in the north of Sweden and in the Mediterranean area. There is an evident different degree of cultural features presented by the two areas, but it is important to note that both of them have been, and still are, influenced by man, although at different levels. If the problem in northern Sweden may be the one of saving some of the remaining cultural evidence of the indigenous populations that once lived there, the matter for Tuscany and Spain is rather to preserve the cultural values of an entire region threatened by a whole set of changes induced by socio-economic development and by current political decisions.

The chapter on Tuscany is the result of an attempt to develop a comparative approach, analysing several study areas according to the same methodology in order to have a better understanding of the trends and also to develop evaluating tools suited to cultural landscapes. The focus on 'spaces' was chosen, among other possible ones, to stress the lack of attention to the dramatic reduction of diversity of spaces (48%) that has occurred in the Tuscan territory in the last 200 years. By far the most important change occurred at regional level, but was not addressed in specific policies in rural development or nature conservation, which also affects other landscapes in the world.

Every cultural landscape has its own spatial structure, but it is important to study this feature in relation to socio-economic influences in order to identify the cultural identity represented by the spatial structure. In this respect, Chapter 4, focusing on the application of sophisticated tools and techniques for assessing land use, understanding of land use history and its linkage to eco-biological and socio-cultural processes, presents a wide number of methods specifically designed for this purpose. When carefully calibrated, these are powerful tools in the hands of planners and managers. A very interesting and effective way of analysing cultural landscapes and their role is presented in the study about the Catalan area of Spain (Chapter 3). The chapter analyses the anatomy of the landscape by revealing its structure, referring to its physiology rather than its appearance, using the social metabolism approach. Agriculture models that could attain the highest energy yields without relying on a large amount of external input are therefore presented as a very good example of sustainability, an approach in line with a very modern way of assessing traditional cultures and their landscapes, facing the ecological footprint of globalization.

1 The Development of a Historical and Cultural Evaluation Approach in Landscape Assessment: the Dynamic of Tuscan Landscape between 1832 and 2004

M. Agnoletti[1]

Department of Environmental Forestry Science and Technology, Università di Firenze, Florence, Italy

Introduction

The study of landscape has been largely influenced by two main streams of scientific thinking, the first one dominated by historical studies, mostly concentrating on the role of man as a cultural agent, but with a reduced interest in the structure and functions of landscape patterns, the second affected by the ecologic approach, interested in explaining landscapes at an ecosystem level. These different views have also developed different methodologies, although both of them, especially the ecological one, have been largely affected by the theory of 'degradationism', emphasizing the negative role of man in the environment, as an agent depleting the ideal state of 'naturalness'.

As already stressed by several investigations carried out in the field of forest history and ecological history, but today generally included in the wider framework of environmental history, there are a wide number of cases where the theory of degradation due to human influence cannot be applied, where man has created valuable landscapes, from both a cultural and an ecological point of view, enhancing biodiversity and improving the condition of the environment.

In the last four decades ecological planning, by far the leading approach at world level, has clearly taken human abuse of landscape as the driving philosophical concept, while bringing human actions into tune with natural processes has been the common strategy of almost all the approaches, with little success in bringing culture as a main issue into planning.

On the other hand, ecology has traditionally tried to obtain laws regarding ecosystems, investigating environments relatively unaffected by man (McHarg, 1981). This resulted in an over-emphasis on natural processes, not only in planning, but also in ecosystem management, although no systems today are unaffected by man (Vogt *et al.*, 1997).

Even 'applied human ecology', considered an alternative to ecological planning, did not succeed in the attempt to successfully include the role of man and the role of time in planning, not even with the development of landscape ecology (Ndubisi, 2002). Also in countries like Italy, much of the emphasis is still put on geomorphology or ecological patterns in explaining landscape structure and planning (Pignatti, 1994; Romani, 1994; Farina, 1998), with a relatively reduced interest in human influence, often resulting in an artificial division between natural features and anthropogenic features of the territory. Therefore, there is still the need to develop specific methodologies to assess human influence, including modern historical research into landscape analysis, a history no more limited to the use of written or printed sources but able to combine different tools and techniques (Agnoletti, 2000). The case presented here proposes an approach taking culture and history as the central philosophical paradigm to understand landscape changes and develop a planning approach.

The Background: is Landscape in Good Shape?

Tuscany is known all over the world for the quality of its cultural landscape, a heritage built through centuries of human influence, but today also an important economic resource. Although being a region where sustainability is the concept on which the most important regional law is based[2] and environmental directives and landscape protection have greatly developed, there was a growing feeling shared by several scholars and administrators that the quality of landscape had degraded in the past decades. This feeling was not only based on the visual effects of urban development, but mostly concerned with the quality of rural landscape, even in protected areas. This impression was not in tune with all the reports on quality of air, water, soil and biodiversity, as well as on certification standards, which showed a fairly good degree of fitness of the regional territory according to the way sustainable development is perceived and applied (Calistri, 2002).

In order to have a clearer view of this problem, the DISTAF of the University of Florence,[3] in collaboration with the regional government, promoted a research project putting together several research institutions coordinated by the author. The

research team involved scholars from the fields of history, agriculture, forestry, economy, ecology and geography. The project had the main goal of developing a methodology based on the evaluation of landscape dynamics, selecting an appropriate spatial and temporal scale, with special attention to factors and processes originating landscape change and the quality of the changes. The expected result was the production of a state-of-the-knowledge report enabling identification of the dynamics of landscape, developing a monitoring system for landscape quality.

Materials and methods

Considering the difficulties of understanding a cultural landscape even in its ecological components without a historical perspective (Motzkin *et al.*, 1996), especially in the Mediterranean region (Naveh, 1991; Grove and Rackham 2001; Agnoletti, 2005a,b), history was not considered an option, but the central part of the method, aiming to understand the trajectory of landscape systems, indicating values, criticalities, degradations and threats. The methodology developed (see Fig. 1.1) did not have as its main goal to express an ecological evaluation, considering the role and the action of man in the ecosystem. In other words, man was not considered one of the elements usually listed in the models used to explain the relationships among the various biotic and abiotic elements across the landscape, but the 'main' actor in the hierarchy of factors and processes affecting and directing evolution and biodiversity.

Inside the working group there was an agreement that it would not be useful to concentrate on past experiences giving a strong emphasis on geomorphological features, or vegetation models used to describe landscape, but to focus on what was already appearing relevant in some of the most interesting studies, although not fully addressed (Vos and Stortelder, 1992). Therefore, most of the attention was given to 'spaces' linked to land uses and their changes through time, by far the most rele-

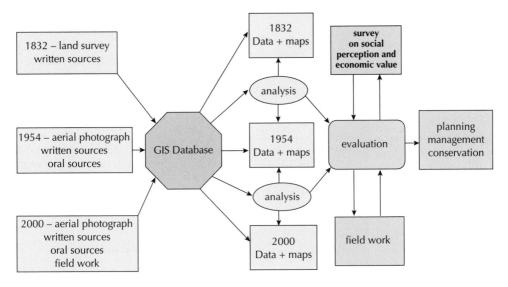

Fig. 1.1. Scheme of the historical and cultural evaluation approach applied in the project. The development of the project also took into account the land use and cover change (LUCC, Lambin and Geist, 2001).

vant issue in the historical dynamic of the landscape of Tuscany, also concerning biodiversity, but not addressed by official investigation (Calistri, 2002). There is in fact a clear relationship between biodiversity and land uses shaping traditional landscapes, often representing valuable 'habitats' for flora and fauna (Wagner *et al.*, 2000; Atauri and De Lucio, 2001; Ortega *et al.*, 2004). Another important aspect of the methodology was the intention to produce a comparative study, analysing many different areas with the same method, meeting also the recent recommendations of UNESCO for cultural landscapes (Fowler, 2003), in order to match the great variety of landscapes existing in Tuscany.

The research meant to cover the following points:

1. Identification of the natural and human factors responsible for landscape changes.
2. Definition of structural typologies and evolutionary patterns.
3. Definition of the historical and cultural value.
4. Determination of the economic value of landscape resources.

5. Definition of management and protection criteria.

To achieve these goals during a 5-year period starting in 2000, the project selected and analysed 13 study areas covering 23,753 ha, approximately 1% of the regional territory. The selection of the areas was made according to the following criteria:

1. To cover the main geographical areas of Tuscany: Apennine mountains, central hills, coastal strip.
2. To include territories with ongoing agricultural and forest activities, abandoned areas and areas placed inside the regional network of protected areas.
3. To provide evidence and sources to justify landscape changes.

The areas selected (Fig. 1.2) represent the geography of the region quite well. In fact, nine areas are located in the hilly region, representing 65% of the whole territory; two are located in the mountains (25%), and two on plains along the coast. Some of the areas also include plains, resulting from the selection of limits including portions of hills and plains. The

- 1 - Moscheta
- 2 - Gargonza
- 3 - Spannocchia
- 4 - Barbialla
- 5 - Castagneto C.
- 6 - Donoratico
- 7 - Bolgheri
- 8 - Montepaldi
- 9 - Paganico
- 10 - Cardoso
- 11 - Migliarino
- 12 - Castiglione Garf.
- 13 - Mensola

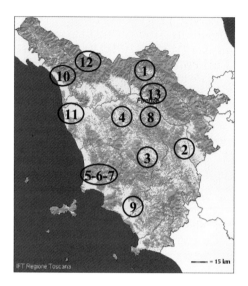

Fig. 1.2. Location of study areas.

choice of making study areas was preferred to methods applied in monitoring or inventories, using remote sensing techniques and statistical grids (Farina, 1998; Köhl, 2003). This method allowed the selection of locations and sizes favouring a good understanding of the structure of the landscape at the level of farming units, facilitating the analysis of the diversity of landscape mosaics. This meant the selection of areas including one or more farming estates, according to the traditional sharecrop system featured in the Tuscan rural economy, with an average size of 1000–2000 ha, facilitated by the availability of written sources preserved in public and private archives and oral sources for recent periods.

In order to develop a dynamic picture of landscape changes, a fairly extended time scale was chosen, selecting three historical moments with different kinds of documents available: 1832, 1954, 2000 (see Fig. 1.1). There is an obviously much longer period that could be analysed, but there was the need to stay away from descriptions already existing in literature (Greppi, 1990; Carandini and Cambi, 2002), but with no detailed descriptions of land uses and

their origins. The year 1832 was chosen because of the availability of a detailed survey represented by the Tuscan Land Register, the cadastre describing almost all the territory on a scale of 1:5000, started by the French at the end of the 18th century and continued by the Lorena Grand Duke after the restoration of the Grand Duchy of Tuscany. This period probably also represents the age with the highest complexity of landscape patterns, due to the strong development of agriculture and demographic growth, although an even higher complexity could be registered in the second part of the 19th century (Agnoletti, 2002). The black and white aerial photographs of 1954 are instead considered the last pictures of the traditional rural Tuscan landscape, documenting the years before the development of mechanization, the use of chemical fertilizers and the abandonment of many farms due to industrial development, the so-called Italian 'economic miracle'. The use of colour digital aerial orthophotos of the year 2000 allowed us to analyse the present landscape, but their interpretation was accompanied by field work to check the data collected.[4] All the material was digitalized and included in a GIS database.

Analyses based on historical photos and cadastral maps have already been developed not only in Tuscany (Vos and Stortelder, 1992; Agnoletti and Paci, 1998), but also by Iseh for Sweden (1988), Foster *et al.* in the USA (1998) and Knowles (2002). However, in this project a systematic methodology was applied to all the study areas, developing tools to compare different years and specific indices to evaluate the historical value of the territory analysed, also described in the chapter about Moscheta. Different sources were used and integrated (oral interviews, written sources, sampling plots) while specific investigations on economic value and social perceptions were carried out by means of interviews with residents and tourists (see Fig. 1.3). The features of the landscape mosaic were studied in each area detected. Surveys of the individual stands were also undertaken to study the

distribution mechanisms of the vegetation. These studies covered the identification of the structural and evolutionary types in abandoned fields, pastures and forests. Some transects were also made to understand the floristic diversities. Synchronic comparisons verified the level of floristic diversity between identical types of crops, especially in the presence of secondary successions. In some areas specific studies on soil and geology were carried out to support the interpretation of changes in land use types. Investigations were also extended to supply further data in case of specific trends or issues, such as the extension of conifer forests for afforestation or of vineyards due to recent market developments.

Different levels of analysis were developed. The main dynamics were synthesized in graphs, while a more detailed evaluation used a cross tabulation

Fig. 1.3. The author during an interview with a lumberman discussing the features of a traditional billhook used for pruning (thanks to Gil Latz).

terracing allows the creation of strips of land for growing cereals and mixed cultivation, with vines bound to olive, maple, poplar, and, more rarely, elm trees. This is the typical landscape created by the sharecrop system, where every farmer shares 50% of his crops with the owner, but must also grow all he needs in his own piece of land, with the help of large families to provide the labour needed. Considering the richness of flora and fauna and the diversity of spaces created by this system, strongly affecting the overall biodiversity (Baudry and Baudry-Burel, 1982), it is perhaps more correct to use the word 'diversity' rather than 'heterogeneity', as landscapes are often described (Farina, 1998).

Woodlands

In 1832, woodlands are the main land use in seven study areas: five in the hills and two on the plains. As described for other European countries (Watkins and Kirby, 1998; Agnoletti, 2000) they are at a moment in their history when they are at their lowest extent, due to a great demographic development that will result in a doubling of the population during the century, with the extension of agriculture even onto high mountain slopes. In Tuscany there is a strict relationship between farming activities and woodlands. Woodlands are managed using a very wide number of techniques to provide products ranging from leaves to nuts, bark and sap, as well as timber for building or fuelwood. The wood patches in the landscape mosaic are often surrounded by pastures and fields, although some large wooded plots can still be found in some areas in the south of Tuscany.

Unfortunately, not all management forms are described in the cadastre, especially the different kinds of coppice woods, the most common management form, as well as the different forms of pollard trees and other forms of culturally modified trees, so widespread in many traditional rural societies of the world (Arnold and Deewes, 1995; Rackham, 1995; Austad, 1988; Sereni, 1997; Östlund et al., 2001).

High forests of oak used for acorn production (often in the form of pastured woods and therefore low density forests shaped to allow the maximum expansion of the canopy to increase acorn production) are described as very common landscapes, a technique still present in Spain (Fuentes Gonzales, 1994). These woodlands were already the most important landscape form on the hills and along the coast in the 18th century (Agnoletti and Innocenti, 2000). Many of the archival documents analysed show how the economy of the farms, more often concerned with raising livestock than cereal production, depended on the production of acorns from oaks to feed pigs grazing freely in the woods. They reported the impossibility to sell the pigs in very dry seasons, due to difficulties in weight increase caused by the scarcity of acorns. Most of woodlands are described as comprising chestnuts, oaks, beech or even shrubs, but chestnut and shrubs (mostly heather) are usually indicated as separate woodland categories. Shrublands (2%), mostly heather, are described in all the study areas, confirming their fundamental role in the economy of farms, providing fuel for domestic ovens and brick kilns, the raw material for making roofs, charcoal, and drainage systems for vineyards. They were cultivated as short rotation coppice (4–5 years) and were often maintained and created with fire. The different types of woodland categories reported in the cadastre can represent up to 40% of the total land use diversity of several study areas, although woodlands rarely represent the main land use type. In the Gargonza study area, on the hills, there are 11 different categories of woodlands listed in the cadastre, out of a total of 27 land use types. In Moscheta, in the mountains there are 15 categories out of 59.

Great attention has been given to chestnut orchards making up 4.3% of the total woodlands surveyed. They are located in only five study areas (Cardoso, Moscheta, Castagneto, Gargonza, Spannocchia), but here they represent a distinctive feature of those landscapes, with an extent close to those of pastureland and cultivated areas.

They played a fundamental role in the production of flour to feed the population and counter-balance the scarcity of wheat (Pitte, 1986) and in the production of timber assortments for many different purposes. In fact we find these trees in all the different geographic regions: mountains, hills and coast, even at 50 m above sea level. This confirms the little relevance of soil and climate conditions in understanding the distribution of this species compared to that of the role of man. Almost all farmhouses had at least a small chestnut wood. Their management included allowing grazing of animals, so that the ground would be cleared of bushes or leaves to facilitate the harvesting of nuts. It is very significant that the account books of the farms had specific sections dedicated to chestnuts involving both timber and nut production, dividing plots among different farmers. Similar occurrences, mostly for umbrella pines producing nuts or firs producing timber, are quite rare.

Pastures

Pasturelands in the reclassified data included also meadows and characterize 28% of the total land use, representing the main landscape in three areas – one in the hills and two in the mountains. An important role was played by wood pasture representing 11.2% of the total land use, but 44.5% of the total pastureland. They are a typical feature of Mediterranean landscapes, offering shelter to animals during the hot season, reducing the temperature of the soil, and producing nuts, leaves and wood (Grove and Rackham, 2001). They made up most of the coastal forest landscape in the 18th century in Tuscany (Agnoletti and Innocenti, 2000), but are still widespread in countries like Spain (Fuentes Gonzales, 1994; Gil et al., 2003). In some areas, the different types of wood pasture may represent up to 95% of total pastureland diversity, due to the presence of different trees in the pasture, and up to 23–27% of total land uses, not including meadows. Trees in pastures are often

chestnut and beech on the mountains, but oaks, walnut, mulberry and even vines and olive trees are found at lower altitudes. Once again it seems that chestnut is the tree most often found in wood pastures, contributing to a large variety of landscapes from the coast to the Apennine ridge. On the other hand, most of the farms studied based their economy more on livestock than on crop production, confirming the importance of a type of landscape created according to economic needs.

Fields

The agricultural cultivations are prevailing only in two hilly areas, a clear symptom of the bad conditions of many plains, still covered with swamps and often flooded, but also of the prevalence of livestock. They are characterized by a larger extent of bare fields (72%) compared to mixed cultivations (28%); however, the latter present many different qualities due to the presence of several tree species and cultivation patterns, according to the distribution of terraces, hedges, single trees, tree rows etc. Specialized cultivations, such as vineyards and olive orchards, play a limited role in terms of extent (0.3%), the latter prevailing on vineyards. The most common pattern where vines and olive trees are found is surely the mixed cultivation technique, inherited from the Etruscans and extended by the Romans (Sereni, 1997). These cultivations have been detailed for each study area, where different types may represent up to 95% of the land uses classified inside cultivated areas, and 26% of all land use types, contributing greatly to the diversity of landscape in terms of habitats and aesthetic values (Fig. 1.5). One typical pattern, especially on terraced slopes, shows the presence of a row at the edge of the field or the terrace, including two or three vines bound to olive trees and maples, more often poplar in the plains, but even oaks or alder are used for this purpose. Sometimes we can find only olive or maple (Acer campestris) to hold the vines. It is certainly the rural part of the landscape creating a

Fig. 1.5. A photograph from the late 19th century showing the richness of mixed cultivation shaping the landscape around Bibbiena. The density of trees could be more than 100/ha.

higher diversity, in terms of patches and their internal features, compared to the woodlands. In fact, fruit trees also contribute to enrich this landscape, although the density of trees in the fields never reaches the level of 200 trees per hectare noted for the Padana valley (Cazzola, 1996). During these years terraces, with dry walls or not, have been greatly extended in the agrarian landscape, up to a point that agronomists argue that they have also been placed in unfavourable geological conditions. Certainly they represent an important expenditure in the account books and are the most common technique to extend cultivation on the hills (Agnoletti and Paci, 1998).

The Landscape in 1954

The landscape of 1954 shows the strong increase of woodlands (+60%) and cultivated areas (+30%), while the importance of pasture land has greatly reduced, covering only 4.3%. The analysis of this year presented some difficulties due to the bad quality of aerial photographs; therefore, the

identification of the internal qualities of each patch could not be as accurate as the description of the cadastre of 1832. There is a significant reduction of the number of land uses (−49%) and landscape patches (−17%), as well as in Hill's diversity number. In this year, more study areas than in 1832 are characterized by the prevalence of woodlands (9 out of 13). Woods are prevalent in five areas on the hills, three on the mountains and one on the plain along the coast, while four areas on the hills and one on the plains show more cultivated fields; but pastures no longer prevail in any of the areas studied. Reforestation is triggered by abandonment of farming, while the vegetation types are determined by the previous land uses and the ecological conditions (Agnoletti and Paci, 1998; Monser *et al.*, 2003). This situation shows that the transformation of the rural economy has already started before the coming of the great innovation affecting Italian society in the following years, although there is still a traditional form of agriculture dominating most of the areas, despite the agrarian reform.

The analysis of the general dynamics of

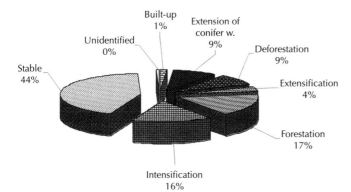

Built-up
1%

Extension of
conifer w.
9%

Unidentified
0%

Deforestation
9%

Stable
44%

Extensification
4%

Forestation
17%

Intensification
16%

Fig. 1.6. Main landscape
dynamics 1832–1954.

the period between 1832 and 1954 (Fig. 1.6) shows that 44% of land uses remained unchanged, while the most important processes are forestation (17%) and intensification (16%). The growth of forest occurs mostly on abandoned pasture and wood pastures, the quality of land use showing the strongest reduction in the landscape, because of the interruption of the practice of letting animals graze freely on the land (they are now kept in stables). The extension of new agricultural forms occurs mostly on former wood pastures, mixed cultivations, pastures and woodlands. New agricultural techniques are substituting old mixed cultivations and new cultivated areas extend on former woodlands or on wetlands in the plains. This is a clear trend along the coast, in Donoratico and Castiglioncello, where the centre of the farming activities moved from the hills to the plains after land reclamation, with the total abandonment of farms on the hills (Bezzini, 1996). Table 1.1 aids the understanding of some of the changes that occurred at regional level in this long period of time,

coming from an elaboration of the original publication by Sereni (1997), but probably not taking into account the different territories included in the different surveys.

Demography played a very important role in this long period, as the population of Italy increased from 22,000,000 to 47,000,000 inhabitants from 1861 to 1955, while the insufficient production of cereals led to a huge increase in the importation of this commodity, making it the second largest import between 1800 and 1900. Also in Tuscany the population grew from 1,303,000 inhabitants in 1810 to 2,317,004 in 1889 (Agnoletti, 2002), so the need for new land extended cultivation towards the high hills and the mountain slopes, favouring the growth of population in mountain areas until 1920–1930, with an increase of 150% (Fig 1.7). The first industrialization of Italy occurred at the end of the 19th century (Castronovo, 1995) and this, together with the law of 1877, favoured deforestation of almost 1,000,000 ha within 50 years, causing the greatest reduction of Italian forests ever seen in modern times.

Table 1.1. Evolution of the main cultivation in Tuscany 1832–1929 (ha ×1000). Modified from Sereni (1997).

Cultivation type	Year			
	1832	1860	1910	1929
Simple fields	681	722	553	454
Mixed cultivation	Not available	–	661	554
Pastures and meadows	583	480	455	134
Woodlands	630	697	909	813
Fallow	448	243	135	255

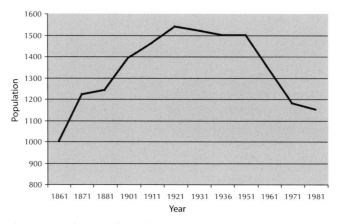

Fig. 1.7. Population in the northern Apennine mountains 1861–1981.

Woodlands

Among woodlands the reduction of shrub-lands (–40%) is very clear. Once cultivated as short rotation coppices and now mostly turned into high forests, they suffered from the interruption of their management due to the abandonment of farming. Coppice is now the most important management form, by far the most useful for farming activities and also for the production of charcoal, comprising 25% of the entire Italian production. Pastured woods no longer exist because the practice of feeding animals with acorns falls into disuse; they are now kept in stables. Chestnut orchards are also reduced in their extent (–84%) due to the changes in agriculture and the abandonment of farms on mountains and high hills, slowly reducing the importance of a method of cultivation more than 2000 years old. Almost 30% of them are turned into mixed woods and coppice, while 40% are woods where chestnut is clearly prevailing, but slowly evolving towards a mixed stand. Chestnut coppice provides poles for vine-yards and the widest number of timber assortments available for agriculture and building on the market, but many new coppices are created after debarking of high trees for the production of tannin. The abandonment of chestnut cultivation is also favouring pests affecting these species, often occurring after the abandonment of the management of a species planted outside its ecological optimum (Vos and Sortelder, 1992).

Conifer forests resulting from afforestation are a new element appearing in the landscape. They represent almost 16% of the woodlands and 10% of the entire landscape, due to the activity of the Italian state in this sector and to private activity. After the unification of Italy in 1861, the state developed a large programme of afforestation affecting all Italy, but until World War II there was little success in this policy, with only 197,000 new forests planted (Agnoletti, 2002). The greatest obstacle to afforestation was not only money, but also the conflicts with shepherds burning plantations to keep pastureland. This conflict was overcome only with the abandonment of mountain areas that occurred after World War II, when the population of the mountains went back down to that of 1861.

Not all the conifer woods have the same significance. The umbrella pines planted along the coast have produced a valuable landscape, useful for protecting fields from sea wind, but also producing edible nuts. The conservation of these forests has been opposed by environmentalists preferring a more natural landscape, especially in protected areas (Agnoletti, 2005b). The afforestation on the mountains, mostly with black pine (*Pinus nigra*), has introduced a degradation in the aesthetic quality of the landscape because of the use of conifers planted in squared plots, typical of artificial plantations, in an area dominated by pastures and broadleaved species, and was never really included in the local culture. Many pasturelands existing in 1832 (43%) have been turned into woodlands, 6% due to afforestation, and 27% into cultivated areas, especially on the plains; but wooded pastures have reduced in extent by almost 80%.

Fields

On the agricultural side there is an increase of cultivated land (+44%), with a very strong expansion of specialized olive orchards, almost 32 times more than in 1832. Specialized vineyards have also increased from 0.23 to 40 ha (see Table 1.2). However, the coming of these specialized cultivations is not yet deleting the old mixed cultivation, which does not show a significant decrease, but are simply adding new elements to the landscape, showing a slight growth. New specialized olive orchards, added to the traditional patterns with sparse olive trees in the fields, are substituting most sowable lands and woodlands, but also new specialized vineyards are replacing sowable lands, making Tuscany one of the Italian regions where the 'wine landscape' is most extensive (Sereni, 1997). Mixed cultivation still remains an important feature of Tuscany, placing the region somewhere between the larger extension of mixed cultivations occurring in the north and the much lower use of trees in the fields that characterizes the southern regions. These changes are slowly introducing the new trends of rural economy in Tuscany, helped by mechanization and chemical fertilizers, which will concentrate agriculture on the best areas and lead to the abandonment of marginal lands on high hills and mountains – a general trend affecting many other countries in the world in the years after the war (McNeill, 2000).

The Landscape in 2004

The years between 1954 and 2000 are a crucial period for Italy and Tuscany. The end of the 1950s and the beginning of the 1960s marks the transformation of Italy from a rural into an industrialized economy, with millions of people moving from the countryside to industrial urban areas. Agriculture and forestry will be strongly affected by these changes from all viewpoints. After an initial period of abandonment, the last decades of the century see a return of people to the land, not as farmers, but as residents interested in the quality of life provided by the Tuscan countryside. Furthermore, many foreigners are buying properties in Tuscany – in some areas in the Chianti region they are approaching the number of local residents. This new interest in the landscape is rapidly increasing the role of services like agritourism (Cox et al., 1994; Casini, 2000), often replacing production as the main source of income, and the role of landscape resources, as more and more people in Italy and abroad are buying wine or come to Tuscany for a holiday. Concerning forestry, the reduced pressure on forest territory has opened the door to the rise of environmentalism in society, and now forests and woodlands are mostly seen as an expression of 'nature', with an interesting and rapid deletion from the memory of the public of their cultural origin. It is the new urban society replacing the rural one which is developing these concepts and creating environmental ideas that will affect the way forests are seen by policy-makers.

The landscape in the areas surveyed shows a small increase in the number of land uses (+10%) and the number of patches, a result probably of the more detailed aerial photographs and field work available in these years, but also due to the new owners buying some of the farms surveyed. New capital is now put into the countryside, similar to what happened in the Renaissance when families of merchants like the Medici invested money made from trade into big farms (Sereni, 1997). However, the further increase in the average surface area of patches and in the average value of the dominance index clearly indicates a simplification of landscape occurring both in forest and agricultural areas, as the new agricultural techniques are not creating valuable landscapes as in the Medicean times.

Table 1.2. Cross tabulation 1832–1954. Columns and lines allow checking of the evolution of the land uses in the period analysed.

1832 \ Extent in ha (1954)	Waters	Anthropic	Shrubs	Sparse vegetation	Woodland	Woods with prevalence of chestnut	Chestnut orchards	Quarry	Fallow	Olive orchards	Pasture	Wooded pasture	Meadow	Afforestation	Sowable	Mixed cultivation	Beach	Vineyard	Total
Waters	123.71	5.32	39.40	0.11	30.58								8.53	32.10	67.85		15.05		322.54
Anthropic		17.26	0.61		5.99	0.38			0.03	1.48	0.69	0.04		0.37	3.97	0.76		0.01	31.71
Shrubs		0.48			201.80	0.22	3.71			0.11	1.01	0.49		27.46	15.06	2.26			252.61
Sparse vegetation					0.36														0.36
Woodland	2.29	42.56	4.19	3.51	3056.21	15.74	14.90	0.21	20.59	269.15	48.85	88.08	0.58	589.06	178.30	221.14		0.43	4555.79
Chestnut orchards		2.44	0.62	16.73	142.12	199.67	77.90	0.60	0.87	7.00	0.95	6.00	0.40	20.84	2.69	9.69			488.50
Fallow		0.61	0.53	0.02	19.64	1.02				0.71	0.15				25.45	1.21			49.33
Not defined		1.91			3.03					10.75	2.03				34.09				51.80
Olive orchards		1.95			0.26	0.16				21.93						0.76			25.06
Pasture		19.30	14.03	114.54	653.03	26.67	4.96	0.06	11.18	53.75	88.71	45.41	0.43	136.30	194.59	169.11	5.86	10.42	1548.35
Wooded pasture		11.50	53.42	1.59	562.58	15.47	6.78		1.81	32.64	54.01	51.77	5.68	43.85	389.15	16.39			1246.65
Meadow		2.16	5.74	1.65	16.67	6.02	1.64	1.22		2.76	7.62	4.95	0.59		27.92	2.24		0.35	81.52
Wooded meadow		2.74		0.00	20.07	1.32	0.65	0.11	0.07	22.45	0.38	5.64	3.66		9.65	0.22		0.00	66.97
Afforestation		5.90												141.32					147.22
Sowable	2.41	58.53	24.26	1.41	243.86	37.28	0.71	0.34	10.64	314.75	10.12	12.47	1.97	29.82	614.98	119.96		28.62	1512.11
Mixed cultivation	2.60	38.49	1.20	7.94	92.17	31.21	2.17	2.06	1.78	78.59	1.97	6.14	0.67	30.97	267.45	33.30		0.36	599.06
Beach		6.55	5.84		10.12									89.29			9.17		120.97
Vineyard		0.05								0.12					0.06			0.00	0.23
Total	131.01	217.76	149.85	147.49	5058.48	335.14	113.41	4.61	46.97	816.19	216.48	220.99	22.51	1141.37	1831.20	577.04	30.07	40.19	11,100.77

Woodlands

Woodlands no longer play a strategic role in terms of charcoal and timber production, showing a further increase of their extension (Fig. 1.8), covering 47% of the region and 55% of the productive land, making Tuscany the most forested region of Italy (Regione Toscana, 1998). However, after a decrease lasting until the oil crisis of 1978, there is a new growth of the importance of fuelwood production for domestic use, allowing the management of coppice woods, representing 75% of all management forms. Woodlands now cover 66% of the landscape in the area surveyed, followed by cultivated fields (20%), while pasturelands have slowly increased (9%). Woods are the most important land use in ten areas; only three areas – one along the coast and two on the hills in the Florence district – show the prevalence of cultivated land, a clear indication of the importance of the abandonment that occurred.

The features of forest landscape are characterized by the prevalence of mixed stands and coppice, but the general patterns show simplified typologies, mostly made of dense, homogeneous forest covers, where diversity is mostly due to the presence of mixed species, that do not contribute much to the diversity of landscape mosaic (Agnoletti, 2002). In terms of management forms, simple coppice has now replaced mixed coppice with standards, a clear indication of the reduced amount of timber assort-ments produced, due to the change in farming systems, once deeply linked to coppice. Many coppices are quite aged, following the interruption of cuttings in areas far from roads, especially after the end of charcoal production in the late 1950s due to the introduction of new energy sources, while pollard trees for fodder or nut production no longer exist. Most coppice woods on the highest part of the mountains have been turned into high stands by foresters in order to develop a protective role, but terraced charcoal kilns on mountain slopes are still there, to testify to the former use of those woodlands. In the high stands there is now a higher proportion of conifers due not only to afforestation (15.5%), but also to secondary successions on abandoned fields and pastures. The new law of 1952, financing afforestation with the goal of creating new jobs, was clearly trying to create new forests not only for protection, but also to produce more timber. About 800,000 ha of new afforestations were registered in Italy in the following decades. Between 1947 and 1997 conifer woods have doubled in area in Tuscany and broadleaved woods have reduced, continuing a trend already observed in many European countries (Agnoletti, 2000; Johann et al. 2004; Brandl, 1992). More than 50% of afforestation since 1954 is in fact occurring on former woodlands, while the rest is occurring on fields, chestnut woods, pastures and old olive orchards. Even today, afforestation continues with no regard for

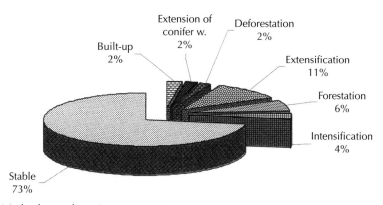

Fig. 1.8. Main landscape dynamics 1954–2004.

the bad quality of landscapes created by these plantations, as the art historian, Mario Salmi, already pointed out to foresters in 1965.

However, the elimination of chestnut orchards by planting conifers is only one of the causes of the dramatic reduction of these woods, already noted in the period 1832–1954, but further reduced by 50% in area by the year 2000. Often, monumental chestnut trees 300–400 years old can still be found (Fig. 1.9), surrounded by beech and hornbeam at higher elevations, or by oak at lower altitudes, and sometimes by olive orchards at sea level. The abandonment has also been favoured by the management policies carried out in some protected areas, considering them an artificial plantation and a cause of hydro-geologic risk, supported also by botanical interpretation (Cavalli, 1990). In 1996, a dramatic flood in the area of Cardoso caused several landslides. The study promoted by the Park suggested that some of the reasons for this was the heavy weight of chestnuts. The research carried out in a portion of this area has demonstrated that 76% of landslides occurred with 'abandoned chestnuts' located on terraces, collapsing because of lack of maintenance (Plate 2). An extension of the research has shown evidence of a strict relationship between hydraulic risk and land use changes, especially abandon-

ment, a problem receiving little attention by local authorities (Agnoletti, 2005a). Chestnut coppice is still managed to produce poles to support vines, but the use of steel poles and the little or no support given in rural policies to the use of wooden ones is not favouring their management. In this respect the analyses made inside two protected areas (Cardoso and Migliarino) are both showing the lack of any effective policies to reduce the loss of traditional landscapes in protected areas. Shrublands have seen an increase of 20% in their extent as the result of abandonment of fields and pasture (Table 1.3), but also as a result of fire. In any case they have completely changed their role in the landscape, passing from a fundamental management form of rural economy, to an aspect of natural evolution. It is very significant that the forest inventory of Tuscany (Regione Toscana, 1998) describes them as an effect of degradation caused by fire or pasture, with no consideration of their cultural origin or their meaning in the landscapes.

Fields

Concerning agriculture, during the 1960s a large portion of farmland was abandoned and modern cultivations developed on the most favourable areas, while mechanization

Fig. 1.9. Some monumental chestnut trees are still surviving but will soon be invaded by natural regeneration and turned into mixed stands if not submitted to silviculture.

Table 1.3. Cross tabulation 1954–2004. Columns and lines allow checking of the evolution of the land uses in the period analysed.

1954 / Extent in ha	Waters	Anthropic	Wood plantation	Shrubs	Sparse vegetation	Woodland	Woods with prevalence of chestnut	Chestnut orchards	Quarry	Fallow	Olive orchards	Pasture	Wooded pasture	Meadow	Afforestation	Sowable	Mixed cultivation	Beach	Vineyard	Total
Waters	128.17					2.84														131.01
Anthropic	0.26	269.65	0.63	1.07		8.53	0.28	0.23		0.02	1.32	1.95	0.01		0.29	0.99	4.55			289.78
Shrubs	0.23	2.49		86.55	1.33	60.26	1.55				0.37	19.66	0.10		6.86		6.65			186.04
Sparse vegetation				3.39	108.62	52.26	1.76			1.87	0.22				8.21					176.33
Woodland	3.11	14.54	4.40	30.98	0.65	5652.29		13.41		3.23	44.19	28.79	26.54	1.18	123.52	3.12	39.49		0.08	5989.51
Woods with prevalence of chestnut		0.65			6.76	1.53	326.20													335.14
Chestnut orchards				0.26	0.01	46.42		45.06				0.54	0.43		20.73					113.44
Quarry									4.61											4.61
Fallow		4.19		3.76		19.17				1.18	6.26	0.48	0.59		0.55	3.42	0.38		7.00	46.97
Olive orchards		175.82	3.52	10.59		137.95				5.85	762.20	1.48	1.26		16.63	31.75	145.15		33.84	1326.38
Pasture				7.00		48.00	3.36	0.36				190.16	11.24		11.89		20.62			292.84
Wooded pasture		0.62		13.74	2.11	88.34	2.30	3.47			0.81	22.90	94.64		19.83		0.56			249.31
Meadow		0.22		2.95		4.28	0.77	0.62					1.79	11.28	0.28		0.32			22.51
Afforestation	0.28	0.14		9.66		30.51		0.61			8.14			2.33	1084.04	5.01	0.81			1141.52
Sowable	7.82	26.77	134.99	65.06		139.53		1.24		3.21	67.93	318.67	79.15	2.37	34.90	124.21	784.59		42.55	1833.00
Mixed cultivation		99.72	4.52			14.44	5.54			11.88	53.95	4.52	0.56	1.40	5.82	202.38	167.47		51.53	623.73
Beach																		30.07		30.07
Vineyard		0.11	9.12	0.05		6.05					5.51	1.31	0.25		0.00	0.14	20.68		2.21	45.43
Total	140.20	595.15	157.18	235.04	119.49	6312.38	341.75	65.00	4.61	27.23	950.92	590.46	216.55	18.56	1333.54	371.00	1191.26	30.07	137.21	12,837.62

and chemical fertilizer are rapidly increasing the amounts produced per hectare. Mechanization favoured the abandonment of terracing and the elimination of trees and hedges, especially in marginal areas, creating large fields and extended monocultures. In many cases, like in Gargonza and Cardoso (Fig 1.10), the forest is today covering terraces once shaping hills and mountains (Agnoletti and Paci, 1998). New vineyards have been planted on mixed cultivations (45%), on fields (30%) and olive orchards (24%), but technical evolution has concentrated its efforts on making large regular plots cultivated uphill, even on steep slopes, often causing erosion and degrading the quality of the landscape. In some areas the extension of the maximum concentration of adjacent vineyards plots has increased from 26 to 253 ha, with a strong simplification of landscape patterns forming large subsystems with only vineyards (Fig 1.11). The development of new vineyards is occurring only on hills and plains, while on the mountains they have been deleted from the landscape.

The decrease of mixed cultivations is quite significant in the areas studied (–66%) – generally diminished by 75% between 1955 and 1974 in the whole region (Agnoletti, 2002) with a great loss also in the wood species included, ranging from fruit trees to woody species. It is worth noting the creation of a new form of 'European' agrarian landscape, due to plantations favoured by the EEC directive 2080/92. The idea of favouring the designation of large portions of farmland with subsidies given to replace existing crops with tree species suited for timber production on

(a)   (b)

Fig. 1.10. Photograph (a) shows a view of Cardoso (Regional Park of the Apuane mountains) at the beginning of the 20th century. It is possible to note that the slopes behind the village were all terraced and planted with trees. Photograph (b) shows the same view of Cardoso today. The old landscape is now covered by forest (see Plates 1 and 2).

Fig. 1.11. Wine monocultures covering entire hill slopes are making landscape more homogeneous, creating a sort of 'globalscape' typical of many wine regions in the world.

a 25-year cycle has often contributed to a further degradation of the cultural features of the landscape, speeding up the abandonment of traditional forms. At the same time this policy has very little chance of affecting the timber market in any way. The landscape quality of these plantations is not necessarily always bad, but it should be asked why, when decisions like these are made about valuable landscapes, there is no evaluation of their impact or any study to adapt them to the local context.

Public perception and economic value

Besides the sampling plots made to analyse the features of vegetation changes, some of the most interesting investigations relate to the perception of the landscape by residents and tourists. There is not enough space here to present these results (Casini and Ferrini, 2002), as will be done in Chapter 5 about Moscheta, but, confirming what has already been noted, there is a strong feeling of cultural identity among the people represented by the landscape (Bacci, 2002). It is very relevant that in areas where our investigations have indicated the most significant landscape forms, characterizing

them from a scientific point of view, the interviews have confirmed our results. This is true even for a very interesting aspect regarding the fact that the same element (e.g. afforestation) may have a totally different meaning in two different areas. Umbrella pine plantations along the coast have created historical landscapes appreciated for their beauty and recreational features, while the afforestation of black pine made on the hills and mountains seems to be valuable only for the foresters who planted them. Another aspect of this investigation is the willingness of most of the people to accept a tax to preserve landscape, while farmers do not accept it. This is an interesting indication of how rural policies have failed to address some of the needs of society (Agnoletti, 2002).

A General View of the 1832–2004 Changes

Landscape changes occurring in Tuscany over the last two centuries are due to direct socio-economic factors. Their size and features are not comparable to any ecological or climate change that occurred in the past two centuries, or to those foreseen for the next. The quality of landscape resources reflects how society develops, especially in the way landscape is perceived by the public, both when development is very much based upon local resources, as in the past, or when this is no longer occurring.

Most of the changes analysed occurred in the period 1832–1954. The following decades confirm a trend initiated before, although some processes such as demographic fall in the mountains are surely very fast, but comparable to the growth between 1861 and 1920. Woodlands and trees in the fields are both central elements involved in the landscape dynamics. Forestation is the most important process occurring (21% of changes), followed by intensification in agriculture (11%) and again by afforestation due to conifer plantations (10%). The increase of woodlands (55% since 1832, 8.7% between 1954 and 2000) is taking place mostly on abandoned pastures and wood pastures (53%), less on cultivated land (20%). Woodlands have increased their area in ten study areas, sometimes by more than double where abandonment has been stronger, as in the mountains and even in the hilly areas. These trends are similar to those reported in several rural areas submitted to abandonment for comparable periods of time (Foster, 1992; Foster *et al.*, 1998), but in our case we have also measured overwhelming changes in the landscape mosaic (see Plate 3). The expansion of woodlands is a process comparable to the general trend reported for Italy in the last 100 years, showing that forests have more than doubled their area, with the evident absence of any real threat for them in the last 50 years (Fig. 1.12). It is worth noting that new plantations with conifers are mostly occurring on territories previously presenting different woodlands or shrubs (65%), but not denuded land.

As in the rest of Italy, the success of afforestation is not only due to the money spent, but to the decrease of population in mountain areas and the reduction of livestock farming, or the cessation of the eco-nomic role of former forests, such as chest-nut orchards. Finally a 'state landscape' came through, replacing the former social landscape of the past, and this new land-scape is the real legacy of afforestation. In fact, these new conifer forests had no influ-ence on the timber market. They probably had a role in reducing risk on former pas-tures and fields on steep mountain slopes, but they definitely had an impact on the landscape, leaving large squared plots, with little relevance to local culture or ecological conditions (see Plate 7).

Pasturelands show a very significant reduction, decreasing to only 25% of their former extent, as do wood pastures (15.5%) (Fig. 1.13); a certain amount of them have been turned not only into forest, but also into specialized cultivations and sowable land. Cultivated fields have slightly increased, but with a strong growth of sowable land (+ 407%). A very substantial increase is the one shown by olive orchards, which have increased 25-fold, and vineyards, almost non-existent as spe-cialized cultivations in the past. It is worth noting the increase of specialized olive

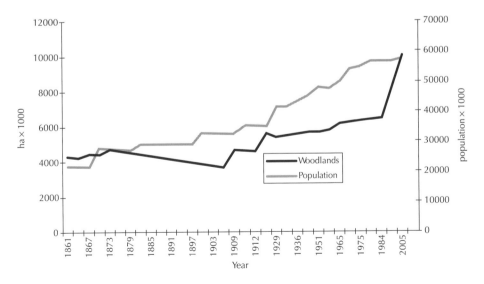

Fig. 1.12. Extent of the Italian forests and population growth between 1861 and 2005. The substantial increase after 1985 is mostly due to the different ways of considering what is 'woodland' by the National Forest Inventory made in 1985 and 2005. Nevertheless, there is undoubtedly a steady growth after World War I due to the changing relationship between socio-economic development and forest resources.

Fig. 1.13. A wood pasture with beech in the study area of Moscheta (see Chapter 5). Wood pastures have almost disappeared because of the reduction of grazing and the advancement of woodlands.

orchards and vineyards. Both of them also show a strong increase in the size of patches, as described for 1954–2004, with large portions of land with a repeated monoculture pattern presented as 'typical' elements of traditional landscapes by rural developers, but clearly presenting the fea- tures of industrial cultivation (Fig. 1.14). The data also show the strong reduction of all the categories of mixed cultivations described for 1832, now reduced to almost one third. Unfortunately, only field work is able to show that most of those still existing are modified and simplified forms, usually

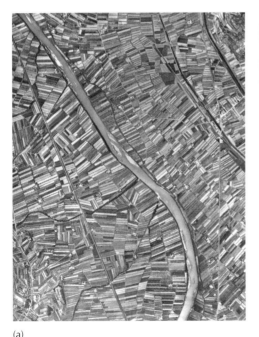

(a)   (b)

Fig. 1.14. The increasing size of the fields in a rural mosaic along the Arno River. Left: a photograph taken in 1954. Right: the same area on a photograph of 1996.

linked to fruit orchards or tree rows at the sides of fields.

One of the most dramatic trends, starting even before 1954, is the great loss of diversity in the landscape, with the decrease of many land uses linked to pastures, fields and woods and a consequent dramatic reduction of the diversity of spaces due to land uses, by almost 45% between 1832 and 2004 (see Plates 1, 2 and 3). The biodiversity of spaces is a part of general biodiversity and a fundamental aspect of the quality of the cultural landscape existing in Tuscany, as well as in the entire Mediterranean region (Baudry and Baudry-Burel, 1982; Naveh, 1998). This reduction, related to the number of land uses, is fully expressive of a reduced biological diversity; very diverse landscapes are more species-rich than individual habitat components. The reduction of this diversity is supported by the trends analysed in other indices. The number of patches is 84% of that existing in 1832, while their average size has also increased by 11% (Table 1.4). These data together with the decrease by 36% of Hill's diversity number confirm the simplification of the landscape mosaic. This reduction in diversity of patches in relatively small areas, according to the size of our study areas, makes the present diversity of the regional landscape mostly based on the features of larger subsystems inside the main geographical areas, confirming the change from a fine-grained to a coarse-grained landscape (Angelstam, 1997).

This loss of diversity is particularly significant from many points of view. First of all, the diversity of species is probably not the only important feature of Italy (although quite significant in the European context), but rather its diversity of spaces created by man in centuries of rural and forest practices, also introducing many species originally not present in the Italian peninsula before Roman times. It is also well known that specific forest types, like chestnut orchards, often present a higher floristic diversity compared to abandoned orchards (Romane and Valerino, 1997). Another problem is the loss of cultural values related to rural and forest practices existent since Etruscan times, connected to small-scale productions creating many small patches in the landscape mosaic, although this feature cannot be generalized for the entire region and Italy. From this point of view, the world-famous pictures of Tuscany showing denuded landscapes with rare cypress trees cannot be considered a common pattern, but an element of the general diversity of the landscapes in the region.

Conclusions

This project has produced a large amount of information on the landscape dynamics occurring in the Tuscan territory in the last 180 years, only partially presented in this text, mostly focused on the factors and processes affecting the main dynamics and the reduction of landscape diversity. The decreased diversity can be clearly ascribed to some main trends. One is the advancement of a continuous forest layer covering

Table 1.4. Indices detailing the changes in landscape diversity for all areas studied.

	Year		
	1832	1954	2004
Number of land uses	310	158	173
Number of patches	1838	1521	1549
Average surface area of patches (ha)	11.66	12.40	13.00
Average number of patches × ha	0.17	0.14	0.15
Shannon index (mean value)	1.09	0.90	1.11
Hill's diversity number (mean value)	8.30	5.00	5.30

the former landscape mosaic like a mantle (see Plate 3). Another is the increased size of fields in agricultural areas. A third is the simplification of the internal structure of landscape patches. In many cases this trend is not sustainable, not only for biodiversity but also for the conservation of landscape resources, not to mention the disappearance of specific woodlands, like chestnut orchards, and a wide number of traditional management practices. In the areas more affected by this process landscape diversity may be only 9% of what it was in 1832, while the number of patches may be only 14% of that in 1832 (see Plates 1 and 2). In the areas where the farmers are still present instead, this tendency can even be inverted, although the internal quality of new landscape patches is not always good. The interruption of traditional techniques like mixed cultivations, terraces, wood pastures, tree rows, and hedges that characterized farming until the 1950s has been replaced by extended monocultures created with mechanization, allowing uphill cultivation even on steep slopes, as in the case of vineyards. This has often created a landscape where diversity and uniqueness, according to UNESCO criteria (Fowler, 2003), are often given mostly by morphological features, while the mosaic can be compared to other regions in the world. The interruption of traditional rural practices also has a strong impact on the hydro-geological risk, causing erosions and landslides affecting especially mountain and hilly areas. In this respect the landscape existing in the area studied until the early 1950s can probably be compared with the ones still surviving in places like Eastern European countries, where industrialization will soon induce the same process. These tendencies are also degrading the economic role of landscape, as an added value for typical products and tourism, and decreasing the quality of life for citizens who prefer a more diverse landscape and feel a very strong cultural relationship with their historical landscapes.

It seems that no real policies have been enhanced to change these trends – on the contrary, many European Union directives concerning rural development and nature conservation are speeding up these trends. The lack of attention given to the role of landscape resources as an added value for rural economy, protecting and favouring the upkeep of traditional practices (Fig. 1.15), creating markets for typical products linking them to their landscapes, promoting the role of landscape for agritourism, have been neglected in favour of other choices. Initiatives promoting setaside, industrial plantations, and technological innovation have denied the fact that development should also care for the conservation of landscape resources as a cultural value, for the quality of life, as a factor of competitiveness. From this point of view, the subsidies given by the EU favouring setaside and plantations have contributed to the disappearance of the typical elements of cultural landscape. Very probably these trends will have the same effect in new eastern EU countries.

Another threat comes from the nature conservation strategies and a paradigmatic way of considering the role of nature and the concept of sustainability. The network of

Fig. 1.15. An old farmer tying vines to maple trees (circa 1930) in a mixed cultivation according to a tradition existing in Tuscany at least since Etruscan times, 1000 BC. Saving traditional knowledge is one of the most important issues in the conservation of cultural landscapes.

protected areas and the habitat directive, acknowledged by Tuscany with a regional law, is enclosing a list of habitats according to the EU habitat list, presenting a peculiar reading of the territory where the protected areas have been established. Areas clearly having a human origin have been described by presenting them as natural or semi-natural. This operation can sometimes be seen as an attempt to offer another 'reading' of the European territory by denying its cultural origin, a factor regarding not only Mediterranean woodlands but also boreal forests (Axelsson and Östlund, 2000).

Another reason for the reduced attention to landscape quality is the perception of managers and a large portion of the public and media that the extension of forest land and renaturalization is a very favourable process, increasing the value of the territory. This vision is supported also by the current way certification standards are developed, enabling agencies to give a label of 'sustainability' to a new forest growing on an ancient rural pattern. In Tuscany, this is also helped by forest legislation saying that an abandoned field covered by a new forest after 15 years is to be considered untouchable, unless the owner pays for a new afforestation of the same size. The fact that a forest is 'untouchable' is very representative of a hierarchy of values paying little atten-

Table 1.5. Criteria and indicators proposed for the sustainable management of cultural landscapes in Tuscany.

Criteria	Indicators	Description
1. Significance	Uniqueness	In the local, regional, national or international context
	Matrix	Internal feature in terms of complexity
	Persistence of mosaic	Historical persistence of the structure of landscape mosaic
	Socioeconomic activities	Practices, traditional knowledge, productions, to maintain landscapes
	Persistence of land uses	Historical persistence of single land uses
	Extension of land uses	Maintenance of the extension of each land use
	Internal features of patches	Maintenance of the internal structure of patches
	Material evidence	Persistence of material evidence in the landscape
	Social perception	Social awareness of landscape values
2. Integrity	Extension	Maintenance of an extension sufficient to ensure functionality
	Geomorphologic features	Maintenance of specific geomorphologic structures
	Aesthetic	Maintenance of aesthetic values
	Management practices	Conservation of traditional knowledge and management forms
	Structure of the matrix	Maintenance of the structure of the landscape matrix
	Structure of the mosaic	State of conservation of landscape mosaic
	Structure of patches	State of conservation of single patches
	Cultural heritage	State of conservation of architectural assets and material evidence
	Natural heritage	State of conservation of flora and fauna
	Conservation and research	Research and conservation activity related to the area
3. Vulnerability	Fragility	Intrinsic fragility of landscape structure
	Farming	Farming activity affecting landscape
	Forestry	Forest activities affecting landscape
	Industrial activity, infrastructure, urbanization	Direct and related activities influencing landscape
	Natural evolution	Fragility of landscape to natural dynamics
	Tourism	Tourist activities influencing landscape
	Social structure	Social features affecting landscape

tion to cultural landscape and denying the role of spaces in biodiversity. Under these circumstances it would be worth reflecting on the way in which sustainability is conceived and applied, and how paradigmatic visions can reduce the chance to develop an approach more adapted to local situations. The methodology developed in this project has been shown to be particularly suited for the development of criteria and indicators to assess significance, integrity and vulnerability of these landscapes, helping to develop management, monitoring and restoration, but also new ways of managing the network of protected areas (Table 1.5). So far, the HCEA methodology has been applied in a wide number of regional proj-

ects ranging from environmental impact assessment to the management of protected areas and urban and landscape planning, but it is currently discussed in the National Strategic Plan for Rural Development and presented also during the work of the Ministerial Conference for the Protection of Forest in Europe, dedicated to the promotion of historical and cultural values in sustainable forest management.[8] However, some signs of new trends appearing in rural development, in the European forest strategy and important events like the European Landscape Convention, suggest a possible change, hopefully taking effect before cultural landscapes are completely lost (see Plate 4).

Notes

1. The author is writing as coordinator of the group of researchers involved in the survey. A team of 20 researchers was involved in the project. The international institutions included: the International Union of Forest Research Organizations, the American Forest History Society, Portland State University and the American Science Foundation.
2. Law number one, the Management of the Territory, n. 5 of 1996, now revised as n. 1 of 2005.
3. Department of Environmental Forestry Science and Technology, Faculty of Agriculture.
4. The project lasted for 5 years; the interpretation of the photographs of the year 2000 was accompanied by field work carried out in 2001, 2002, 2003, 2004 and 2005, according to the different areas surveyed.
5. For two areas photographs of 1981 were used. However, the result was incorporated in the data regarding 1954, since the comparison between the two years did not show significant changes in the trends.

References

Agnoletti, M. (2000) Introduction: factors and process in the history of forest research. In: Agnoletti, M. and Anderson, S. (eds) *Forest History: International Studies on Socioeconomic and Forest Ecosystem Change.* CAB International, Wallingford, pp.1–19.

Agnoletti, M., (2002) *Il paesaggio agro-forestale toscano, strumenti per l'analisi la gestione e la conservazione.* ARSIA, Firenze.

Agnoletti, M. (2005a) *Landscape changes, biodiversity and hydrogeological risk in the area of Cardoso between 1832 and 2002.* Regione Toscana, Tipografia Regionale, Firenze.

Agnoletti, M. (2005b) *The evolution of the landscape in the Migliarino Estate between the 19th and the 20th century.* Regione Toscana, Tipografia Regionale, Firenze.

Agnoletti, M. and Innocenti, M. (2000) Caratteristiche di alcuni popolamenti di farnia e rovere presenti lungo la costa toscana alla metà del settecento. In: Bucci, G., Minotta, G. and Borghetti, M. (eds) *Applicazioni e prospettive per la ricerca forestale italiana.* Atti del II congresso SISEF, Bologna.

Agnoletti, M. and Paci, M. (1998) Landscape evolution on a central Tuscan estate between the eighteenth and twentieth centuries. In: Kirby, K.J. and Watkins, C. (eds) *The Ecological History of European Forests.* CAB International, Wallingford, pp.117–127.

Angelstam, P. (1997) Landscape analysis as a tool for the scientific management of biodiversity. *Ecological Bulletins* 46, 140–170.

Arnold, J.E. and Deewes, P.E. (1995) *Tree management in farmer strategies.* Oxford University Press, Oxford.

Atauri, J.A. and De Lucio, J.V. (2001) The role of landscape structure in species richness distribution of birds, amphibians, reptiles and lepidopterans in Mediterranean landscapes. *Landscape Ecology* 16,147–159.

Austad, I. (1988) Tree pollarding in western Norway. In: Birks, H., Birks, H.J.B., Kaland, P.E. and Moe, D. (eds) *The cultural landscapes. Past present and future.* Cambridge University Press, Cambridge, pp. 11–30.

Axelsson, A. and Östlund, L. (2000) Retrospective gap analysis in a Swedish boreal forest landscape using historical data. *Forest Ecology and Management* 5229, 1–14.

Axelsson, A., Östlund, L. and Hellberg, E. (2001) Use of retrospective analysis of historical records to asses changes in deciduous forests of boreal Sweden 1870s–1999. *Acta Universitatis Agriculturae Sueciae, Silvestria* 183.

Bacci, L. (2002) *L'impatto del turismo nell'economia regionale e locale della Toscana.* IRPET, Firenze.

Baudry, J. and Baudry-Burel, F. (1982) La mesure de la diversité spatiale. Relation avec la diversité spécifique. Utilisation dans les évaluations d'Impact. *Acta Ecologica, Oecol. Applic.* 3, 177–90.

Bezzini, L. (1996) *Storia di Castagneto Bolgheri e Donoratico dalle origini al 1945.* Bandecchi & Vivaldi Editori.

Brandl, H. (1992) Entwicklungen und tendenzen in der Forstgeschichte seit ende des 18. Jahrhundert. *Mitteilungen der Forstlichen Versuchs und Forschungsanstalt Baden-Wuerttemberg,* Freiburg.

Calistri, E. (ed.) (2002) *Environmental Signals in Tuscany.* Edifir, Firenze.

Carandini, A. and Cambi, F. (eds) (2002) *Paesaggi d'Etruria: valle dell'Albegna, valle d'Oro, valle del Chiarone, val le del Tafone : progetto di ricerca italo-britannico seguito allo scavo di Sette finestre.* Edizioni di storia e letteratura, Roma.

Casini, L. (2000) *Nuove prospettive per uno sviluppo sostenibile del territorio.* Studio Editoriale Fiorentino, Firenze.

Casini, L. and Ferrini, S. (2002) Le indagini economiche. In: Agnoletti, M. (ed.) *Il paesaggio agro-forestale toscano, strumenti per l'analisi la gestione e la conservazione.* ARSIA, Firenze, pp. 49–68.

Castronovo, V. (1995) *Storia economica dell'Italia.Dall'Ottocento ai giorni nostri.* Einaudi, Torino.

Cavalli, S. (1990) Costruzione della natura. In: Greppi, C. (ed.) *Quadri ambientali della Toscana,* Vol. I, II, III. Marsilio Editori, Venezia, pp. 101–118.

Cazzola, F. (1996) Disboscamento e riforestazione ordinata nella pianura del Po: la piantata di alberi nell'economia agraria padana, secoli XV-XIX. *Storia Urbana,* XX, n. 76–77, 35–64.

Cox, L.J., Hollyer, J.R. and Leones, J. (1994) Landscape services: an urban agricultural sector. *Agribusiness* 10, 13–26.

Di Berenger, A. (1859–1863) *Dell'antica storia e giurisprudenza forestale in Italia.* Treviso e Venezia.

Farina, A. (1998) *Principles and methods in landscape ecology.* Chapman & Hall, London.

Foster, R.F. (1992) Land-use history (1730–1990) and vegetation dynamics in central New England, USA. *Journal of Ecology* 80, 753–772.

Foster, D.R, Motzkin, G. and Slater, B. (1998) Land-use history as long-term broad scale disturbance: regional forest dynamics in central New England. *Ecosystems* 1, 96–119.

Fowler, P.J. (2003) *World Heritage Cultural Landscapes 1992–2002.* UNESCO, Paris.

Fuentes Gonzales, C. (1994) *La encina en el centro y suroeste de Espana.* Servantes, Salamanca.

Gil, L., Manuel, C. and Diaz Fernandez, P. (2003) *La transformation historica del paisaje forestale en las islas baleares.* Egraf, Madrid.

Greppi, C. (ed.) (1990) *Quadri ambientali della Toscana,* Vol. I, II, III. Marsilio Editori, Venezia.

Grove, A.T. and Rackham, O. (2001) *The Nature of Mediterranean Europe. An Ecological History.* Yale University Press, Ehrhardt.

Hughes, D. (2003) Europe as consumer of exotic biodiversity: Greek and Roman times. *Landscape Research* 28(1), 21–31.

Iseh, M. (1988) Air photo interpretation and computer cartography-tool for studying the cultural landscape. In: Birks, H., Birks, H.J.B., Kaland, P.E. and Moe, D. (eds) *The Cultural Landscapes. Past Present and Future.* Cambridge University Press, Cambridge, pp.153–164.

Johann, E., Agnoletti, M., Axelsson, A.L. *et al.* (2004) History of secondary Norway spruce in Europe. In: Spiecker, H., Hansen J., Klimo, E., Skovsgaard, J.P, Sterba, H. and von Teuffel, K. (eds) *Norway Spruce conversion. Option and consequences.* EFI research report 18. Brill, Leiden, pp. 25–62.

Knowles, A.K. (ed.) (2002) *Past Time, Past Place. GIS for History.* ESRI Press, Redlands.

Köhl, M. (2003) New approaches for multi resource forest inventories. In: Corona, P., Köhl, M. and Marchetti, M. (eds) *Advances in Forest Inventories for Sustainable Forest Management and Biodivesity Monitoring.* Kluwer Academic Publisher, Dordrecht, pp. 1–18.

Lamdin, E.F. and Geist, H.G. (2001) Global land use and cover change: What have we learned so far? *Global Change Newsletter* 46, 27–30.

McHarg, I. (1981) Human ecology planning at Pennsylvania. *Landscape Planning* 8, 109–120.

McNeill, J. (2000) *Something New Under the Sun.* Penguin Books, London.

Monser, U., Albani, M. and Piussi, P. (2003) Woodland recolonization of abandoned farmland in the julian pre-Alps (Friuli, Italy). *Gortania* 25, 207–231.

Motzkin, G., Foster, D., Allen, A., Harrod, J. and Boone, R. (1996) Controlling site to evaluate history: vegetation patterns of a New England sand plain. *Ecological Monographs* 66(3), 345–365.

Naveh, Z. (1991) Mediterranean uplands as anthropogenic perturbation dependent systems and their dynamic conservation management. In: Ravera, O.A. (ed.) *Terrestrial and Aquatic Ecosystems, Perturbation and Recovery*. Ellis Horwood, New York, pp. 544–556.

Naveh, Z. (1998) Culture and landscape conservation: A landscape-ecological perspective. In: Gopal, B. *et al.* (eds) *Ecology Today: An Anthology of Contemporary Ecological Research*. International Scientific Publications, New Delhi, pp. 19–48.

Ndubisi, F. (2002) *Ecological Planning*. Johns Hopkins, Baltimore.

Ortega, M., Elena Rossello, E. and Garcia del Barrio, J.M. (2004) Estimation of plant diversity at landscape level: a methodology approach applied to three Spanish rural areas. *Environmental Monitoring and Assessment* 95, 97–116.

Östlund, L., Zackrisson, O. and Hörnberg, G. (2002a) Trees on the border between nature and culture – culturally modified trees in boreal Scandinavia. *Environmental History* 7(1), 48–68.

Östlund, L., Zackrisson, O. and Hörnberg, G. (2002b) Trees on the border between nature and culture – Culturally modified trees in boreal Scandinavia. *Environmental History* 7(1), 48–68.

Pignatti, A. (1994) *Ecologia del Paesaggio*. Utet, Torino.

Pitte, J.R. (1986) *Terres de Castanide*. Fayard, Évreux.

Rackham, O. (1995) *Trees and Woodlands in the British Landscape*. Weidenfeld and Nicholson, London.

Regione Toscana (1998) *Inventario Forestale*. Tipografia Regionale, Firenze.

Romane, F. and Valerino, L. (1997) Changements du paysage et biodiversité dans les châtegnairaies cévenoles (sud de la France). *Ecologia Mediterranea* 23 (1/2), 121–129.

Romani, V. (1994) *Il Paesaggio: teoria e pianificazione*. Milano, Franco Angeli.

Sereni, E. (1997) (reprint of first edition 1961) *History of the Italian Agricultural Landscape*. Princeton University Press, Princeton.

Vogt, K.A., Gordon, J.C. and Wargo, J.P. (1997) *Ecosystems*. Springer, New York.

Vos, W. and Stortelder, A. (1992) *Vanishing Tuscan Landscapes*. Regione Toscana, Firenze.

Wagner, H.H., Wildi, O. and Ewald, K.C. (2000) Additive partitioning of plant species diversity in an agricultural mosaic landscape. *Landscape Ecology* 15, 219–227.

Watkins, C. and Kirby, K.J. (1998) Introduction – historical ecology and European woodland. In: Kirby, K.J. and Watkins, C. (eds) *The Ecological History of European Forests*. CAB International, Wallingford, pp. ix–xv.

influences biological processes; tree growth and decomposition of woody material are typically slow, or very slow, in these areas. Scots pines can reach ages exceeding 800 years and Norway spruces can reach ages of up to 500 years (Fig. 2.2).

People, Subsistence and Landscapes

Hunting, fishing and collecting edible plants formed the economic basis of northern Fennoscandian societies throughout prehistory and into historical times. Since the first hunter-gatherer communities established soon after deglaciation (about 10,000 BP), subsistence and settlement patterns have undergone significant changes over time. Within the hunter-gatherer framework, different forms of resource utilization evolved, all setting characteristic marks on the landscape. During the first

Fig. 2.2. Detail of old-growth pine forest in Tjeggelvas forest reserve. Trees with ages exceeding 400 years are common in this forest. Forest structure is shaped by recurrent forest fires and low intensity human use. Photo by Lars Östlund.

millenium AD reindeer pastoralism developed among the indigenous Sami population, leading to new forms of land use. Although the archaeological records give evidence of stock-breeding and cultivation in the Atlantic and Bothnian coastal regions during the first millenium BP (Johanssen, 1990; Baudou, 1992; Liedgren, 1992; Edgren and Törnblom, 1993), hunting and fishing remained crucial to subsistence. In northern Sweden, farming was firmly established during the first centuries AD, although exclusively restricted to coastal areas (Baudou, 1992).

In the 18th and 19th centuries farming was slowly and successively established in the interior regions of northern Sweden. The so-called 'settlers' were either Swedish farmers immigrating from the coastal areas, or of Sami origin. To the Sami, stock-breeding and cultivation, mainly of potatoes and barley, presented either alternative or complementary subsistence activities to reindeer herding. Hunting and fishing remained equally important as subsidiary activities to both farmers and reindeer herders until the 1950s. Today, extensive reindeer herding still provides a livelihood for many Sami, while the economic significance of hunting and fishing has fallen. Farming has dramatically declined since the 1960s and very few farmsteads are still active.

Recent types of landscape use include hydro-power exploitation,ʳ industrial forestry and mining. The very first mining enterprises in the Swedish mountain region started as early as 1600 AD. Although of short duration, mining had a large-scale environmental impact, mainly due to the extensive logging of wood to produce the charcoal required by the smelting-houses. Although the central points of different ecological niches may have varied between different economies, resource areas overlapped to a great extent. Hunter-gatherer and pastoralist economies, as well as farming, were ultimately delimited by the seasonal changes characterizing boreal areas. Consequently, exploitation strategies and settlement patterns were logistic in character, focussing on the seasonal occurrences of resources and the storage of supplies.

Trade has also made a continuous economic contribution to northern pre-industrial societies. Trade and the exchange of gifts occurred throughout their history, cutting across ethnic, linguistic, social and religious boundaries and connecting societies through extensive networks. Trade relations formed one of many arenas where values and conceptions of the world met, prompting reflections and altering perceptions of landscape disposition, use and meaning. The economic and cultural dimensions structuring northern landscapes are to a large extent intact and distinguishable. Recent subsistence enterprises have not consistently altered or erased traces of ancient economies and, consequently, northern Swedish landscapes present optimum conditions for studying the interactions between people and their physical environment.

Apart from providing the bases for economic interests and resources to exploit, landscapes constitute ideological constructions where culturally conditioned values materialize in the form of settlements, migration routes, sacrificial sites, etc., and cultural differences are expressed in the significance and meaning attributed to the landscape. To the Sami, cultural identity is deeply rooted in the landscape. Mountains, boulders, lakes, brooks and other natural formations constitute significant and substantial elements that tell of subsistence and settlement history, social and religious spaces, important events, family history and prominent persons. Until about 50 years ago, movements between settlements were conducted either by foot, boat or on skis. Family members moved together and the pace of movement facilitated the transmission of experiences and knowledge from one generation to the other. By ascribing identity to places they became important points of reference. It was crucial to the people's survival to obtain detailed knowledge of the landscape. Also, information embodied in oral traditions and place-names served as an instrument of enculturation and reinforced cultural identity (Bergman and Mulk, 1996). The media whereby landscape information is transferred have changed during the past 50 years, mainly due to changes in the mode

of reindeer herding. The continuous journeys by foot or ski between seasonal settlements have been replaced by quick trips from one settlement to another by helicopter, snowmobile or other motorized vehicles. Consequently, landscape perception has dramatically altered.

Methods and Sources to Interpret Past and Present Cultural Landscapes

A variety of methods can be applied in studying and interpreting past human impact on boreal forest landscapes. However, more general approaches have often proven to provide data that are too limited to be of great value, while detailed case studies with carefully chosen methods are far more fruitful (Hellberg, 2004). In order to uncover the complex and intrinsic structures of the dialectics between ecosystem processes and human land use, an inter-disciplinary approach is required. Such an approach offers the best possibilities to understand landscape change and to explain the driving forces behind changes (Östlund and Zackrisson, 2000). For instance, written information on traditional Sami land-use is sparse, and detailed maps are totally lacking. The only possible means of analysing features such as settlement patterns and the subsistence history of sites older than 100 years is by a combination of historical, ecological and archaeological records and methods. In our experience, productive case studies necessitate close cooperation between scholars in the field. Inter-disciplinary discussion in the field allows discipline-specific perspectives to be confronted and to converge in theory and practice, and thus realize the full potential of landscape analysis.

The archaeological record

In northern Sweden, the deglaciation of the Weichselian ice forms the *terminus post quem* of the archaeological and palaeoecological records. Recent inter-disciplinary archaeological and ecological research

focusing on early post-glacial pioneer colonization in interior northern Sweden, has
verified the establishment of a hunter-
gatherer society in 8600 BP (uncalibrated
date) (Bergman *et al.*, 2003). Excavations
have revealed a diverse settlement pattern,
including different types of sites, and indications of logistic procurement strategies.
Palaeoecological data from close canopy
sites (in mires) have verified an early
human environmental impact (Hörnberg *et
al.*, 2005). The settlement sites, including
the immediate surroundings, were deliber-
ately deforested to open up a pleasant
living space.

Fig. 2.3. Partly excavated hearth of Sami *árran*
type, Adamvalta, Arjeplog, Sweden, with stone
construction uncovered. The hearth was placed in
the centre of an (11th century) hut dwelling, now
completely decomposed. Photo by Lars Liedgren.

Although procurement strategies and
settlement patterns changed through time,
hunting, fishing and gathering still
remained the main economic activities
throughout a period spanning almost 8000
years. Settlement sites are distinguished by
their location close to the shores of lakes
and watercourses. The composition and
quantity of artefacts and features, as well as
the spatial extent of activity areas, are
indicative of the function and status of a
site within the overall settlement pattern
(Bergman, 1995). The same repertoire of
features, including cooking pits, pit hearths
and heaps of fire-cracked stones occur on
sites dating from 8600 BP to 1 AD.

During the first centuries AD, the
hunter-gatherer societies of interior northern Sweden underwent substantial
changes. Settlements were no longer
located close to shores, but in forested areas
at a distance from larger lakes and watercourses. The archaeological record dating
from 500 AD onwards is distinguished by
large numbers of hearths (i.e. fireplaces
made of stones) forming the most common
elements of the prehistoric Sami landscape.
The hearths are the only remaining structures of Sami type hut dwellings (Figs 2.3
and 2.4) and generally appear in pairs or
groups of 3–5. Sometimes hearths display a
significant spatial pattern, including rows
of up to 10–20 hearths spaced at a regular
distance (5–10 m).

Although there is generally great variation in the size of the hearths, those
arranged in rows are very homogeneous

Fig. 2.4. Sami hut dwelling, Rebraure, Arjeplog
parish, Sweden. The hut was probably built in the
19th century and abandoned in the early 20th
century. This type of log construction is typical of
the forest Sami in the Arjeplog, Jokkmokk and
Arvidsjaur areas. Photo by Lars Liedgren.

(Bergman, 1991). Hearth rows are interpreted as reflecting a social structure
including related families, each hearth representing a family household. In a corresponding manner, hut foundations in the
high mountain regions, dating to 500–1200
AD, are arranged in rows of 3–7. The huts
represent the expansion of settlements to an
ecological niche close to the tree-limit that
had not previously been systematically
exploited. Altogether, the observed changes
in settlement location reflect a profound
shift in the subsistence basis, probably due
to the emergence of reindeer pastoralism.

During the following period, around 1200–1500 AD, reindeer herding underwent changes with regard to management and land use, reflecting changes in social organization and settlement patterns. Herding was conducted with a high degree of mobility between pastures, and the nomadism recognized from later periods evolved. At the same time, large cooperative groups split into smaller social units.

Sami settlement sites dating to 1500 AD and later are characterized by huts in various degrees of decomposition, the oldest being completely decomposed, but with remaining hearths (Figs 2.3 and 2.4). Site structure and composition vary with respect to season and subsistence. Generally, two or three dwelling huts occur at most sites. In addition, there may be other huts with various functions – for example huts for keeping goats, or for smoking meat and fish. Furthermore, the sites host the remains of a number of storage buildings of various types of construction. The oldest standing storage buildings date to the end of the 16th century, but at sites where storage buildings have either decomposed, or have been removed for secondary use at other places, their original locations can be discerned from remaining sills and post-holes. Settlement sites included profane as well as religious spaces. Seventeenth and 18th century sources mention wooden sacrificial platforms as regular elements at Sami settlement sites (Högström, 1747; Schefferus, 1956; Rheen, 1983; Tornaeus, 1983). Also, trees in the vicinity of settlements could be subjected to religious practices (Rheen, 1983; Graan, 1983; Högström, 1747; Niurenius, 1983). No standing platforms have been preserved, but in old forest stands sacrificial trees may still be found.

Biological archives

Since forests cover entire landscapes and trees can reach great ages, dendrochronology is an important tool for analysis of past land use (Niklasson, 1998). Dendrochronological studies related to landscape history have primarily focused on fire history (Niklasson and Granström, 2000), forest structure (Östlund and Lindersson, 1995; Axelsson et al., 2002) and native peoples' impact on forest ecosystems (Zackrisson et al., 2000; Östlund et al., 2002, 2004). The potentially great age of the dominant tree species provides opportunities to analyse events in the forests related to human activity far back in time (see Fig. 2.2). It is also possible to connect chronologies from live trees with sub-fossil wood from snags, down-logs and trees submerged in lakes or peat. In drier habitats, snags can stand for many centuries and down-logs can also reach very great ages. Changes in water levels of lakes and peat accumulation in mires have often preserved submerged trees for thousands of years (Östlund et al., 2004). Marks or blazes on trees have received particular attention recently (see Mobley and Eldridge, 1992; Östlund et al., 2002). Such trees, often called culturally modified trees, can be used to analyse forest use far back in time at their actual location and with precise dating. Typical culturally modified trees include border trees, trees marking paths and bark-peeled trees (Fig. 2.5).

The major limitation affecting this resource is the fact that most of the forests have been logged since the end of the 19th century, and most of the old trees have been removed since that time (Ericsson et al., 2003; Andersson and Östlund, 2004). This highlights the importance of forest reserves and remaining old-growth forests along the Caledonian mountain range. In these forests there are better opportunities to study past land use than in the forest landscape affected by modern forestry. This creates a problem, however, since the protected areas are mostly located in the western areas at relatively high-altitude, low-productivity sites, thus producing a skewed overall picture of past land use.

Traditionally, reconstructions of long-term vegetation history and agricultural history have depended on pollen analysis. Vegetation changes in relation to human activities and climate change have been studied in many areas in northern Scandinavia (see, for instance, Engelmark, 1976;

Fig. 2.5. Scots pine with typical Sami bark-peeling scar. Photo by Lars Östlund.

Segerström, 1990; Emanuelsson, 2001). While these studies provide good examples of large-scale changes in vegetation and land use, they seldom give detailed information about individual forest stands or about how they were used. Recently, more refined methods have been promoted, including palaeoecological studies of 'closed canopy sites' (Bradshaw, 1988). Using these techniques it is possible to get a detailed record of a particular forest or forest stand (Bradshaw and Zackrisson, 1990; Hörnberg *et al.*, 1999). This approach is of particular interest in northern Sweden, where it is possible not only to use lake-sediments and peat, but also, due to the slow decomposition of organic matter, mor humus in the forests.

Historical records

In northern Sweden, forest history and agricultural history can be studied by the use of historical records dating back to the 13th

century. Some of the oldest records are the law-rolls from the different counties of Sweden, several of which were written in the period 1250–1350. The general law that applied to the coastal and sparse inland settlements in northern Sweden was 'Hälsingelagen' (Holmbäck and Wessén, 1979). Other medieval sources include deeds of gifts, ecclesiastical tax books, tax records (notably the tax record of 1413) and records of tithes. The records can be used to confirm the year of establishment and land area as well as the development of the villages in the coastal region of northern Sweden during medieval times (see, for instance, Lundholm, 1987). The creation of a strong centralized government during the 16th century, under the reign of Gustavus Vasa, led to administrative measures designed to register all landed property and to establish a system of individual taxation as opposed to the previous collective system. From that time the tax records give more detailed and coherent information about the agricultural activities in northern Sweden (see Myrdal and Söderberg, 1991). During the 17th century many new records were created, which are of great importance for the interpretation of the land-use history of agricultural land as well as forest land adjacent to the older villages in northern Sweden. In 1628 the National Land Survey Board came into being. One of the first tasks for this authority was to produce geometric maps of villages and parishes in the country (Peterson-Berger, 1928).

Most of the available records prior to the 19th century deal with the agriculturally developed coastal parts of northern Sweden, and only a few give information about the much more sparsely inhabited inland parts of this region. In the 17th and 18th centuries most of the inland region was used by Sami communities, which paid taxes for their exclusive use of the natural resources. The territory was organized into large taxation units, in Swedish 'lappskattelands', within which the families moved around on a yearly basis, utilizing different parts at different times. Usually the borders between the 'lands' were marked in the forest by blazed trees (Hultblad, 1968).

The 18th century was also a period of exploration of the less familiar northern parts of the country. Although people had travelled earlier in this part of the country and described their experiences, for example Olaus Magnus in 1555, Schefferus in the 1670s (Schefferus, 1956) and Rudbeck in the 1690s (von Sydow, 1968–1969), a new wave of travel accounts were written down and in some cases published during the 18th century. Most famous was of course the account by the young Linneaus travelling in the far northern parts in 1732 (von Platen and von Sydow, 1977). His description of the province of Lapland is of great value from an ecological perspective. Not only was he interested in botany, but also in different kinds of natural history as well as local customs and economic activities. The travels and work of Linnaeus stimulated followers and many important descriptions from different parts of northern Sweden were published in the 18th century.

While many important records emerged during the 18th century, the rapid economic development in the following century produced an immense number of detailed records. By the first decades of the 19th century many lumber companies were established in northern Sweden, these later became very large and have dominated the lumber market up to the present time. Especially important resources in the forest company archives are the forest surveys. The first surveys in northern Sweden were carried out during the late 19th century, usually as simple stem counts, including only the number and diameter of the large, valuable trees (Linder and Östlund, 1998). Later, smaller trees, dead trees and other variables were included in the surveys, and management plans were also produced. The exploitation and the following debate and concern over forest resources resulted, inter alia, in new forest legislation in 1903 (Stjernquist, 1973), more detailed company surveys and a national forest survey starting in 1923 (Nordström, 1959).

In order to understand the major changes that have taken place in the boreal forest landscape, it is possible to use maps and corresponding descriptions from the delineation process, which was carried out primarily during the 18th and 19th centuries. The purpose of the delineation was to confirm the property rights over the forest land of the Swedish state and the northern farmers. The maps and descriptions from the period following 1873 are valuable since the work was carried out extensively over the two northernmost counties, Västerbotten and Norrbotten, and because they were produced according to detailed specifications (Almquist, 1928). For these reasons, they provide interesting details of the forest landscape over large areas and allow regions which at that time were influenced by human activities to varying degrees to be compared. The detailed scale of the maps and the accompanying descriptions make it possible to study landscape elements as small as 0.5 ha. The information provided includes, inter alia, data on tree species composition, time since forest fires and grazing quality (which can be used to interpret the density of the forest).

Interpretation of Past and Present Landscapes – Spatial Patterns and Cognitive Aspects

The long-term relationship between people and forest ecosystems has created dynamic landscape patterns, which can be interpreted with appropriate tools. Different resources were utilized at different locations, with differing intensities and at different intervals in the landscape. This in turn created spatial patterns in the forest, affecting variables such as forest age structure, the openness of the forests and species composition. On a larger scale, patterns of 'cultural islands' in a forest matrix can be recognized (Ericsson, 2001). One such example from the southern boreal zone is the system of summer farms scattered in the forests around villages. The distance between a summer farm and the corresponding permanent settlement could exceed 50 km. This system facilitated the use of grazing resources over large areas and allowed surplus production. Each part of the forest landscape in such a system, where some areas were used for cattle grazing, others for winter-fodder production, and still others for hunting and fishing, generally exhibits typical characteristics. Fire was used specifically in grazing areas to improve grazing conditions, which in turn created large areas of open, pine-dominated forests in the areas between permanent settlements (Ericsson, 1997; Ericsson *et al.*, 2000). There were also gradients with increasing numbers of large, dead and old trees from the permanent settlements towards areas used less intensively (Östlund, 1993).

Semi-permanent settlements used by the native Sami in northern Sweden represent another example of cultural islands in a forest matrix. Each Sami family practising intensive reindeer herding in the forest area used 8–10 settlements during different parts of the year, each with specific grazing lands. A typical pattern around a settlement included a small area (covering a few hectares) that was almost deforested or covered by younger forest which was used for firewood (Östlund *et al.*, 2003). Within this area fences for penning reindeers were constructed, and within the corrals very large and old pines were saved in order to provide shade for the animals. In the next, exterior, zone the trees were generally older and some Scots pine trees were used as sources of inner bark, which was an important staple food for the Sami people (Zackrisson *et al.*, 2000). However, not all trees of a suitable age were used. Other factors appear to have influenced the number of trees being peeled for food in any given area. Dead trees were also characteristic features of this exterior forest. Old trees with and without the characteristic bark-peeling scars died of natural causes, but the Sami also intentionally killed some trees by girdling to increase the availability of firewood. The resulting exterior forest around Sami settlements, an uneven, sparse old-growth mixed coniferous forest, was thus partly a natural forest and partly influenced by people. At a further distance, trees were blazed to show borders around the settlement. Paths through the forest led to the next settlement a few or several kilometres

away. This pattern was repeated across the forest landscape, creating a landscape mosaic of more intensively used forest around settlements interspersed with lightly affected forest covering most of the land (Östlund *et al.*, 2002).

Yet another aspect of Sami land use that added to the environmental impact was their utilization of seasonal settlements. Each site represented a short period of habitation, covering a time-span of 1–4 weeks, and had limited effects on forest stands and grazing. However, there was a successive and cumulative environmental impact, leading eventually to the abandonment of sites. In general, access to firewood was the limiting factor setting the lifetime for each site. Consequently, a consistent feature of Sami settlement patterns is the succession of sites. At any given point in time the landscape encompassed settlement sites and site catchment areas in various stages of abandonment, decomposition and environmental recovery.

Settlement history, as represented by the concrete remains of dwelling huts, storage buildings and visible traces of land use, formed a key element in the cognition of landscape meaning and served as a medium for transmitting traditions and family history. Today, place names (if still remembered) reflect aspects of ancient ways of attaching importance to landscape elements. Forest history, and specifically culturally modified trees, provide other means of interpreting past cognitive landscapes, however fragmentary. The collection of inner bark was one of the decisive elements structuring time and space, and bark-peelings provide data on site location, the spatial extent of resource areas, social organization and seasonal logistics involved in the peeling. Quantitative estimates of exploitation pressure reflect the importance of inner bark in the regular diet and the weighing of landscape qualities with regard to access to inner bark. The size, form and direction of peeling scars reveal underlying norms and standards to which the Sami adhered in ancient times. Also, peeling was conducted with attention to ritual practice and strict behavioural rules, thereby expressing religious aspects of the relationship between people and nature (Bergman *et al.*, 2004).

The economic importance of pine inner bark, as well as the social and ideological implications of peeling, clearly illustrates the fact that affiliation to land cannot be defined purely by rational choices and foraging efficiency. Consequently, the interpretation of past and present landscapes should include a truly contextual and dialectic approach, recognizing landscapes as complex mosaics that exceed the sum of their parts.

Preservation and Protection of Cultural Landscapes in Northern Europe

The land use practised by farmers and native people shaped the forests and landscapes in characteristic ways for millennia up until the late 19th century. The methods used shifted, and the intensity of the land use grew successively over time, but many landscape elements were constant over long periods of time. During the 19th century a major transition in land use occurred. Large-scale industrial forest exploitation and, more recently, modern forest management was introduced across the entire region and carried out regardless of earlier land use. The first wave of forest exploitation targeted the oldest and largest Scots pine trees. In successive waves of the timber-frontier, smaller trees of all species were logged for use as wood pulp. This dramatic change in land use has erased earlier landscape patterns and specific traces of earlier land use. The landscape pattern in the north-western part of the boreal forest created by semi-permanent Sami settlements in a forest matrix has been totally eradicated over most of the forest landscape. Single-species forest stands in different age classes with no really old forest have replaced the old mixed coniferous forests with dead trees and many culturally modified trees. Mechanized soil scarification on clear-cuts has destroyed hearths that showed the locations of historic and pre-historic Sami settlements. The same

developments in other regions in the boreal forest have destroyed other cultural landscapes. However, some areas still retain the cultural heritage within the landscape. These include forest reserves and national parks, which are almost invariably protected for their natural qualities and for biodiversity reasons, rather than for their cultural values. Inaccessible forests at high elevations and small patches of old forest that have not yet been logged may also contain important traces of past land use.

Conclusion

For the future we believe that remaining traces of the cultural heritage in the forests must be protected from the impact of forestry operations. Such traces include archaeological remains, adjacent forest stands and culturally modified trees. The most important action to take is to educate people in the forestry sector and make them aware of the value of these resources. Modified forestry practices, with less impact, must also be developed. New laws protecting specific features, such as culturally modified trees, may also be needed. A second important measure is to include protection for valuable cultural sites together with nature conservation efforts. High cultural values in a forest add unique qualities, and make forest reserves more, not less, valuable. This idea must permeate future conservation work and requires new cooperation between different authorities.

Recent inter-disciplinary research into ancient land use in northern boreal landscapes stresses the importance of detailed case studies in attempts to attain a general understanding of the intrinsic and complex processes involved in landscape development. In analysing Sami landscape qualities, factors other than measurable environmental data have been taken into account. Cognitive aspects have also been considered, reinforcing the interpretative and explanatory frameworks. The results facilitate the development of new preservation strategies, and data on the specific characteristics of each investigated area can be immediately applied. Detailed case studies not only add to our general theoretical knowledge of landscape genesis, but also constitute methodological examples of preservation in practice. Such studies will provide tools for selecting landscapes for conservation, and provide methods for preserving and restoring landscape functions.

References

Almquist, J.E. (1928) Det norrländska avvittringsverket. In: Anon (ed.) *Svenska Lantmäteriet 1628–1928*. Stockholm, pp. 365–494.

Andersson, R. and Östlund, L. (2004) Spatial patterns, density changes and management implications for old trees in the boreal landscape of northern Sweden. *Biological Conservation* 118, 443–453.

Axelsson, A-L., Östlund, L. and Hellberg, E. (2002) Changing deciduous tree distributions in a Swedish boreal landscape, 1820–1999 – Implications for restoration strategies. *Landscape Ecology* 17(5), 403–418.

Baudou, E. (1992) *Norrlands forntid – ett historiskt perspektiv*. Förlags AB Wiken, Höganäs.

Bergman, I. (1991) Spatial structures in Saami cultural landscapes. *Readings in Saami history, culture and language II. Center for Arctic Cultural Research* 12, 59–68.

Bergman, I. (1995) Från Döudden till Varghalsen. En studie av kontinuitet och förändring inom ett fångstsamhälle i övre Norrlands inland, 5200 f. Kr. -400 e. Kr. *Studia Archaeologica Universitatis Umensis 7*.

Bergman, I. and Mulk, I.M. (1996) Det samiska landskapet. *Västerbotten* 3, 12–15.

Bergman, I., Påsse, T., Olofsson, A., Zackrisson, O., Hörnberg, G., Hellberg, E. and Bohlin, E. (2003) Isostatic land uplift and mesolithic landscapes: lake-tilting, a key to the discovery of mesolithic sites in the interior of northern Sweden. *Journal of Archaeological Science* 30(11), 1451–1458.

Bergman, I., Östlund, L. and Zackrisson, O. (2004) The use of plants as regular food sources in ancient subarctic economies – a case study based on the Sami use of Scots pine inner bark in northern Fennoscandia. *Arctic Anthropology* 41(1), 1–13.

Bradshaw, R.H.W. (1988) Spatially-precise studies of forest dynamics. In: Huntley, B. and Webb III, T. (eds) *Vegetation History*. Kluwer Academic Publishers, Dordrecht, Germany, pp. 725–751.

Bradshaw, R.H.W. and Zackrisson, O. (1990) A two thousand year history of a northern Swedish boreal forest stand. *Journal of Vegetation Science*, 1, 519–528.

Edgren, T. and Törnblom, L. (1993) *Finlands Historia 1*. Schilds, Ekenäs.

Emanuelsson, M. (2001) Settlement and land-use history in the central Swedish forest region. The use of pollen analysis in interdisciplinary studies. *Silvestria 223*. Swedish University of Agricultural Sciences.

Engelmark, R. (1976) The vegetation history of the Umeå area during the past 4000 years. *Early Norrland 9*.

Ericsson, T.S. (1997) *Alla vill beta men ingen vill bränna. Rapporter och uppsatser 8*. Institutionen for skoglig vegetationsekologi. SLU, Umeå, Sweden.

Ericsson, T.S. (2001) Culture within nature – key areas for interpreting forest history in boreal Sweden. *Silvestria 227*. Swedish University of Agricultural Sciences.

Ericsson, S., Östlund, L. and Axelsson, A.-L. (2000) A forest of grazing and logging: Deforestation and reforestation history in a central boreal Swedish landscape. *New Forests* 19(3), 227–240.

Ericsson, S., Östlund, L. and Andersson, R. (2003) Destroying a path to the past, culturally modified trees along Allmunvägen, N Sweden. *Silva Fennica*, 37(2), 283–298.

Graan, O. (1983) *Relation Eller En Fulkomblig Beskrifning om Lapparnas Vrsprung, så wähl som om heela dheras Lefwernes Förehållande. Berättelser om samerna i 1600-talets Sverige*. Kungl. Skytteanska Samfundets Handlingar, Nr 27, Umeå, Sweden.

Hellberg, E. (2004) Historical variability of deciduous trees and deciduous forest in northern Sweden. Effects of forest fires, land-use and climate. *Silvestria 308*. Swedish University of Agricultural Sciences.

Helmfrid, S. (1996) *The Geography of Sweden. National Atlas of Sweden*. Almqvist & Wiksell, Stockholm.

Högström, P. (1747) *Beskrifning öfwer de til Sweriges krona lydande Lapmarker*. Stockholm.

Holmbäck, Å., and Wessén, E. (1979) *Svenska landskapslagar. Södermannalagen och Hälsingelagen*. Almqvist & Wicksell, Uppsala, Sweden.

Hörnberg, G., Bohlin, E., Hellberg, E., Bergman, I., Zackrisson, O., Olofsson, A., Wallin, J.E. and Påsse, T. (2005) Effects of Mesolithic hunter-gatherers on local vegetation in a non-uniform glacio-isostatic land uplift area, northern Sweden. *Vegetation History and Archaeobotany* 15, 13–26.

Hörnberg, G., Östlund, L. and Zackrisson, O. (1999) The genesis of two Picea-Cladina forests in northern Sweden. *Journal of Ecology* 87, 800–814.

Hultblad, F. (1968) Övergång från nomadism till agrar bosättning i Jokkmokks socken. *Acta Lapponica* XIV. Almqvist & Wiksell/Gebers., Stockholm.

Johanssen, O.S. (1990) *Synspunkter på jernalderens jordbrukssamfunn i Nord-Norge*. Stencilserie B, nr. 29, Institutt for samfunnsvitenskap, Universitetet i Tromsø, Norway.

Kempe, G., Toet, H., Magnusson, P.-H. and Bergstedt, J. (1992) *Rikskogstaxeringen 1983–87 – Skogstillstånd, tillväxt och avverkning*. No. 51. Institutionen för skogstaxering, Sveriges lantbruksuniversitet, Umeå, Sweden.

Liedgren, L. (1992) Hus och gård i Hälsingland. *Studia Archaeologica Universitatis Umensis* 2. Umeå universitet, Umeå, Sweden.

Linder, P. and Östlund, L. (1998) Structural changes in three mid-boreal Swedish forest landscapes, 1885–1996. *Biological Conservation* 85, 9–19.

Lundholm, K. (1987) Staten blir till – ett nordskandinaviskt exempel. *Bebyggelsehistorisk Tidskrift*, 14, 137–150.

Mobley, C.M. and Eldridge, M. (1992) Culturally modified trees in the Pacific Northwest. *Arctic Anthropology* 29, 91–110.

Myrdal, J. and Söderberg, J. (1991) *Kontinuitetens dynamik. Agrar ekonomi i 1500-talets Sverige*. Almqvist & Wiksell International, Stockholm.

Niklasson, M. (1998) *Dendroecological Studies in Forest and Fire History*. Swedish University of Agricultural Sciences, Silvestria.

Niklasson, M. and Granström, A. (2000) Numbers and sizes of fires: Long-term spatially explicit fire history in a Swedish boreal forest landscape. *Ecology* 81, 1484–1499.

Niurenius, O.P. (1983) *Lappland eller beskrivning över den nordiska trakt, som lapparne bebo i de avlägsnaste delarne av Sskandien eller Sverge. Berättelser om samerna i 1600-talets Sverige*. Kungl. Skytteanska Samfundets Handlingar, Nr 27, Umeå, Sweden.

Nordström, L. (1959) Skogsskötselmetoder och skogslagstiftning. In: Arpi, G. (ed.) *Sveriges skogar under 100 år*. Kungl Domänstyrelsen, Stockholm, pp. 241–262.

Östlund, L. (1993) Exploitation and structural changes in the north Swedish forest 1800–1992. PhD thesis, Department of Forest Vegetation Ecology, Swedish University of Agricultural Sciences, Umeå, Sweden.

Östlund, L. and Lindersson H. (1995) A dendroecological study of the exploitation and transformation of a boreal forest stand. *Scandinavian Journal of Forest Research* 10, 56–64.

Östlund, L. and Zackrisson, O. (2000) The history of the boreal forest in Sweden – and the sources to prove it! In: Agnoletti, M. and Andersson, R. (eds) *Methods and Approaches in Forest History*. CAB International, Wallingford, UK.

Östlund, L., Zackrisson, O. and Hörnberg, G. (2002) Trees on the border between nature and culture – Culturally modified trees in boreal Scandinavia. *Environmental History* 7(1), 48–68.

Östlund, L., Ericsson, S., Zackrisson, O. and Andersson, R. (2003) Traces of past Saami forest use – an ecological study of culturally modified trees and earlier land-use within a boreal forest reserve. *Scandinavian Journal of Forest Research* 18, 78–89.

Östlund, L., Bergman, I. and Zackrisson, O. (2004) Trees for food – a 3000 year record of subarctic plant use. *Antiquity* 78, 278–286.

Peterson-Berger, E. (1928) Lantmäteriets kartografiska verksamhet. In: *Svenska Lantmäteriet 1628–1928*. Stockholm, pp. 259–297.

Rheen, S. (1983) *En kortt Relation om Lapparnes Lefwerne och Sedher, wijdSkiepellsser sampt i många Stycken Grofwe wildfarellsser. Berättelser om samerna i 1600-talets Sverige*. Kungl. Skytteanska Samfundets Handlingar, Nr 27, Umeå, Sweden.

Schefferus, J. (1956) Lappland. *Acta Lapponica VIII. Nordiska Museet*. Almqvist & Wiksells, Uppsala, Sweden.

Segerström, U. (1990) The natural holocene vegetation development and the introduction of agriculture in northern Norrland, Sweden. Studies of soil, peat and especially varved lake sediments. PhD thesis, University of Umeå, Sweden.

Stjernquist, P. (1973) *Laws in the Forest*. Blom, Lund, Sweden.

Tornaeus, J. (1983) *Berättelse om Lapmarckerna och Deras Tillstånd. Berättelser om samerna i 1600-talets Sverige*. Kungl. Skytteanska Samfundets Handlingar, Nr 27, Umeå, Sweden.

Von Platen, M. and Von Sydow, C.-O. (eds) (1977) *Carl Linnaeus Lapplandsresa år 1732*. Wahlström & Widstrand, Stockholm.

Von Sydow, C.-O. (1968–1969). *Rudbeck d.y.:s dagbok från Lapplandsresan:* Med inledning och anmärkningar I. Svenska Linnésällskapets årsskrift. Almqvist & Wicksell, Uppsala, Sweden.

Zackrisson, O., Östlund, L., Korhonen, O. and Bergman, I. (2000) Ancient use of Scots pine innerbark by Saami in N. Sweden related to cultural and ecological factors. *Journal of Vegetation History and Archaeobotany* 9, 99–109.

3 Energy Balance and Land Use: the Making of an Agrarian Landscape from the Vantage Point of Social Metabolism (the Catalan Vallès County in 1860/1870)[1]

E. Tello,[a] R. Garrabou[b] and X. Cussó[b]

[a]Departament d'Història i Institucions Ecònomiques, Facultat de Ciences Econòmiques i Empresarials, Universitat de Barcelona, Barcelona, Spain, and [b]Departament d'Economia i d'Història Econòmica, Universitat Autònoma de Barcelona, Bellaterra, Spain

Landscape can be seen as the territorial expression of the metabolism that any given society maintains with the natural systems sustaining it. One way of understanding when and why the human shape of the territory changes consists of analysing the path of social metabolism that leaves its ecological footprint on its surroundings. Approaching this analysis of the exchange of society's energy, materials or waste with its sustaining sources requires, at the same time, a broadening of the window used for observing reality. And this broadening of our field of vision can only be achieved through a trans-disciplinary dialogue among different areas of knowledge within the social and natural sciences, all of which are capable of adopting a common historical perspective.

Social Metabolism

Karl Marx was the first to introduce the concept of social metabolism in the realm of economics and history. Based on the notion of metabolic exchange developed in his day by the field of biology, Marx characterized human labour as the intentional modulation of that metabolism, and in one of the few occasions on which he categorically specified what he viewed as socialism, he defined it as the conscious organization of an exchange between human beings and nature 'in a form conducive to full human development' (Marx, 1976 [1867], p. 141). Nevertheless, just as Joan Martínez Alier has explained, Marx and Engels rejected the proposition set forth by Sergei Podolinsky to analyse the social metabolism in an operational way via the calculation of energy flows (Martínez Alier and Schlüpmann, 1991; Martínez Alier, 1995; Fischer-Kowalski, 1998). On the one hand, the theory of value-labour polemically tied them to the liberal economists of the time. On the other hand, the rigid Hegelian dialectic schema led them to blindly believe in the 'growth of productive forces' as the fulcrum of social change. Marx believed in an inexorable historical process: 'the destruction of the purely spontaneous original conditions of that exchange between human beings and nature' (see

Sacristán, 1992; Foster, 2000). This short-circuited consideration of the environmental question in the 20th century Marxist traditions, while the ecological leanings of other authors such as Herbert Spencer, Stanley Jevons, Wilhelm Ostwald, Leopold Pfaundler, Eduard Sacher, Patrick Geddes and Frederick Soddy suffered the same fate in the mainstream of economic thinking.

The need to confront the social–environmental crisis of our time has enabled the concept of social metabolism to be rescued. The development of economic ecology, based on the work by Nicholas Georgescu-Roegen, has given it a new energy and materials to account for the biophysical flows of human societies (Georgescu-Roegen, 1971; Fischer-Kowalski and Hüttler, 1999; Martínez Alier and Roca Jusmet, 2000; Haberl, 2001a,b). This emerging approach revives the task initiated years ago by pioneering authors such as David Pimentel, Gerald Leach, Vaclav Smil, José Manuel Naredo, Pablo Campos and Mario Giampietro, who constructed the energy balances of diverse agrarian systems (Pimentel and Pimentel, 1979; Campos and Naredo, 1980a, b; Leach, 1981; Giampietro and Pimentel, 1991; Giampietro et al., 1994; Smil, 1994, 2001, 2003). Secondly, ecological economics research is generating specific alternatives to the exclusive use of macro-economic indicators used in national accounting via the development of a parallel system of national biophysical accounts. Thirdly, William Rees and Mathis Wackernagel have proposed the territorial translation of the most significant headings of those biophysical flows via the estimate of their *ecological footprint* (Rees and Wackernagel, 1996a,b; Fernández, 1999; Costanza et al., 2000; Haberl et al., 2001; Carpintero, 2002, 2005).

Ecological Footprints: From Local to Global

This opens up a very interesting bridge spanning the study of socio-ecological flows and the evolution of the territory, on both a local and global scale.[2] Combining the different approaches and their respective methods or tools, we can relate the geographical and historical study of the landscape with the analysis of the path of the social metabolism that has led to the replacement of multiple local ecological footprints, imprinted on the territory by the land requirements corresponding to each particular way of using the resources, by an ever more global, uniform ecological footprint which is further removed from the perceptions of those who originate it (Norgaard, 1997; Fischer-Kowalski and Amann, 2001).

Our research project aims to study the transformation of the agricultural landscapes in the northwest Mediterranean, and lies precisely on this bridge spanning two major avenues of research. First, it is inspired by the famous inter-disciplinary symposia held in 1955 and 1987 on the transformation of the Earth by human actions, which have paved the way in recent years for the international project, Land Cover – Land Use Change (Thomas et al., 1956; Turner, 1990, 1995; Boada and Saurí, 2002). Second, and coinciding with a majority of the innovative approaches emerging from the Department of Social Ecology at the Institute for Interdisciplinary Studies at the University of Vienna (IFF), we believe that in the long term, the driving forces of change in land uses obey the transformations experienced by the social metabolism of human activity with the natural environment. Following the approach spearheaded by the Austrian researchers from the IFF, a methodological key to understanding the evolution of the territory lies in a cross between the analysis of the energy and material flows which underlie a certain pattern of consumption, the forms of land use that shape the landscape, and the balances of the use of time or the working capacity of the population that consumes those products and inhabits the same land in order to meet its needs (Fischer-Kowalski, 1998; Fischer-Kowalski and Hüttler, 1999; Haberl, 2001a, b; Krausmann, 2001; Schandl and Schultz, 2002).

We start from the very simply formulation that is similar to the working method of agrarian economic historians: the *land requirement by unit of product and inhabitant*. Using the conventional yields or productivity rates, we can estimate how much agricultural, forested or livestock grazing land was needed to obtain each unit of food and energy consumption. Then, comparing the territorial capacity that was truly available with the historical increase of land requirements, we can identify the situations or moments of rupture which led different human societies to change the shape of the cultural landscapes they had inherited. This dynamic approach implies taking into account the different aptitudes of the soils, the social norms of territorial management, property rights and other regulations on land access, and the availability of networks for more far-reaching trade exchanges, in order to identify those moments of crisis and transformation that led people to modify agricultural uses through the activation of human labour using the range of tools and knowledge at their disposal.

This places some of the main issues traditionally analysed by economic historians – such as the role of the demographic dynamic, technological change and market networks – into a broader frame of reference which enables the corresponding biophysical flows, as well as their ecological footprint, to be included in the analysis.

Case Study: the Vallès Oriental County in Catalonia (Spain) in 1860/1870

The first study area where we have applied this methodology is located in the region known as Vallès Oriental, a small plain situated in a tectonic basin between Catalonia's littoral and pre-littoral mountain ranges, whose diversity of geological substrata, together with above-average rainfall for the Mediterranean, have led to the development of a considerable variety of soils with a greater range of agricultural possibilities than in more arid areas (Rodríguez Valle, 2002). The proximity to Barcelona – between 5 and 12 h on horseback according to a timetable map from 1808–1809 – meant that the Vallès was connected very early on with the commercial dynamics of Catalonia's demographic and urban centre of gravity (Vilar, 1962). Starting from the end of the late medieval agrarian struggles, the landowners who held the poly-cultural agricultural farms that were characteristic of the mid-northeast of Catalonia organized themselves into compact units around a rural dwelling (called *masies*), and gradually gained control of the access rights to the main resources through a complex and conflict-ridden transition from feudalism to agrarian capitalism (Serra, 1988; Garrabou and Tello, 2004).

Figure 3.1 shows the distribution of the soil textures described in the *Estudio Agrícola del Vallès*, an anonymously written manuscript submitted in 1874 to the Barcelona Economic Society of Friends of the Country, which is the main source of our analysis (Garrabou and Planas, 1998).

The *Estudio* established seven kinds of soil characterized by their texture (Table 3.1):

1. Those composed of slate and shale, situated in mountainous areas and slopes generally covered in woodland or scrub.
2. Clayey-calcareous, with a certain proportion of limestone and fine matter, suitable for cereals.
3. Clayey-sandy, with a large proportion of clay, very heavy to work with due to its compacted nature and water retention, but which could be made into good fertile land with a certain amount of investment in fertilizer and labour.
4. Compact calcareous, situated on steeply sloping land and only exploitable for forestry or for planting vines with a great deal of terracing work.
5. Sandy-calcareous, thin soil on slopes located in areas where the tectonic basin had contact with calcareous outcrops, also exploited for vines and forestry.
6. Those described as sandy-clayey, made up of sands and flood silts which give them good drainage and good water-retaining capacity, easy to work and very fertile for growing cereals or legumes.

1

LEGEND

study area
farmyard
woodland
wood with chestnut
hemp field with trees and walnuts
hemp field and walnuts
hemp field and trees
house
house and farmyard
chestnut grove
chestnut grove and shrubs
church
rocks
arable with trees
arable with chestnut
arable with fruit trees
arable with mulberry trees
arable with walnuts and trees
arable with walnuts and chestnuts
arable and shrubs
arable and best quality chestnuts
arable and walnuts
arable and pasture
arable and meadow
arable and meadow with vines
arable with vines
arable with vines and walnuts
arable with vines and chestnut grove
arable with vines and fruit trees
shrubland
mill
olive grove
vegetable garden
vegetable garden and walnuts
village
pasture

pasture with trees
pasture with chestnuts
pasture with chestnuts and shrubs
pasture with chestnuts and walnuts
pasture with rocks
pasture with walnuts
pasture with vines
pasture and woodland
pasture and hemp fields
pasture and stone quarry
pasture and bundles
pasture and shrubs
pasture with olive trees
pasture, meadow and hemp fields
meadow
meadow hemp fields and walnuts
meadow with chestnuts
meadow and hemp fields
meadow and mulberry trees
meadow and shrubs
meadow and walnuts
meadow and pasture
meadow and vines
meadow with vines
heath
fallow
fallow and chestnuts
fallow and walnuts
hydric surface

Plate 1. The 3D image shows the land use of 1832 in the study area of Cardoso, in the Regional Park of the Apuane Mountains in Tuscany. In this year there are 65 land uses in an area of 1000 ha, forming a mosaic of 618 patches. The green colour refers to woodlands, the orange to pastures, the pink to meadows and the yellow to arable land. The violet colour represents vineyards. All the slopes are terraced as shown in the photograph of the area in 1900. (See Chapter 1.)

2

LEGEND:
- Study area boundary
- rock outcrops
- anthropic area
- shrubland
- area with sparse vegetation
- wood with prevalence of hornbeam
- wood with prevalence of chestnut
- wood with prevalence of beech
- mixed wood
- mixed wood with chestnut
- quarry
- terraced cultivations
- ex-terraced cultivations
- landslide
- shrubland
- pasture
- wooded pasture
- meadow
- riparian

Plate 2. The 3D image shows the land use of 2002. In this year there are only 18 land uses, forming a mosaic of only 84 patches. Woodlands have more than doubled their extension, replacing the former mosaic of land uses. The red areas show landslides occurring on the abandoned terraced chestnut orchards. (See Chapter 1.)

3a

3b

Plate 3. The obliteration of a traditional landscape operated by natural forestation. Image (a) shows a portion of a territory managed under the share-crop system, with small-scale production involving woodlands, cultivated fields and pastures. We can note chestnut orchards, pastured woods, wood pastures, terraces, fields, pollard trees and dry stone walls. Image (b) shows the same landscape after several decades of total abandonment; the complexity due to cultural influence has disappeared. (See Chapter 1.)

4

Plate 4. Fragmented landscape mosaics, made of many patches linked to small-scale production and mixed cultivation are becoming quite rare. The diffusion of large monocultures and the extension of woodland on abandoned fields and pastures are removing them from the landscape. (See Chapter 1.)

5

Plate 5. Ortophoto of the year 2000 on a DEM showing the boundaries of the Moscheta study area. (See Chapter 5.)

6

Plate 6. View of the area of the Moscheta project from the slopes of Monte Fellone facing west. (See Chapter 5.)

7

Plate 7. Afforestation near the abbey of Moscheta. It is evident that this plantation is not integrated into the landscape of the area, even from an aesthetic point of view. (See Chapter 5.)

8

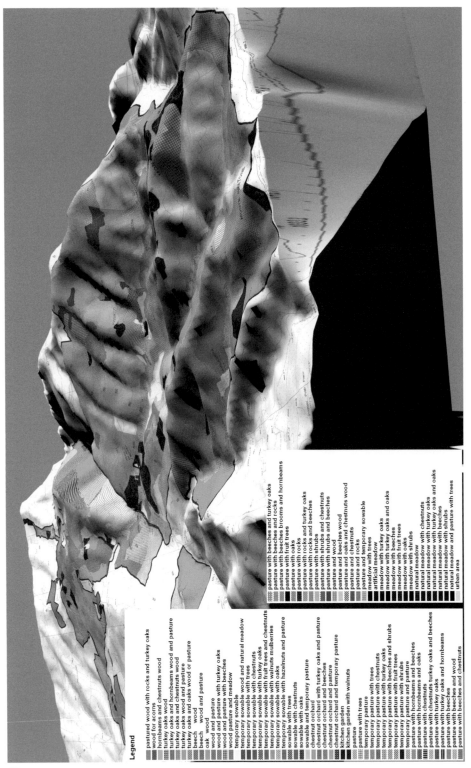

Legend

pastured wood with rocks and turkey oaks
wood for acorns
hornbeams and chestnuts wood
turkey oaks wood
turkey oaks and hornbeams wood and pasture
turkey oaks and chestnuts wood
turkey oaks wood and pasture
turkey oaks and oaks wood or pasture
beech wood
beech wood and pasture
oak wood
wood and pasture
wood and pasture with turkey oaks
wood and pasture with beeches
wood pasture and meadow
temporary sowable
temporary sowable wood and natural meadow
temporary sowable with trees
temporary sowable with chestnuts
temporary sowable with oaks
temporary sowable with hazelnuts and pasture
temporary sowable with fruit trees and chestnuts
temporary sowable with walnuts mulberries
temporary sowable with turkey oaks
sowable with oaks
sowable with chestnuts
sowable with oaks
sowable and temporary pasture
chestnut orchard
chestnut orchard with turkey oaks and pasture
chestnut orchard and beeches
chestnut orchard and pasture
chestnut orchard and temporary pasture
kitchen garden
kitchen garden with walnuts
pasture
pasture with trees
temporary pasture
temporary pasture with trees
temporary pasture with chestnuts
temporary pasture with turkey oaks
temporary pasture with beeches and shrubs
temporary pasture with fruit trees
temporary pasture with shrubs
temporary pasture and pasture
pasture with hornbeams and beeches
pasture with hornbeams and oaks
pasture with chestnuts
pasture with chestnuts turkey oaks and beeches
pasture with turkey oaks
pasture with turkey oaks and hornbeams
pasture with beeches
pasture with beeches and wood
pasture with beeches and chestnuts

pasture with beeches and turkey oaks
pasture with beeches and rocks
pasture with beeches brooms and hornbeams
pasture with fruit trees
pasture with oaks
pasture with rocks and turkey oaks
pasture with rocks and beeches
pasture with shrubs
pasture with shrubs and chestnuts
pasture with shrubs and beeches
pasture and wood
pasture and beeches wood
pasture and oaks and chestnuts wood
pasture and chestnuts
pasture and rocks
pasture and temporary sowable
meadow with trees
artificial meadow
meadow with turkey oaks
meadow with turkey oaks and oaks
meadow with beeches
meadow with fruit trees
meadow with oaks
meadow with shrubs
natural meadow
natural meadow with chestnuts
natural meadow with turkey oaks
natural meadow with turkey oaks and oaks
natural meadow with beeches
natural meadow with shrubs
natural meadow and pasture with trees
urban area

Plate 8. 3D map of the land use of Moscheta in the year 1832. (See Chapter 5.)

9

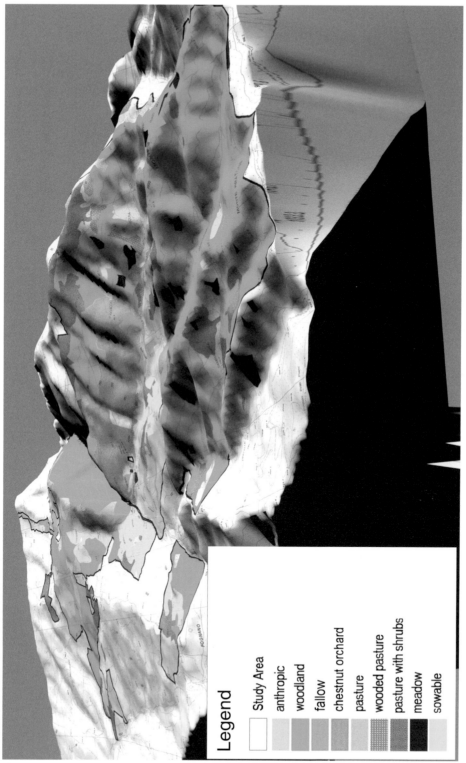

Legend

Study Area
anthropic
woodland
fallow
chestnut orchard
pasture
wooded pasture
pasture with shrubs
meadow
sowable

Plate 9. 3D map of the land use of Moscheta in the year 1954. (See Chapter 5.)

10

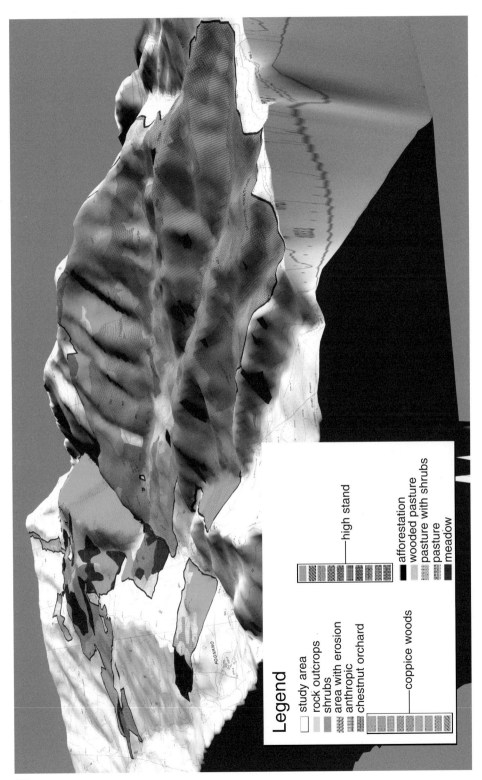

Legend

- study area
- rock outcrops
- shrubs
- area with erosion
- anthropic
- chestnut orchard

coppice woods

- high stand

- afforestation
- wooded pasture
- pasture with shrubs
- pasture
- meadow

Plate 10. 3D map of the land use of Moscheta in the year 2000. (See Chapter 5.)

11

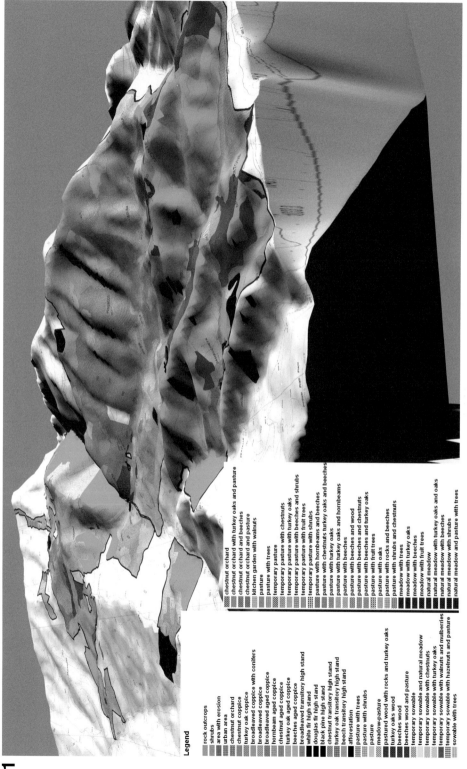

Legend

rock outcrops
shrubs
area with erosion
urban area
chestnut orchard
chestnut coppice
turkey oak coppice
broadleaved coppice with conifers
broadleaved coppice
broadleaved aged coppice
hornbeam aged coppice
chestnut aged coppice
turkey oak aged coppice
beeches aged coppice
broadleaved transitory high stand
white fir high stand
black pine high stand
douglas fir high stand
chestnut transitory high stand
turkey oak transitory high stand
beech transitory high stand
afforestation
pasture with trees
pasture with shrubs
pasture
meadow-pasture
pastured wood with rocks and turkey oaks
turkey oaks wood
beeches wood
beeches wood and pasture
temporary sowable
temporary sowable and natural meadow
temporary sowable with chestnuts
temporary sowable with turkey oaks
temporary sowable with walnuts and mulberries
temporary sowable with hazelnuts and pasture
sowable with trees

chestnut orchard
chestnut orchard with turkey oaks and pasture
chestnut orchard and beeches
chestnut orchard and pasture
kitchen garden with walnuts
pasture
pasture with trees
temporary pasture
temporary pasture with chestnuts
temporary pasture with turkey oaks
temporary pasture with beeches and shrubs
temporary pasture with fruit trees
temporary pasture with shrubs
pasture with hornbeams and beeches
pasture with chestnuts turkey oaks and beeches
pasture with turkey oaks
pasture with turkey oaks and hornbeams
pasture with beeches
pasture with beeches and wood
pasture with beeches and chestnuts
pasture with beeches and turkey oaks
pasture with fruit trees
pasture with oaks
pasture with rocks and beeches
pasture with shrubs and chestnuts
meadow with trees
meadow with turkey oaks
meadow with beeches
meadow with fruit trees
natural meadow
natural meadow with turkey oaks and oaks
natural meadow with beeches
natural meadow with shrubs
natural meadow and pasture with trees

Plate 11. 3D map showing the future land use of Moscheta after the restoration project. (See Chapter 5.)

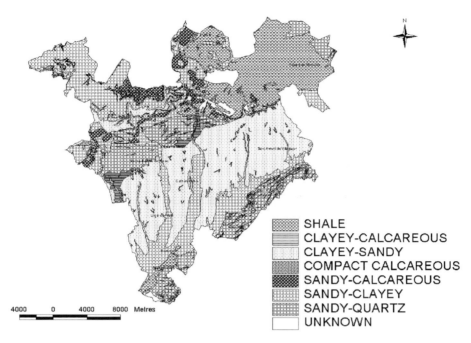

4000   0   4000   8000 Metres

Fig. 3.1. Area of study and main types of soil. Source: Rodríguez Valle (2002).

7. Sandy-quartz, formed from the decomposition of granite which emerges along both sides of the fault, shallow and with a low water-retaining capacity, largely devoted to vines or forestry exploitation.

The 'unknown' category corresponds to areas with such a variety of outcrops that it proves difficult to fit them into the 1874 classification, located in areas that are scarcely or not at all cultivatable.

The *Estudio Agrícola del Vallès* reproduced the summary of the use of land and livestock for each municipality in the 1860–1865 tax assessments (Tables 3.2, 3.4 and 3.5). From this, and by assigning each crop rotation to the type of soil, we have estimated the total production by applying the average yields per hectare. Although some local variations can be seen, for the overall study area the surface area estimated by the 1860s assessments varies little

Table 3.1. Classification of soils and types of exploitation.

Type of texture	Ha	%	Main soil use around 1860–1870
Shale	11,396.7	19.1	Woodland, scrub or pasture
Clayey-calcareous	1,000.7	1.7	Cereals and fodder (without fallowing)
Clayey-sandy	17,199.4	28.9	Cereals and legumes (without fallowing)
Compact calcareous	309.2	0.5	Vines or scrub and pasture
Sandy-calcareous	2,864.4	4.8	Vines or woodland, scrub and pasture
Sandy-clayey	16,379.3	27.5	Legumes, potatoes, cereals, fodder
Sandy-quartz	8,636.8	14.5	Vines or woodland, scrub or pasture
Unknown	1,813.4	3.0	(Woodland, scrub or pasture)
Total	59,600.0	100.0	

Source: Garrabou and Planas ([1874]1998); Rodríguez Valle (2002).

from the 59,600 officially registered at present: 52,020 ha, 12.7% less. A large part of the difference lay in mountainous municipalities and corresponded to extensive barren or forested areas. The reliability of the agricultural surface area measurement seems to be reasonably good.[3]

Subtracting the 3404 ha declared to be unproductive – some 6–7% of the total expanse used – our analysis will refer to the 48,616 ha that in around 1860 made up the useful agrarian area, with 36% of it arable. The then-cultivated area of 18,476 ha was substantially more than what is currently under cultivation. Once areas given over to housing development have been subtracted, and especially taking into account the abandonment of the countryside, which has left a great deal of land which is undergoing an extensive process of reforestation unexploited, the useful agricultural land area registered in the last agricultural census from 1999 comes to less than 30,000 ha, less than 10,000 of which are cultivated.

Between 1718 and 1860, the population multiplied by a factor of 2.6, going from 25 to 64 inhabitants/km². The available land area was reduced from 4 to 1.6 ha/inhabitant, and the cultivated land from 1.23 to 0.5 ha/inhabitant (Table 3.2). In an organic-based economy in which the people still lived mainly from the land, satisfying their basic needs obliged them to increase the land use to a maximum. At the same time as agricultural, livestock and forestry exploitation reached their maximum extent, the lower availability of land per inhabitant required an increase in the amount of energy from each land unit directed towards human use. This challenge led to the development of a specific Mediterranean form of advanced organic agriculture. It is significant that the average population density between 1860 and 1900 remained at the same level of 63 inhabitants/km², despite the existence of significant local variations. Mountain areas became depopulated as their role as suppliers of timber and firewood was reduced, while the population came to be concentrated in the lowland industrial centres.

Land Cover and Land Use

As a result of demographic growth and the intensification of exchanges, there was transference of woodland and pasture land into cultivated land, particularly vineyards. According to the assessments from the mid-19th century, the 48,616 ha of agricultural land used were distributed in quite similar proportions between wooded areas and cultivated land, with a lesser area of uncultivated land set aside for pastures.

Fifty-five percent of the cultivated area was being used for cereals and legumes, with 28% for vines, 4% for olive trees and other fruit trees, and 13% for vegetable gardens and irrigation. The 2300 irrigated hectares represented little more than one-ninth of the total area, yet they corresponded to one-fifth of the *available*

Table 3.2. Demographic growth and land available.

	Inhabitants	Population density inhab./km²	Area available per inhabitant (ha/inhab.)	
			Total	Cultivatable
1718	14,993	25.16	3.98	1.23
1787	20,051	33.64	2.97	0.92
1860	38,342	64.41	1.55	0.48
1900	38,390	63.30	1.55	0.48
1930	56,111	94.15	1.06	0.33
1960	77,039	129.26	0.77	0.24

Source: Authors' own table, taken from data in published censuses and from www.idescat.es. The estimate for *cultivatable* land area is based on that actually cultivated in 1860–1870.

Table 3.3. Land use and available land.

Uses	Land classification in hectares			Total existing hectares	%	Total available hectares
	1st class	2nd class	3st class			
Crops	2,824.12	5,427.51	10,312.79	18,475.54	35.5	25,392.65
Woodland	2,586.63	5,156.20	10,229.60	18,564.43	35.7	18,564.43
Pasture	2,743.68	3,843.24	4,989.07	1,575.99	22.3	11,575.99
Unproductive	–	–	–	3,404.35	6.5	–
Total	8,154.43	14,426.95	25,531.46	52,020.31	100.0	55,533.07

Source: 'Resumen del Estado demostrativo de la riqueza rústica del Vallés', Garrabou and Planas (1998).

cultivated area, taking into account multiple annual crops. According to the 1874 *Estudio* and the *Cartillas Evaluatorias* (Evaluation Handbooks) published by the Catalan Agricultural Institute of San Isidro (IACSI), the practice of leaving land fallow had practically disappeared, and the predominant rotations in the non-irrigated areas combined wheat and maize with runner beans, broad beans and peas, fodder or potatoes.[4] On irrigated lands, two or three crops were obtained annually, alternating between growing hemp and legumes or potatoes with wheat, other late-ripening fruit and fodder. On the best non-irrigated lands, wheat was sown, except in the very cold areas at the foot of the Montseny mountain, where it was replaced by rye, while on the worst lands, various mixed bread-making wheats were obtained. Average cereal yields fluctuated between 10 and 18 hectolitres per ha, thus multiplying the seed by between 5.5 and 10 times. The increasing importance of vines should be stressed, with their presence continuing to rise until the grapevine fever of the final decades of the 19th century, spurred on by the high wine prices caused by French vineyards being affected by the phylloxera plague.

Taking into account the agronomic capacities evaluated in Table 3.3, we have estimated the area where more than one annual crop was obtained as 15% of the cultivated land. Given these intensive crop rotations without fallowing, and the fact that in irrigated areas or a certain percentage of non-irrigated areas several annual crops were obtained, the cultivated area that was truly available came to around 25,000 ha. With a population of 38,342 inhabitants in 1860, the useful agricultural area fluctuated between 1.27 ha/inhabitant, not counting double crops, and 1.36 ha/inhabitant if these are included. If we take current measurements and do not discount unproductive lands, this level would rise to 1.5 ha/inhabitant. The land area cultivated per person in around 1860 was between 0.48 and 0.65 ha, depending on whether we take into account double crops or not.

These are values that were lower than the 2 ha/inhabitant estimated by Paolo Malanima for more densely populated parts of Europe towards the middle of the 18th century, or the 2.4 that Fridolin Krausmann calculates for the whole of Austria in 1830–1850. While the woodland area per inhabitant was still close to Malanima's levels, the land set aside for pasture was much less (0.3 ha against a range of between 0.5 and 1 ha per inhabitant in Western Europe, and 0.7 in Austria in around 1830–1850). The amount of available agricultural land was also less (0.5–0.7 in the Vallès, against 0.8–1 ha per person in Europe on the eve of the Industrial Revolution).[5]

The highest energy output per hectare (Table 3.4) was obtained in the forested area, although 17% of woodland production was leaves, grasses or woodland undergrowth species that were of limited human use. Timber, firewood and other human extractions came to 28.7 GJ/ha of woodland, slightly below the agricultural 29.9 GJ/ha. Adding other extractions, the overall forest final output amounted to 29 GJ/ha and accounted for 42.6% of the primary

Table 3.4. Area used and energy value of the crops, woodland and pasture.

	Area available (ha)	Total production (kg)	Energy value of the product (TJ)	Output per ha/year (GJ)	Output per inhabitant/year (GJ)
Crops	18,475.54	75,297,024	551.9	29.9	14.4
Woods	18,564.43	56,103,978	638.5	34.4	16.7
Pasture	11,575.99	20,107,541	72.6	6.3	1.9
Total	48,615.96	151,508,543	1,263.0	26.0	32.9

Sources: Cussó *et al.* (2006). We have used the yields in the 1874 *Estudio* and the 1879 IACSI *Evaluation Handbooks*, completed with the data on pasture land and woodland from Naredo and Campos (1980a, b) and the *Ecological and Forestry Inventory of Catalonia* (CREAF, 2000). The caloric transformation was performed from Mataix Verdú (2003⁴), Moreiras-Varela *et al.* (1997³), and Naredo and Campos (1980a, b). Conversion: 1 Kcal = 4186.8 J; 1 MJ = 10^6 J; 1 GJ = 10^9 J; 1 TJ = 10^{12} J. An average human being needed an energy intake of 3.5 GJ/year.

agro-forestry conversion. A comparison with some initial estimates put forward by classic authors such as Sacher and Podolinsky leads to certain divergences, yet our data is consistent with that obtained by Pablo Campos and José Manuel Naredo, Fridolin Krausmann, and Manuel González de Molina (Campos and Naredo, 1980a, González de Molina *et al.*, 2002; Krausmann, in press).[6]

Need, and also the opportunities arising from the intensification of exchanges, led to the development of a specific Mediterranean form of advanced organic economy (Wrigley, 1988, 2004) between 1860 and 1900. Taking into account the fact that the Vallès Oriental was a rural area close to Catalonia's main urban system, it may be surprising that the final agricultural production was not greater; but our results are consistent with the agricultural statistics for production and consumption from the Granollers administrative area in 1862–1864, which endorse the existence of a small deficit of cereals compensated for by wine exportation (Planas, 2003). In caloric terms, the high water content of vineyard production involved a low energy production per surface area: only 4.4 GJ/ha, against 11.5 for cereals or 12.6 for potatoes (Table 3.5). Nevertheless, the higher relative prices for wine allowed the acquisition of a greater caloric equivalent through imported cereals.

Vines were planted in poor soil, and except at the time of initial planting, no manure was applied. A partial wine-growing specialization allowed cultivators to concentrate manure on the better land devoted to vegetable gardens, cereals, legumes or hemp. Vineyard pruning and green shoots even went towards fertilizing other crops, either directly as compost or indirectly as fodder. These were responses to the challenge of feeding a population that had doubled between 1787 and 1860 using an organically based intensive agriculture on a land subject to the water restrictions typical of the Mediterranean environment, where keeping livestock and obtaining fertilizer became severely limiting factors (González de Molina, 2001a, b). Market insertion allowed more to be made of the area's ecological possibilities, and paved the way for a reduction in the bottleneck when obtaining organic fertilizers.

Energy Balance and Land Use

The energy balance described in Fig. 3.2 allows us to note the intensive, and at the same time integrated, nature of the agro-ecosystem (Cussó *et al.*, 2006a,b). The integration between cropping, livestock breeding and forestry explains why the caloric equivalent of the final agrarian product still represented 59% of the solar energy fixed by photosynthesis, although losses from livestock conversion and reuse consume 41%.

In 1950–1951, the final Spanish agrarian production only accounted for 38% of

Table 3.5. Land area, product and energy value of the main crop component.

	Area available (ha)	Physical yield (kg or l/ha)	Total production (kg or l/year)	Unit energy value (MJ/kg)	Energy value of the total production		
					TJ/year	GJ/ha	GJ/inhab/year
Cereals (kg)	8,082.81	925.79	7,482,984.7	12.46	93.2	11.5	2.4
Legumes (kg)	5,401.84	662.09	3,576,504.2	12.96	46.3	8.6	1.2
Wine (l)	4,996.89	1,345.21	6,721,866.4	3.27	22.0	4.4	0.6
Oil (l)	771.12	195.90	151,062.4	37.01	5.6	7.3	0.1
Potatoes (kg)	1,429.52	4,000.00	5,718,080.0	3.14	18.0	12.6	0.5
Fodder (kg)	3,289.59	3,478.32	11,442,246.7	3.55	40.6	12.3	1.1
Others (kg)	1,970.81	—	—	—	27.8	14.1	0.7
Total	25,392.65	—	—	—	253.5	13.7	6.6
Seeds					25.8	1.4	0.7
Fodder					40.6	12.3	1.1
Final product for human consumption					187.1	10.1	4.9

Source: Cussó *et al.* (2006a)

the energy contained in the biomass, while the other 62% went toward reuses and transformation losses. In 1977–1978, in spite of the replacement of animal traction and manure by tractors and chemical products, 68% of the primary agro-forestry conversion continued to be devoted to livestock conversion. The energy value of the final agrarian product accounted for 47% of the primary output, due in reality to imported foodstuffs whose caloric total was equivalent to 11% of the solar energy fixed by photosynthesis in Spain (Naredo, 1996; Simón Fernández, 1999; Carpintero, 2005; Carpintero and Naredo, in press).

From Social Metabolism to the Making of an Agrarian Landscape

The analysis of social metabolism aids us in understanding the anatomy of the landscape by revealing its structure. However, based on that we cannot deduce the functional logic of its various parts or elements – what we could refer to as its physiology – nor the appearance of its specific availability in the territory (its phenology). Cultural landscape is always a social construction. In order to understand it beyond its basic functional shape, any explanation must also include the role of institutional factors,

the economic logic behind the functioning of its builders, and the resolution of social conflicts related to access rights to resources (Tello, 1999; Garrabou and Tello, 2004). As an example, we shall see the GIS-based reconstruction of the cadastral plot map dating from 1853 from the municipality of Caldes de Montbui (Fig. 3.3).

We have also evaluated through GIS the suitability of land or soil characteristics for the main agricultural uses (grain, fodder, vineyards, olive and almond orchards) considering the then-prevailing different land uses and agrarian techniques in the Vallès county. Although the best soils tended to be used for cereal crops, while on mediocre soils grapevines were grown and inferior lands were covered with uncultivated woodland for grazing, the correspondence between land use and the quality of the soil was anything but linear. In spite of the expected high correspondence between the main agrarian uses and the agrological soil capabilities, we have found that 34% of the vineyards and 23% of the area with cereal crops were located on non-suitable or poorly suitable land for these uses. This means that to understand that land-use pattern we need to analyse the prevailing entitlement rules, and the economic goals or options of socially different farm units.

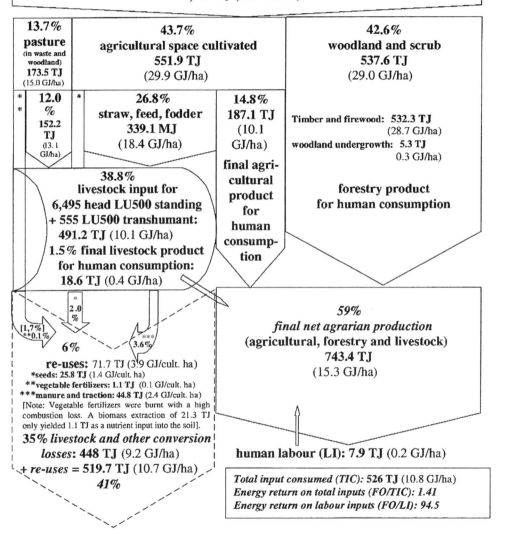

total surface area: 59,600 ha   woodland area: 18,564 ha (38.2%)
inhabitants in 1860/70: 38,342   cultivated area: 18,476 ha (38.0%)
useful agrarian area: 48,616 ha   pasture area: 11,576 ha (23.8%)

direct solar energy:
(54.4 TJ/ha × 59,600 ha = 3,243,933 TJ/year)

agro-forestry conversion in the useful agrarian area:
1,263 TJ (26.0 GJ/ha)

| 13.7% | 43.7% | 42.6% |
| pasture (in waste and woodland) 173.5 TJ (15.0 GJ/ha) | agricultural space cultivated 551.9 TJ (29.9 GJ/ha) | woodland and scrub 537.6 TJ (29.0 GJ/ha) |

* 12.0 %**   *** 26.8%**   **14.8%**
152.2 TJ   **straw, feed, fodder**   **187.1 TJ**   Timber and firewood: 532.3 TJ
(13.1 GJ/ha)   **339.1 MJ**   (10.1   (28.7 GJ/ha)
                   (18.4 GJ/ha)   GJ/ha)   woodland undergrowth: 5.3 TJ
                                                                              0.3 GJ/ha)
                                            final agri-
                   38.8%   cultural
                   livestock input for **product**   **forestry product**
                   6,495 head LU500 standing   **for**   **for human consumption**
                   + 555 LU500 transhumant: **human**
                   491.2 TJ (10.1 GJ/ha)   **consump-**
                   1.5% final livestock product   **tion**
                   for human consumption:
                   18.6 TJ (0.4 GJ/ha)

                        *
                   2.0
                   %   **59%**
[1,7%]   *final net agrarian production*
0.1%   *   *(agricultural, forestry and livestock)*
          6%   3.6%   **743.4 TJ**
   re-uses: 71.7 TJ (3.9 GJ/cult. ha)   **(15.3 GJ/ha)**
  *seeds: 25.8 TJ (1.4 GJ/cult. ha)
 **vegetable fertilizers: 1.1 TJ (0.1 GJ/cult. ha)
***manure and traction: 44.8 TJ (2.4 GJ/cult. ha)
[Note: Vegetable fertilizers were burnt with a high
combustion loss. A biomass extraction of 21.3 TJ
only yielded 1.1 TJ as a nutrient input into the soil].
35% livestock and other conversion
    losses: 448 TJ (9.2 GJ/ha)   **human labour (LI): 7.9 TJ (0.2 GJ/ha)**
+ re-uses = 519.7 TJ (10.7 GJ/ha)
             41%   *Total input consumed (TIC): 526 TJ (10.8 GJ/ha)*
                                                Energy return on total inputs (FO/TIC): 1.41
                                                Energy return on labour inputs (FO/LI): 94.5

Fig. 3.2. Summary of the annual energy flows of the agrarian system. (Originally published in *Ecological Economics* 58, 49–65).

Fig. 3.3. Original cadastral map from 1853 and organization of plots. Source: authors' compilation via GIS by Marc Badia for our research project SEC2003–08449–C04–03.

The owners of the *masies* that exerted control over the territory usually opted for poly-cultivation in the organization of their farms, which enabled them to extract the maximum market advantages without having to depend on it for their reproduction (Pascual, 1990, 2000). For the *masies* located on flatlands one of those inputs was firewood, and this explains why they maintained forested tracts or coppices on prime quality soils within the poly-cultivation agrarian mosaic that was still predominant in the Vallès in the mid-19th century. At the same time, this land-owning class tried to take economic advantage of their least suitable land for maize by leasing it in small plots to a new sub-peasant class of wine-growers (called *rabassers* in Catalan, or 'slash and burn growers'). These new specialized sub-peasants were mostly labouring immigrants coming from the Pyrenees mountains or non-inheriting descendants.

By leasing them small plots, the owners of the *masies* also tried to prevent further social conflict with them.

This may explain why in 1853, in the village of Caldes, 85% of the cereal fields were set on slopes of less than 20%, while 30% of the vineyards were placed on slopes of more than 30%, and sometimes even 60–70%. Slopes protected with stone terraces occupied about 700 ha, 43% of the overall cultivated land, and 80% of that area was then used for vineyards. Building of those terraces, which were generally small and were mainly undertaken on relatively soft geological materials (mudstone and sandstone), may have needed 120,000 work days. Most of those vineyards grown in the poorest soils of Vallès County were abandoned after the phylloxera crisis, and the terraces are now left behind covered by woods. Ironically, they still help to prevent deeper soil degradation when this abandoned forest or brushwood catches fire, but there are no public or private initiatives to preserve them for the future.

These results show that land characteristics influence land-use decision making and long-term historical landscape changes. The extension of agriculture to non-suitable land in 1853 also reflects the existence of other economic goals and deeper social conflicts which eventfully led to a vast labour investment in soil conservation to grow vineyards by the poorest rural classes.

Driving Forces towards Non-sustainability

Towards the end of the 18th century, several reports were written advocating coal imports in order to preserve the Montseny forests that were considered overexploited at that time (Caresmar (1780) and Comes (1786), quoted in Martí Escayol, 2002). Although we do not have enough historical data on timber and firewood cutting to confirm that trend, photographs taken in the Vallès county during the first half of the 20th century did show a picture of intense deforestation and rejuvenation apparent in the surviving forest.[7]

We also suspect that over-grazing or exploitation of the undergrowth reinforced the effects of deforestation. The scarcity of fertilizer is yet another clear result of our reconstruction of the energy balance of the agrarian system in the Vallès county during the second half of the 19th century. The 1874 *Estudio* tells us of the existence of many different local enterprises which attempted to recycle all types of human and animal waste matter, and of industrial by-products that could be reused as fertilizer. It also mentions the appearance of small amounts of guano, which we have not been able to include in our balance. In order to meet the incessant demand for nutrients, there were two other possibilities: increasing the transhumant livestock[8] and increasing the compost extracted from forests or scrub to make the vegetable fertilizers used in the crop lands.

Planting grapevines provided a replacement for the firewood and pastures via the reuse of vineyard prunings, olive tree prunings, and the use of green shoots as fodder. However, the vineyard option required a major investment in labour, mainly provided by tenants (*rabassers*) with very little or no land of their own who depended on the planting contracts offered by the *masie* owners. Rises in the relative prices of wine, such as that brought about by the oidium plague in 1840–1850, encouraged planting contracts (*rabassas*) to be hired. Perhaps we should also take into account the value of the by-products such as pruning or fodder in encouraging these plantings, which initially tended to be located over old forested and scrub areas.

A second wave started in 1867 when the phylloxera plague hit French vineyards, causing relative prices for Catalan wine to soar. This grapevine disease suddenly displaced poly-cultivation, provoking the first episode of environmental and economic globalization that linked the fate of that area to the international value of a single export product. This ended abruptly with the arrival of the disease in the Vallès county in 1883. By 1890 it had killed the old vines, and the region's agriculture swung towards the production of fresh milk and vegetables for daily delivery to the nearby cities. The widespread use of coal and petroleum put an end to the role of the Montseny mountains as suppliers of firewood and charcoal, although during the First World War and the Spanish Civil War the demand for forestry products enjoyed a short-lived increase. The new model was consolidated circa 1930, when the available land had been reduced to 1 ha per inhabitant, with only one-third of it arable.

During the first half of the 20th century, the land required for consumption by the population had already outstripped local availability. Since then, the globalization of economic networks has been sustained by the growing globalization of social metabolism and its ecological footprint. It is interesting to note that even in such a globalized context, the ecological footprint of Spanish food in the second half of the 20th century remained practically the same as for the Vallès county circa 1860–1870: 0.5 ha per inhabitant. Since then, the main difference from the earlier advanced organic economy is the considerable energy footprint resulting from the combustion of fossil fuels (Carpintero, 2002, 2005).

Conclusions

Studying past landscapes as organic advanced agricultures that could attain the highest energy yields, without relying on a high amount of external inputs, may help in achieving a better understanding of the two sides that have led to a lower energy performance after the so-called 'green revolution': (i) the injection of an external energy subsidy coming mainly from fossil fuels; and (ii) the functional disconnection between the agricultural, pasture and forestry spaces within the agrarian ecosystem that brought about the human abandonment of a huge portion of our territories and a much more inefficient land use from an ecological standpoint (Agnoletti, 2002; Díaz Pineda et al., 2002). Both sides are worth distinguishing, because the solutions to overcoming their ecological effects are necessarily different. An energy subsidy that replaces animal work power, or helps to make human labour less painful, should not be considered a problem in itself if provided by clean, local and renewable sources.

The true problem is that at present it comes from fossil or nuclear fuels, and overcoming the past dependency on bioconverters has led to a dysfunctional and ecologically unsound land use that is causing serious environmental pathologies (Naredo, 2001; Naredo and Parra, 2002) and also contributing to the ecological degradation of old cultural landscapes. Ongoing research undertaken by several landscape ecologists intends to verify the hypothesis we have put forward that a serious reduction of the territorial efficiency, which is related to a significant landscape transformation, underlies the observed loss of energy efficiency experienced by the agrarian systems during the last 150 years (Marull et al., 2006).

A structural and functional analysis of the changes experienced by the agrarian landscape in five municipalities of the Vallès county has been made through GIS, applying an innovative methodology that combines two indicators from the landscape ecology (cover diversity and ecological fragmentation) with two new socio-environmental indexes recently developed by some of the authors (eco-landscape structure and ecological connectivity). The preliminary results do show an increase in cover diversity, ecological fragmentation and anthropogenic barriers, as well as a growth in landscape heterogeneity which is associated with a substantial reduction of the ecologically functional areas, and its ecological connectivity. These transformations resulted in a severe loss of landscape functionality and territorial efficiency in the study area. All these changes happened at the same time as a sharp reduction in the energy agricultural returns to energy inputs, suggesting the need for a deeper trans-disciplinary dialogue between natural and social sciences, in a common historical background, to develop new criteria and methods in order to undertake a more sustainable type of land use planning (Marull et al., 2006).

Notes

1. This work has comes from the project SEC2003-08449-C04 of the Spanish Ministry of Science and Technology. Marc Badía, Fernando L. Rodríguez and Oscar Miralles have developed the cartography and the GIS-based treatment. We also wish to thank David Molina for his help in the cartography, Óscar Carpintero for his useful comments on energy balances and ecological footprints, and especially José Manuel Naredo and Pablo Campos for their advice on processing environmental accounts and developing indexes.

2. To distinguish between the global 'ecological footprint', calculated according to Rees and Wackernagel's method by examining average agricultural yields on a worldwide scale, and the local footprint of social metabolism calculated based on local historical productivity, in our study we prefer to refer to the latter as the land requirement by unit of product or inhabitant. The meaning of this term is identical to the third method proposed by Haberl, Erb and Krausmann to calculate the biocapacity and the ecological footprint using local yields (2001).

3. During the 1850–1860 years a cadastral plot map had been drawn up in several municipalities of the Vallès Oriental County (see Muro *et al.*, 1996, 2002 and 2003). We have made a GIS study of the 1850s, 1950s and present cadastral map of Caldes de Montbui and other four municipalities.

4. The 1874 *Estudio* describes various rotations of non-irrigated land, adapted to each type of soil, in which there was no fallowing and one or two crops were obtained annually. Thus, for example, in clayey soils, three crops were produced in two years, in the first legumes, and in the second year another two of cereals or late legumes; or alternatively, one of potatoes the first year, and the second year another two of cereals and maize, fodder or beans (Garrabou and Planas, 1998).

5. Paolo Malanima believes between 15 and 20,000 Kcal/inhabitant/day (22.9 to 30.6 GJ/inhabitant/year) were obtained on some 2 ha/inhabitant in the more densely populated parts of Europe, coming from 500–800 g of cereals per person, 2 kg of timber, and the traction provided by one ox, mule or horse for every six inhabitants (Malanima, 1996, 2001).

6. At the end of the 19th century, Eduard Sacher evaluated the energy from the wheat fields and forests of Austria and Prussia at 18×10^6 Kcal, or 75 GJ/ha/year. Sergei Podolinsky reduced this figure to 8.1 and 6.4×10^6 Kcal, or 33.9 to 26.8 GJ/ha/year for wheat and pasture in France (Martínez Alier and Schlüpmann, 1987; Martínez Alier, 1995). Fridolin Krausmann estimated values in around 1830–1850 of between 26 and 39 GJ/ha/year for the crops in various Austrian villages, between 34 and 44 GJ in the forests, and between 17 and 36 in pastureland (Krausmann, in press). For all crops, and including fallow land, González de Molina *et al.* obtain 6.1×10^6 Kcal or 25.5 GJ/ha/year in 1856 for the Granada plain of Santa Fe (2002). In the countryside around the Guadalquivir River, Pablo Campos and José Manuel Naredo obtain a mean value of 5.9×10^6 Kcal or 25 GJ/ha/year (corresponding to 6.3 GJ in the fallow pasture, 21.8 GJ in the seeded fallow land, and 46.1 GJ in that of wheat). The energy value of pastureland used here (1.497×10^6 Kcal or 6.268 GJ/ha/year) comes from Campos and Naredo, 1980a.

7. This explains why we have had to apply a reduction in both firewood and timber extractions to our historical sources in order to get sustainable forestry energy flow.

8. We have deduced the number of transhumant livestock by the difference between the capacity of pastures, fodder and feeds and the requirements of the existing livestock which appears in the 1865 livestock census. All the transhumant contracts were orally settled, and there are no written records.

References

Agnoletti, M. (2002) *Il Paesaggio Agro-forestale Toscano. Strumenti per l'Analisi, la Gestione e la Conservazione*. Arsia/Regione Toscana, Florence, Italy.

Boada, M. and Saurì, D. (2002) *El Cambio Global*. Rubes, Barcelona, Spain.

Campos, P. and Naredo, J.M. (1980a) La energía en los sistemas agrarios. *Agricultura y Sociedad* 15, 17–114.

Campos, P. and Naredo, J.M. (1980b) Los balances energéticos de la agricultura española. *Agricultura y Sociedad* 15, 163–256.

Carpintero, O. (2002) La economía española: el 'dragón europeo' en flujos de energía, materiales y huella ecológica, 1955–1995. *Ecología Política* 23, 85–125.

Carpintero, O. (2005) *El Metabolismo de la Economía Española. Recursos naturales y huella ecológica (1955–2000)*. Fundación César Manrique, Lanzarote/Madrid, Spain.

Carpintero, O. and Naredo, J.M. (in press) Sobre la evolución de los balances energéticos de la agricultura española, 1950–2000. *Historia Agraria*.

Costanza, R., Ayres, R.U., Deutsch, L. *et al.* (2000) Forum: The ecological footprint. *Ecological Economics* 32, 341–394.

CREAF (2000) *Inventari Ecològic i Forestal de Catalunya. Regió Forestal V*. Department of the Environment of the Generalitat de Catalunya, Barcelona, Spain.

Cussó, X., Garrabou, R. and Tello, E. (2006a) Social metabolism in an agrarian region of Catalonia (Spain) in 1860–70: flows, energy balance and land use. *Ecological Economics* 58, 49–65.

Cussó, X., Garrabou, R., Olarieta, J.R. and Tello, E. (2006b) Balances energéticos y usos del suelo en la agricultura catalana: una comparación entre mediados del siglo XIX y finales del siglo XX. *Historia Agraria*.

Díaz Pineda, F., De Miguel, J.M., Casado, M.A. and Montalvo, J. (2002) *La Diversidad Biológica en España*. Prentice Hall, Madrid, Spain.

Fischer-Kowalski, M. (1998) Society's metabolism. The intellectual history of materials flow analysis. Part I, 1860–1970. *Journal of Industrial Ecology* 2(1), 61–78.

Fischer-Kowalski, M. and Amann, C. (2001) Beyond IPAT and Kuznets Curves: globalization as a vital factor in analysing the environmental impact of socio-economic metabolism. *Population and Environment* 23 (1), 7–47.

Fischer-Kowalski, M. and Hüttler, W. (1999) Society's metabolism. The intellectual history of materials flow analysis. Part II, 1970–1998. *Journal of Industrial Ecology* 2(4), 107–136.

Foster, J.B. (2000) *Marx's Ecology. Materialism and Nature*. Monthly Review Press, New York.

Garrabou, R. and Planas, J. (eds) (1998) *Estudio Agrícola del Vallés (1874)*. Museu de Granollers, Granollers.

Garrabou, R. and Tello, E. (2004) Constructors de paisatges: amos de masies, masovers i rabassaires al territori del Vallès (1716–1860). In: *Josep Fontana. Història i projecte social*. Crítica, Barcelona, Spain.

Georgescu-Roegen, N. (1971) *The Entropy Law and the Economic Process*. Harvard University Press, Cambridge, Massachusetts.

Giampietro, M. and Pimentel, D. (1991) Energy efficiency: Assessing the interaction between humans and their environment. *Ecological Economics* 4, 117–144.

Giampietro, M., Bukkens, S.G.F. and Pimentel, D. (1994) Models of energy analysis to assess the performance of food systems. *Agricultural Systems* 45, 19–41.

González Bernáldez, F. (1981) *Ecología y paisaje*. Blume, Barcelona, Spain.

González de Molina, M. (2001a) Condicionamientos ambientales del crecimiento agrario español (siglos XIX y XX). In: Pujol, J., González de Molina, M., Fernández Prieto, L, Gallego, D. and Garrabou. R. *El Pozo de todos los Males. Sobre el Atraso en la Agricultura Española Contemporánea*. Crítica, Barcelona, Spain.

González de Molina, M. (2001b) El modelo de crecimiento agrario del siglo XIX y sus límites ambientales. Un estudio de caso. In: González de Molina, M. and Martínez Alier, J. (eds) *Naturaleza Transformada. Estudios de Historia Ambiental en España*. Icaria, Barcelona, Spain.

González de Molina, M., Guzmán Casado, G. and Ortega Santos, A. (2002) Sobre la sustentabilidad de la agricultura ecológica. Las enseñanzas de la Historia. *Ayer* 46, 155–185.

Haberl, H. (2001a) The energetic metabolism of societies. Part I: Accounting concepts. *Journal of Industrial Ecology* 5(1), 107–136.

Haberl, H. (2001b) The energetic metabolism of societies. Part I: Empirical examples. *Journal of Industrial Ecology* 5(2), 53–70.

Haberl, H., Erb, K.H. and Krausmann, F. (2001) How to calculate and interpret ecological footprints for long periods of time: The case of Austria, 1926–1995. *Ecological Economics* 38, 25–45.

Krausmann, F. (2001) Land use and industrial modernization: An empirical analysis of human influence on the functioning of ecosystems in Austria 1830–1995. *Land Use Policy* 18, 17–26.

Krausmann, F. (2006) La transformacion de los sistemas centroeuropeos de uso del suelo: una perspective biolisica de la modernización agricola en Austria des de 1830. *Historia Agraria*.

Leach, G. (1981) *Energía y Producción de Alimentos*. Ministry of Agriculture and Fisheries, Madrid, Spain.

Malanima, P. (1996) *Energia e crescita nell'Europa pre-industriale*. La Nuova Italia Scientifica, Rome.

Malanima, P. (2001) The energy basis for early modern growth, 1650-1820. In: Prak, M. (ed.) *Early Modern Capitalism. Economic and social change in Europe, 1400–1800*. Routledge, London.

Martí Escayol, M.A. (2002) Indústria, medicina i química a la Barcelona de finals del segle XVIII. El tintatge i la introducció del carbó mineral des d'una perspectiva ambiental. *Recerques* 44, 5–20.

Martínez Alier, J. (ed.) (1995) *Los Principios de la Economía Ecológica. Textos de P. Geddes. S. A. Podolinsky y F. Soddy*. Fundación Argentaria/Visor, Madrid, Spain.

Martínez Alier, J. (ed.) (1998) *La Economía Ecológica como Ecología Humana*. Fundación César Manrique, Madrid.

Martínez Alier, J. and Roca Jusmet, J. (2000) *Economía Ecológica y Política Ambiental*. Fondo de Cultura Económica/United Nations Environmental Programme, Mexico.

Martínez Alier, J. and Schlüpmann, K. (1991) *Ecological Economics. Energy, Environment and Society*. Basil Blackwell, Oxford, UK.

Marull, J., Pino, J., Tello, E. and Mallarach, J.M. (2006) Análisis estructural y funcional de la transformación del paisaje agrario en el Vallès durante los últimos 150 años (1853–2004): relaciones con el uso sostenible del territorio. *Áreas* 25, 105–126.

changes, examine different landscape patterns, analysis of the connection between multi-temporal dynamics and environmental factors and to model and foresee its future evolution.

Methodological Framework for Monitoring Landscape Change

Once the geographical area of interest and the map scale have been defined, the acquisition of the digital land use/cover maps at two different points in time (occasions) is the first step to assess land use/land cover changes. If the study has the final aim of modelling possible future changes, it will be better to consider at least three different occasions. Commonly (but not mandatory), one of the occasions is the current point in time.

Mapping is generally carried out by aerial photo interpretation, by polygon delineation and digitizing or by classification of multi-spectral remotely sensed images (characterized by a geometric resolution congruent with the nominal scale level of the analysis).

A possible alternative is to monitor the development over time of special phenomena using directly multi-temporal remotely sensed images. In these special cases, the analysis is not focused on the development of the entire landscape, but just on a specific event, and a complete land use/cover mapping project is not necessary. For instance, such a methodology is applied to monitor forest fires, coastal erosion, urban development, or desertification processes. For these kinds of projects, advanced methods of elaboration of multi-temporal remotely sensed data have been developed (Gillesond and Kiefer).

The analysis of multi-temporal land use/cover maps is the main methodology used for the historical analysis of a landscape. From a practical point of view, two main approaches can be adopted:

- a serial approach based on the production of successive maps for the different occasions: i.e. at the first occasion, a land use/cover map by the most accurate available procedure is created (for instance, by digitizing a digital orthophoto); maps for successive occasions are created by updating the initial map; or
- a parallel approach based on the creation of maps on independent data.

The first method minimizes the risk that geometrical discordances between maps are wrongly considered as land use/cover changes.

The two land use/cover maps created at two different occasions can be intersected through GIS analysis (see later in this chapter) to create a change map, where the dynamics in land use/cover are reported. The theoretical number of change classes is equal to the product of the numbers of classes at the first occasion and the second occasion. In order to better interpret the result, the change classes are usually aggregated in change types, whose number and features depend on the purposes of the project at hand.

Land use/cover maps can also be analysed by spatial indexes in the GIS environment to understand changes of landscape structures and to investigate if relationships in space and time exist between such trends and other ecological or social factors. In the next paragraphs more details are given regarding both remote sensing and GIS procedures to implement these kinds of procedures.

Earth Observation Tools

The map format derived from remotely sensed images is the main tool that can be used to describe the use/cover type and associated spatial variations. Earth observation (EO) techniques are an ideal tool for extensive surveys, as they provide geo-referenced information at relatively low cost.

Available types of remotely sensed images

Land use/cover data can be analysed by various types of remotely sensed images.

Fig. 4.1. An example of the use of aerial photography for monitoring land use/cover changes. Left: the image of a landscape in central Italy from an aerial photo taken in 1954. Right: the same landscape from an aerial image taken in 1996. The extension of woodland can be easily detected.

Satellite images may be available for the analysed study area and can be used if suitable for the considered map scale and thematic system of nomenclature. Satellite imagery may be used for historical studies too: for instance, Landsat images have been available since 1972, and SPOT images available since 1986. However, in general, historical landscape analysis is based on aerial photographs (Fig. 4.1) or old maps.

EO data can be classified on the basis of several properties:

- vector: the type of platform on which the sensor instrument is mounted, basically satellite or aerial;
- sensor technology: passive if the sensors are scanners sensitive to the solar radiation reflected by the earth's surface or active if the system emits radiation, whose reflection is detected by the sensor (e.g. radar or LIDAR);
- spectral resolution: number of sensors and wavelength of radiation to which they are sensitive;
- geometric resolution: dimension at the ground of the image pixel (Fig. 4.2);
- time resolution: frequency in the acquisition of images over time;
- price: ranging from free to very expensive datasets.

In Table 4.1, characteristics of some of the most frequently used digital sensors are described.

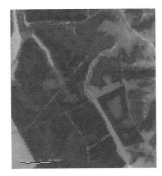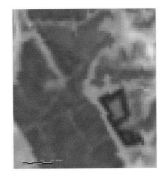

Fig. 4.2. Different resolution of remotely sensed images for the same area. Left: Quick Bird image (pixel of 2.8 m). Centre: SPOT XS (pixel of 10 m). Right: Landsat 7 ETM+ (pixel of 30 m). All images are compositions of RGB 432 in grey-scale palette.

Table 4.1. Characteristics of selected digital sensors most frequently used in land use/cover mapping.

Name of the sensor	Type of vector	Type of sensor	Geometric ground resolution	Spectral resolution	Time frequency
Landsat 7 ETM+	NASA satellite	Multi-spectral	30 m in multi-spectral 15 m in panchromatic	7 bands from visible to far infrared, 1 band panchromatic	19 days
XS	SPOT ESA	Multi-spectral	10 m in multi-spectral 2.5 m in panchromatic	4 bands from visible to near infrared, 1 band panchromatic	26 days
Quick Bird	Satellite	Multi-spectral	2.8 m in multi-spectral 0.7 m in panchromatic	4 bands from visible to near infrared, 1 band panchromatic	On demand
IKONOS	Satellite	Multi-spectral	4 m in multi-spectral 1 m in panchromatic	4 bands from visible to near infrared, 1 band panchromatic	On demand
AVHRR	NOAA satellite	Meteorological	1.1 km	4 bands from visible to near infrared	Daily
MIVIS	Aerial	Hyperspectral	Depending on flight height	106 bands from visible to infrared	On demand

Pre-processing techniques

The use of digital remotely sensed images to produce geocoded thematic maps requires several pre-processing steps. Some of them have to be done in all cases while others are required just in specific situations. To derive a geocoded map of land use/cover, raw images have to be corrected geometrically in order to project them in the desired geographic system of coordinates: this procedure is called georeferencing. Several methods can be implemented but nearly all of them require the acquisition of a number of ground control points (GCP) where both the row and column coordinate of the raw image and the geographical coordinate are known. In such a way it is possible to create a mathematical relationship between the two coordinate systems that is then applied to each pixel of the raw image to create a new resampled geocoded image. Several options are available: if the mathematical relationship is built taking into consideration the geo-

metric distortion induced by orography, then the process is an orthocorrection; if geographical coordinates are acquired on the basis of another geocoded image, the process is called co-registration.

While georeferencing is a mandatory geometric pre-elaboration, other processes implemented to spectrally correct the image are needed just for some specific applications. Topographic normalization is needed in mountain areas to correct the effect of slope and aspect. Atmospheric correction is needed when images have to be transformed into a quantitative dataset of physical meaning (e.g., to measure surface temperature or reflected radiance), reducing the filtering effect of atmospheric haze.

Other simple and very common techniques are applied to transform a set of grey-scale multi-spectral images into colour images. This is achieved by assigning one spectral band to each basic colour channel (red, green and blue) in order to prepare a more understandable image for photo intepreters.

Change-detection methods

As mentioned, two main approaches can be used to transform raw data from remote sensing into useful information for historical landscape studies. The first is to classify images according to a defined reference scale and a defined system of nomenclature of land use/cover classes, and then to intersect maps with GIS analysis. The second is to analyse directly multi-temporal images in order to enhance specific land-cover changes

From remotely sensed images to maps

It is possible to map land use/cover from a digital remotely sensed image with two main conventional approaches:

- photo interpretation: delineating polygons by manual digitization and then assigning each polygon to one of the classes of the selected system of nomenclature;
- classification: analysing by specific algorithms the raw image in order to assign each single pixel or group of pixels to one of the classes of the selected system of nomenclature.

The feasibility of classification can be increased by:

- decreasing the geometric resolution of images;
- increasing the number of spectral bands of the images;
- increasing the amount of ancillary information available to guide the classifier;
- decreasing the landscape complexity in terms of orography and land-cover mosaics.

When the images are not digital, such as traditional aerial photos, then two main procedures are available:

- acquiring aerial photographs by high resolution scanner, transforming them into digital orthophotos by geocoding with the use of DEM (digital elevation model). The orthophoto is then usually imported into a GIS and photointerpreted on screen;

- using traditional manual photointerpretation methods, and then digitizing the polygons delineated on the photos.

Change detection

Change-detection techniques are applied to multi-temporal remotely sensed images to identify and quantify specific phenomena. They are usually applied to map events that have transformed an area under consideration in a very fast way, while the remaining part of the landscape, not subject to any events, remains more or less unchanged. On the basis of such assumptions, at least two images must be acquired: one before and one after the event. A constraint is that the time interval between the acquired images has to be as short as possible. The analysis is based on the hypothesis that areas subject to the phenomena of interest have a very different spectral behaviour in the image space, while unaffected areas show a similar spectral response. Typical applications are mapping floods, forest fires (Fig. 4.3), deforestation and logging.

Advances

Fusion techniques

Promising possibilities are offered by a new generation of satellite-based products, orthoimages and fusion images which are analysed by specific application software (e.g., for change detection), and enable the user to take advantage of the information coming from different sources. Indeed, data fusion may significantly improve the resolution (spectral, spatial) of satellite images. The fusion between multi-spectral and panchromatic data aims at the preservation of the geometric features of panchromatic images and the spectral content of multi-spectral images. Sound data merging may improve the visual interpretability of the images, producing a multi-spectral image with a higher level of geometric detail (Fig. 4.4).

The fusion of multi-spectral and radar

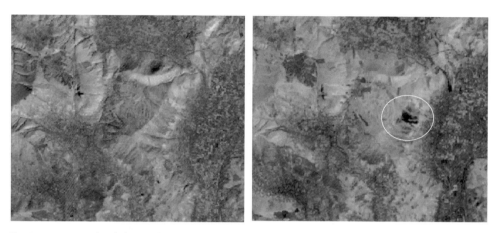

Fig. 4.3. An example of change-detection analysis for forest fire monitoring by Landsat 7 ETM+ imagery. Left: a pre-fire image. Right: a post-fire image. Both images are produced utilizing band 4 (near infra red). The circle in the right-hand image indicates the burnt area.

Fig. 4.4. An example of image fusion of Quick Bird data. Left: an image from multi-spectral bands (RGB 432 in grey-scale palette, resolution 2.8 m). Right: the same image fused with panchromatic band (resolution 0.7 m).

data may have two goals: the preservation of the multi-spectral content of the optical imagery and of the radar-derived information, or the integration of the radar data as an additive band of the multi-spectral image.

Future research effort will focus on testing the performances of available fusion algorithms (based for example on IHS transform multi-resolution analysis and wavelet transform, principal component analysis, arithmetic techniques, colour normalizing technique, simple band substitution, local regressions) and on the design of new ones. Whatever the data-fusion technique, the

best results are expected when using merged data collected during the same day, or with a time interval of a few days. Nevertheless, not all data-fusion methods work properly under each condition. The IHS merger, for example, is based on the assumption that multi-spectral and panchromatic bands cover the same spectral range. If not, the inclusion of the panchromatic band will modify the colour composition results of original multi-spectral bands: such a result can be useful when the image is photo-interpreted, but not where the image has to be automatically classified, due to the fact that its spectral content is altered.

The information content of maps based on remotely sensed data can be increased by combining field assessments and remote sensing imagery. This approach is especially useful when maps are required that have to show the spatial pattern of attributes that are either not directly assessable in remote sensing imagery or that cannot be assessed with appropriate accuracy.

Segmentation

Segmentation is a technique used to limit the original pixel variability of remote sensing imagery. Original images are in fact sets of raw data. To support land management or ecological studies, images have to be considered as base data useful to elaborate geographical datasets, usually in mapped format.

A land use/cover map is a vector model (see p. 67) of the original remotely sensed image. Segmentation is a methodology able to create automatically a vector model of a multi-spectral remotely sensed image (Fig. 4.5). Its distinctive aim is to process image objects separated according to homogeneity criteria. The segmentation procedure separates parts of landscapes as long as they appear significantly to contrast with each other on the analysed image (Giannetti *et al.*,

2003). This procedure substantially reduces the number of units to be handled; the following image classification is then based on the attributes of image objects rather than on the attributes of individual pixels (see below).

In most common segmentation algorithms, pixels of the image are aggregated in polygons on the basis of their spectral similarity (colour). Shape and dimension of desired polygons can be determined by changing driving variables of the segmentation process. Polygons then have to be classified on the basis of the chosen system of nomenclature by manual (photo-interpretation), automatic or semi-automatic systems (object-oriented classification) (Pekkarinen and Tuominen, 2003).

Object-oriented classification

This technique is based on a vector model of a remotely sensed image developed manually by photo-interpretation or automatically by segmentation. In pixel-oriented methods the underlying concept is the possibility to define a 'spectral signature' of each class of land use/cover acquiring a consistent number of pixels representing each class. For such 'training areas' the spectral signature is acquired. Then all the pixels of the image are assigned to the most spectrally similar class. The methods to

Fig. 4.5. Segmentation of remotely sensed data. Left: part of a Landsat 7 ETM+ image (30 m resolution, RGB 432 composition in grey-scale palette). Right: the same image after segmentation – note the black boundary polygons automatically generated.

evaluate such similarity are manifold and under continuous development. Each approach results in a classified raster map.

In object-oriented methods the target of the classification is not a pixel but a polygon (object). There are two benefits of such an approach (Benz *et al.*, 2004):

- the final product of the classification is in vector format instead of the raster format of a pixel-oriented approach; and
- the vector format is already a cartography oriented map and ready to be implemented in LIS (land information system) (Fig. 4.6).

A larger number of descriptive parameters can be associated with an object rather than individual pixels. Only the spectral signature can be developed for a pixel, while a huge number of attributes can be derived for polygons: studying the distribution of spectral values of pixels included in the polygon, the shape of the polygon, the type and number of contacts between polygons, the absolute position of the polygon and its relative position to other polygons, etc. Such properties enable a better characterization of land use/cover classes and thus an easier classification. Finally, after an automatic classification a final manual revision is possible.

kNN and spectral unmixing

Since the 1990s, two methods have been described that facilitate the operational use of remote sensing even at the local level: the kNN method and spectral unmixing.

The 'k nearest neighbour' (kNN) method relates terrestrial samples to the spectral information of pixels (Tomppo, 1993; Chirici *et al.*, 2003). For the entire set of pixels without associated ground assessments, the k nearest neighbours in the spectral image space are determined among those pixels which coincide with the location of field samples. The values of attributes assessed on the ground at the location of the k nearest pixels are weighted by the distances in the spectral image space and assigned to the respective pixels for which no ground information is available. Pixel estimates are plotted to produce maps that show the spatial distribution of attributes assessed on the ground in the resolution of the remote sensing data (Köhl, 2003).

Most ecosystems show small-scale heterogeneity, which results in a large amount of mixed pixels by the currently available multi-spectral sensors and their spatial resolutions. By utilizing hyperspectral remote sensing sensors, image data are collected in an enormous number (i.e., from 30 to more than 200) of narrow and adjacent spectral bands. Despite the fact that hyperspectral imagery is an extension of multi-spectral imagery, the tools applied for image analysis and interpretation differ from the well-known approaches in multi-spectral image analysis. A hyperspectral scene can be seen as an image with a spectrum of grey values, which are available for each pixel. For a

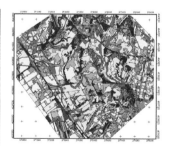

Fig. 4.6. Pixel- vs. object-oriented classifications. Left: an image acquired by Quick Bird satellite (2.7 m resolution, RGB 432 composition in grey-scale palette). Centre: an example of raster pixel-oriented classification, each grey level corresponding to a land use/cover class. Right: vector object-oriented classification, each grey level corresponding to a land use/cover class.

given geographic area the data can be viewed as a cube, having two dimensions that represent the spatial position and one that represents wavelengths. The image spectra can be compared with known spectra from field or laboratory experiments and enable detection and mapping of the spectral signatures of objects. This technique is known as spectral analysis and utilizes the information of the entire spectral image space and searches for characteristics of spectra that are similar to the known spectra of objects (Köhl, 2003).

An alternative technique for analysing hyperspectral data is called spectral unmixing. Spectral unmixing assumes that the reflectance of a pixel in an individual spectral band is a linear combination of the spectral reflectance of different objects or, in the nomenclature of hyperspectral image analysis, endmembers. The resulting spectra are thus a composite of the endmembers or pure spectra of objects in a pixel, weighted by their area proportion. Köhl and Lautner (2001) found for a test site in the Ore Mountains (Germany) that spectral mixture analysis provided good results for the assessment of mixture proportions of deciduous and coniferous trees. Classification of stand types, using the maximum likelihood algorithm, provided good results especially for further differentiation of tree species groups and for mapping age classes.

Analysis Based on Geographic Information Systems

A geographic information system (GIS) is a system for the capture, storage, retrieval, manipulation and visualization of geocoded datasets. A geocoded dataset is a set of data that has, as one of its main descriptive features, its position in the space (Aronof, 1995). Data acquired by remote sensing, pre-elaborated and transformed manually (by photo-interpretation), automatically (by one or more methods described above) or by a mixed approach in thematic maps are finally elaborated by GIS technologies. Geographic information systems provide an efficient way to overlay layers of mapped data too, but their capabilities are by no means limited to formulating and producing maps. Although GISs have their roots in cartography they have evolved into powerful data management, analysis and display tools to analyse, study and model the structural landscape and environmental dynamics.

Spatial data are represented in a GIS in two different ways: the raster format and the vector format. In raster format, a grid is used to represent the area of interest: the location of features is indicated by a designated code in each cell containing that feature. Vector data represent geographic features by coordinates of points, lines and polygons: points designate small features or individuals, such as a den tree, or a surveying benchmark; lines are used to represent linear features such as roads and streams; and polygons are employed to designate areas such as lakes or forest stands and are bounded on all sides by a series of straight-line segments (Avery and Burkhart, 2002).

Data entry

Change detection on classified maps

The base dataset for the analysis of change detection within a geographic information system is a multi-temporal land use/cover database. Such a dataset is usually developed by intersection of two land use/cover maps acquired on two different occasions for the same area. The maps should be consistent both from a geometrical and thematic point of view. They should have the same resolution (same scale and same size of the minimum mapping unit) and should be developed with the same system of nomenclature.

However, frequently such prerequisites are not satisfied because original mapping projects have been developed with independent procedures and objectives. In such cases a standardization operation has to be performed. From a geometrical point of view the most common choice is to rasterize all coverages with a resolution (pixel dimension) consistent with the scale of the least accurate dataset (the dataset mapped at the lowest scale). From a thematic point of view a unique relationship has to be found for all the classes of the system of nomenclature used in order to ensure the possibility of reclassifying the classes, creating a new unique system usable for all the maps. In such an operation, the new common system of nomenclature should be based on international standard definitions, as, for instance, for forest and other wooded lands (FAO, 1995; MCPFE, 2003).

Land use/cover maps are usually derived by remotely sensed images, but in historical studies could also be derived from old paper maps. In such cases, the maps have to be acquired in raster format by cartographic scanners. The images have

to be geocoded just like remotely sensed images and finally digitized to convert them to vector format. An example of change detection on classified maps is shown in Fig. 4.7.

Cross-tabulation matrix

The analysis of land use/cover changes is completed by cross-tabulating multi-temporal raster land use/cover maps. The matrix reports in the rows land uses of the oldest date and in the columns land uses at the current date (an example is given in Table 4.2). The vectors in the matrix have specific meanings. A row vector indicates how old land use developed. A column vector indicates from which class the present land use developed. Every cell of the matrix

Table 4.2. An example of a cross-tabulation matrix derived by the analysis of datasets in Fig. 4.7. Values reported in cells are in hectares.

Old	New				
	B	DG	LG	W	Total
B	6.1		1.1		7.2
DG		2.7			2.7
LG			6.3	2.0	8.3
W			4.9	4.9	
Total new	6.1	2.7	7.4	6.8	23.0

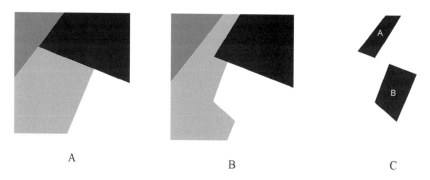

A   B   C

Fig. 4.7. An example of change detection on classified maps. In images A (old) and B (new) the same area (approximately 23 ha) with a classified land cover maps with four land cover classes: black (B), dark grey (DG), light grey (LG), white (W). Image A is taken at the year X and image B at time X+n years. In image C, the change analysis by GIS shows two changes. Polygon A changed from B to LG and polygon B changed from LG to W.

contains the extension of a specific class of land-use change. Matrixes can also be elaborated in order to simplify their reading and interpretation. For this purpose all possible land use changes can be classified in dynamics classes.

Advances

Landscape indexes

In order to examine the landscape patterns and their evolution, it is possible to apply landscape ecology techniques. Many indexes have been specifically developed to analyse certain structural landscape features. Such indexes can be applied at various levels:

- at whole study-area level: the index is calculated on a very wide area. GIS techniques are not necessary but the result cannot be mapped (e.g., the average size of patches);
- at patch level: the index is calculated for every patch. Usually, the calculation is based on vector data. The output can be mapped: for instance, the relationship between patch area and perimeter expresses the level of complexity of the form of the patch (fractal dimension);
- at a specific window level: the index is calculated on a geographical neighbourhood of patches. GIS technology is applied to digital maps in raster form.

This procedure allows the best mapping output to be obtained. For every pixel the index is based on the adjacent pixels. For instance, if the window is 3 × 3 pixels, a simple diversity index will be calculated on the eight adjacent pixels. The value is referred to the target pixel (the central one), then the window shifts to the adjacent pixel and the calculation is repeated (Fig. 4.8).

At each application level, changes of values provide substantial information related to different occasions on the same area.

Landscape ecology focuses on three characteristics of the landscape: structure, function and change (Forman and Godron, 1986). Thus, landscape ecology involves the study of landscape patterns. Landscape indices are a widely used tool to quantify spatial landscape structures. The large number of landscape indices that have been described in the literature can be grouped into the following eight classes (McGarigal and Marks, 1994): area metrics; patch density metrics; edge metrics; shape metrics; core area metrics; nearest neighbour; diversity metrics; contagion and interspersion metrics (Fig. 4.9).

Geographic vs. geometric windows

The calculation of one index for an entire landscape often fails to reflect the structure of the landscape. As mentioned, the

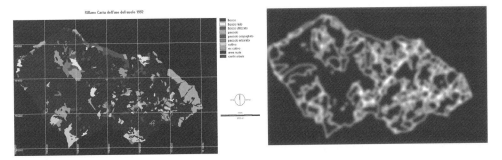

Fig. 4.8. An example of the application of a landscape index (Shannon diversity) on a moving window. Left: a land use/cover map. Right: the result of the application of the Shannon diversity index with a moving window of 30 original pixels; the diversity increases from black to white.

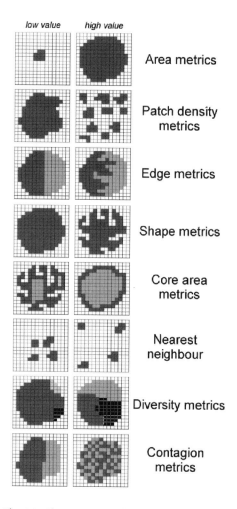

low value   high value

Area metrics

Patch density
metrics

Edge metrics

Shape metrics

Core area
metrics

Nearest
neighbour

Diversity metrics

Contagion
metrics

Fig. 4.9. The main types of landscape indexes (modified from Häusler *et al.*, 2000).

application of the moving windows technology has been suggested in the context of landscape analysis. Windows (synonyms are kernels, masks or filters) have their origin in image analysis and are used to characterize spatial information in a neighbourhood. Sub-areas are separated from the area under concern and analysed.

According to Merchant (1984) and Dillworth *et al.* (1994) windows of fixed size – so-called geometric windows – show disadvantages in situations where not the neighbourhood of pixels, but the neighbourhood of patches has to be analysed. For these situations Merchant (1984) proposed to use geographic windows, where, based on an initial geometric window, the window is expanded until all patches covered by the initial geometric window are included. Ricotta *et al.* (2003) and Köhl and Oehmichen (2003) compare geometric and geographical windows in a landscape analysis context (Fig. 4.10).

Conclusion

New developments of technologies for capturing spatial and temporal information significantly influenced research on the assessment of landscape dynamics. Standard image processing and GIS software are able to transform raw images into complex thematic maps (landscape structures,

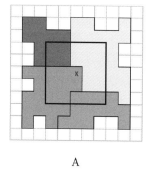

A

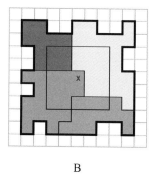

B

Fig. 4.10. Selection of patches by (A) a geometric window (in the outlined square) and (B) a geographic window (the darker outlined area).

landscape dynamics and history, etc.). However, a lack of knowledge still exists concerning the relationships among environmental and anthropogenic factors at landscape level. For instance, most landscape indices do not have clear ecological meaning, and difficulties arise in verifying the extracted relationships, with conventional ecological data intensive and costly to assess in the field. Despite the fact that land-use dynamics over time provide significant information, the debate regarding the integration of this information into decision processes by land-use managers is still going on.

It is widely accepted that the new methodological approaches will become operational in the near future and offer new perspectives to land management-oriented issues. A transition from the reconstruction of landscape to the forecasting of change trends will facilitate the implementation of decision-support systems for land-management planning.

References

Aronof, S. (1995) *Geographic Information Systems: a Management Perspective*. WDL Publications, Ottawa, Canada.

Avery, T.E. and Burkhart, H.E. (2002) *Forest Measurements,* 5th Edition. McGraw Hill, New York.

Benz, U.C., Hofmann, P., Willhauck, G., Lingenfelder, I. and Heynen, M. (2004) Multi-resolution, object-oriented fuzzy analysis of remote sensing data for GIS-ready information. *ISPRS Journal of Photogrammetry and Remote Sensing* 58, 239–258.

Chirici, G., Corona, P., Marchetti, M., Maselli, F. and Bottai, L. (2003) Spatial distibution modelling of forest attributes coupling remotely sensed imagery and GIS techniques. In: Amaro, A., Reed, D. and Soares, P. (eds) *Modelling Forest Systems*. CAB International, Wallingford, UK, pp. 41–50.

Dillworth, M., Whistler, J.L. and Merchant, J.W. (1994) Measuring landscape structure using geographic and geometric windows. *Photogrammetric Engineering & Remote Sensing* 60, 1215–1224.

FAO (1995) *Forest Resource Assessment 1990. Global Synthesis*. FAO Forestry Paper 124. Rome, Italy.

Forman, R.T.T. and Godron, M. (1986) *Landscape Ecology*. John Wiley & Sons, New York.

Giannetti, F., Gottero, F. and Terzuolo, P.G. (2003) Use of high resolution satellite images in the forest inventory and mapping of Piemonte region (Italy). In: Corona, P., Koehl, M. and Marchetti, M. (eds) *Advances in Forest Inventory for Sustainable Forest Management and Biodiversity Monitoring*. Kluwer, Dordrecht, pp. 87–96.

Gustafson, E.J. (1998) Quantifying landscape spatial pattern: what is the state of the art? *Ecosystems* 1, 143–156.

Häusler, T., Akgö, E., Gallaun, H., Schardt, M., Ekstrand, S., Löfmark, M., Lagard, M., Pelz, D.R. and Obergföll, P. (2000) Monitoring changes and indicators for structural diversity of forested areas. In: Zawila-Niedzwiecki, T. and Brach, M. (eds) *Remote Sensing and Forest Monitoring*. EUR 19530, Office for Official Publications of the European Communities, Luxembourg, pp. 392–414.

Köhl, M. (2003) New approaches for multiresource forest inventories. In: Corona, P., Koehl, M. and Marchetti, M. (eds) *Advances in Forest Inventory for Sustainable Forest Management and Biodiversity Monitoring*. Kluwer, Dordrecht, pp. 1–6.

Köhl, M. and Lautner, M. (2001) Erfassung von Waldökosystemen durch Hyperspektraldaten. *Photogrammetrie–Fernerkundung–Geoinformation* 2, 107–117.

Köhl, M. and Oehmichen, K. (2003) Comparison of landscape indices under particular consideration of the geometric and geographic moving window concept. In: Corona, P., Koehl, M. and Marchetti, M. (eds) *Advances in Forest Inventory for Sustainable Forest Management and Biodiversity Monitoring*. Kluwer, Dordrecht, pp. 231–244.

Krummel, J.R., Gardner, R.H., Sugihara, G., O'Neill, R.V. and Coleman, P.R. (1987) Landscape patterns in a disturbed environment. *Oikos* 48, 321–324.

McGarigal, M. and Marks, B.J. (1994) *FRAGSTATS: Spatial Pattern Analysis Program for Quantifying Landscape Structure*. Reference manual for Science Department, Oregon State University, Corvallis, Oregon, USA.

McGarigal, K., Cushman, S.A., Neel, M.C. and Ene, E. (2002) *FRAGSTATS: Spatial Pattern Analysis Program for Categorical Maps*. Computer software program produced by the authors at the University of Massachusetts, Amherst, Massachusetts.

MCPFE (2003) *Vienna Declaration and Vienna Resolutions.* Adopted at the Fourth Ministerial Conference on the Protection of Forests in Europe. 28–30 April, 2003. Vienna, Austria.

Merchant, J.W. (1984) Using spatial logic in classification of Landsat TM data. In: *Proceedings of the Pecora IX Symposium*, Sioux Falls, South Dakota, pp. 378–385.

Pekkarinen, A. and Tuominen, S. (2003) Stratification of a forest area for multisource forest inventory by means of aerial photographs and image segmentation. In: Corona, P., Koehl, M. and Marchetti, M. (eds) *Advances in Forest Inventory for Sustainable Forest Management and Biodiversity Monitoring.* Kluwer, Dordrecht, pp. 111–124.

Ricotta, C., Cecchi, P., Chirici, G., Corona, P., Lamonaca, A. and Marchetti, M. (2003) Assessing forest landscape structure using geographic windows. In: Corona, P., Koehl, M. and Marchetti, M. (eds) *Advances in Forest Inventory for Sustainable Forest Management and Biodiversity Monitoring.* Kluwer, Dordrecht, pp. 221–229.

Romme, W.H. (1982) Fire and landscape diversity in subalpine forests of Yellowstone National Park. *Ecological Monographs* 52, 199–221.

Tomppo, E. (1993) Multi-source national forest inventory of Finland. In: Nyyssonen, A., Poso, S. and Rautala, J. (eds) *Proceedings of Ilvessalo Symposium on National Forest Inventories*, Helsinki, Finland, pp. 52–59.

PART II

Management

There can be different ways of managing landscapes according to cultural or ecological approaches. The ecosystem is usually a fundamental concept used by almost all the approaches to understand landscapes, as a complex system of interacting biological, physical and cultural factors, while ecological planning seeks to maintain the stability of ecosystems and their flows of energy, material and species. However, there are different ways of considering the role of ecological processes in valuable cultural landscapes.

The management of fire proposed by Métailié (Chaper 7), not only to maintain pastures, but also to enhance biodiversity, would be an almost impossible proposal in contexts like Italy, for instance, not only because of the different way of understanding the role of fire from an ecological point of view and the danger represented by fires in densely populated areas, but also because of the perception of fire induced in the public by certain fields of scientific research and media.

On the other hand, the need for a revision of the way sustainability is interpreted is needed. In a continent like Europe with such a long history of cultural influence on the environment, the simple fact of having a large network of areas to protect nature, but no equivalent instruments to protect cultural landscapes is quite symptomatic. Neither is it useless to stress 'cultural' when describing these landscapes, as the association of the concept of landscape to that of nature often creates misleading interpretations regarding management strategies. As a matter of fact, although the aims of the NATURE 2000 network of protected areas are also to include cultural factors in their management, the way this directive is applied shows that it is better suited to protecting endangered species, rather than the habitats created or dominated by cultural influence, that are quite widespread in Europe. It is therefore very useful to propose studies like the one by Bradshaw (Chapter 6), showing the importance of cultural influence in southern Sweden and the progressive incorporation of this into the practical management of biodiversity. There is in fact the need to elaborate a better strategy for biodiversity and also to propose instruments to protect and restore not only biodiversity of species, but also diversity of spaces. The project for the park of the rural landscape in Moscheta (Tuscany) (Chapter 5), indicating management choices to reduce the expansion of woodlands restoring the diversity of spaces, could even be interpreted as a provocation, rather than a scientific proposal, according to how the situation is at a scientific and political level. Although the restoration of an ideal earlier 'natural ecosystem' is widely accepted, the idea of restoring a cultural landscape finds many more difficulties, although the first is a much more undefined goal compared to the second.

This is a view that brings together very different situations across the world as shown by the chapter by Nancy Langston for the USA (Chapter 11), stressing the limited success of such policies, and bringing up the matter of the views that forest managers apply to forest management, creating a direct parallel with EU forest strategies. A rethinking of traditional preservation approaches is however underway, as documented by the experience illustrated by the Marsh–Billings–Rockefeller Park, a small but significant place for management approaches, and also by the chapter by Blank (Chapter 10), addressing the widespread problem of parks in the neighbourhoods of urban areas.

This also brings up the matter of the economic aspect of landscape restoration, an activity proposing an alternative to the globalization also affecting countryside, contributing to the sustainable development also of less favoured areas. The importance of this concept is presented by Angelstam (Chapter 8), describing the crucial moment for Eastern European countries recently joining the EU and waiting to develop their economies. This process risks destroying their landscape, with the help of economic incentives for agriculture and strategies for agriculture and nature conservation. This is the case presented by Montiel Molina for Spain (Chapter 14), but could actually be applied to all the EU countries, where widely used subsidies to carry out afforestation projects often damaged landscapes. This occurred because of the initial absence of specific indications in this respect; therefore, after a couple of decades, the effects and conflicts provoked in regions where landscape is a valuable resource are evident. On the other hand, these trends are also favoured by the way sustainability is interpreted and applied by important international processes like certification standards. Anderson (Chapter 12) is considering them in forestry, but the same problem can be seen for eco-certification in agriculture, denying any important role played by cultural values, or the links between food quality and landscape. The lack of specific criteria and indicators are obviously affecting management approaches based on them, which will affect cultural landscapes when applied. The hope is that initiatives like the European Landscape Convention, described by Weizenegger and Schenk in Chapter 13, will be able to affect this problem, although nothing can be done without a growth of social sensibility about this matter. However, the Convention is willing to promote an approach favouring the recognition of all landscapes, independent of their value.

5   The Project for the Rural Landscape Park in Moscheta (Tuscany, Italy)

M. Agnoletti, V. Marinai and S. Paoletti
*Department of Environmental Forestry Science and Technology,
University of Florence, Florence, Italy*

Introduction

Moscheta is one of the 13 study areas covered during the project for Tuscany reported in Chapter 1. In this case the HCEA methodology has been applied to the problem of defining the management criteria for the creation of a landscape park. The study addresses two main problems referring to the conservation of cultural landscapes and the development of marginal rural areas, but also to the long-term analysis of landscape dynamics. For the first problem we hope to offer a contribution to the definitions of cultural landscapes, according to the criteria indicated by UNESCO for the World Heritage List, making this concept less opaque and vague (Fowler, 2003). The development of rural regions, especially those located in mountain environments, is one of the most important issues at world level, addressed by several international directives. Marginal territories, especially those once intensively cultivated and now abandoned, represent a problem not only in developing countries, but also in Western societies. Also, for this reason the Mountain Community of the Mugello valley, the administrative body managing the mountain districts where the study area of Moscheta is located, promoted and supported this study, trying to tackle the lack of any real policy at regional, national and European Union level favouring the conservation of cultural landscape in rural development or nature conservation. Moreover, none of them offer a chance to create a network of landscape parks, as happened with the NATURA 2000 network of protected areas, creating a situation particularly problematic for countries like those in the Mediterranean area, where cultural landscape represents an important resource.

Site and History

The Moscheta study area is located in a small mountain valley on the right side of the Santerno River basin, flowing from the eastern side of the Tuscan Apennines towards the Emilia-Romagna region. It extends for 901 ha and includes the bottom of the valley and the mountains around it, presenting an average altitude of 680 m above sea level (Plates 5 and 6). The main geological formations include the Marnoso-Arenacea formation and the Caotic Complex of the Ligurian Units. The first has all the characteristics of a deep sedimentary basin: sandstones and silt in rythmic sequences. The second refers to an olistostrome sequence: distorted claystones, marlstones or shales, stratified packets, a few metres to several hundred metres in length, and masses of breccia. From a vegetation point of view Moscheta is situated in the cold and warm zones of Castanetum, according to the phyto-climatic classification of Pavari (1916), between the hill belt of hop horn-beam and deciduous oak woods, and the mountain belt of beech forests, according to Ozenda (1975). Historically, the valley is characterized by the presence of an abbey, founded in the year 1034 by Saint Giovanni Gualberto, a monk of the Benedictine order, on a piece of land donated by the Ubaldini family having the same extension as our study area. The monks started to carry out farming activities and in two centuries the abbey became one of the most important in Tuscany. In a 17th century drawing (Fig. 5.1) the abbey is surrounded by pastures and woods made of beech and fir, the latter reflecting a distinctive attitude of the Benedictine monks, who liked to plant fir around their abbeys, managing them with clear-cuts and selling the timber (Agnoletti and Paci, 2001). In the 18th century the abbey lost its importance and was suppressed during the reforms introduced by the Lorena, the Austrian dynasty replacing the Medici family as the rulers of Tuscany at the beginning of the 19th century. The new owners organized the area as a rural estate, according to the share-crop system, where each farmer shares the crops produced with the owner. This structure is the one we find at the beginning of the 19th century when our investigation begins, when the estate appears to be divided into holdings, 'poderi', each one with a stone house for the farmer.

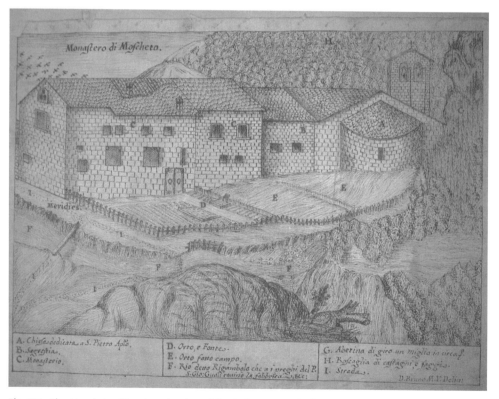

Fig. 5.1. The Moscheta abbey in a drawing of the 17th century. The land uses listed below the abbey describe the presence of fir, which can be clearly seen on the right side of the picture, together with chestnut and beech.

The Landscape in 1832

In 1832 the Moscheta estate was organized into eight farms, each farmhouse hosting more than one family of farmers and owned by the Martini family of Florence. In this period, 254 people were registered as residents in the 'Community of Moscheta', including also neighbouring small communities. Cereals and chestnut flour were the main crops, but the income of the farms, according to the account books, was mostly coming from livestock and timber production. The landscape in 1832 was made up of 89 different land uses (see Plate 8), representing a higher degree of diversity compared to better farming areas at lower altitudes in Tuscany and also had a high level of spatial diversity (Baudry and Baudry-Burel, 1982). More than 40% of this variety of land uses was given by wooded pastures (see Fig. 5.2), as only one land use was described by the cadastre as simple pastureland, confirming the role of wood pasture in the landscape of Tuscany, a technique of crucial importance for cattle breeding in many parts of Europe from north to south (Fuentes Sanchez, 1994; Rotherham and Jones, 2000). Pasture represents the most important land use, covering 55% of the territory, located not only on the high part of the mountains, but also in the bottom of the valley regardless of slope or altitude. Pastures with beech make up 18.6% of the total, while 'pastures with small woods of beech' make up 15%, indicating an interesting feature of the landscape in terms of patterns, but also that beech is the dominating element in 43% of pastureland. About 17% of the land is covered by different qualities of meadows (included pasturelands in our reclassified land use), which are all wooded as well.

Forests represent 22.4% of the landscape with a list of 20 different categories; six of them described as pastured woods, five as chestnut orchards, including also five mixed woods with chestnut, beech and oak. The most common species represented is beech, making up 19% of the total, while oak covers 2.8%. Most of the beech forest was used for charcoal production, as testi-

fied by the terraced charcoal burning sites still existing, but also for fodder and acorn production to feed livestock. Chestnut orchards make up 20% of the woodlands. They represent an essential factor for the survival of farmers integrating low cereal production with chestnut flour. Each family had one plot of chestnuts, even if far away from the house. There is no trace in the cadastre of the fir forest appearing in Fig. 5.1, which indicated an extent of 21 ha. The most extensive wood category is instead 'woodlands with beech and pastures', covering almost 110 ha, showing once again the prevailing role of pastures in the landscape, followed by beech forest. Sowable land represents 16% of the land uses. Even in this case only one category is described as without trees. The most extensive category is simple sowable land, with more than 20 ha, followed by 'mixed sowable land with pastures and trees', making 5 ha. The contribution of the land uses listed, in terms of extension, is not proportional to their number, as pastures are covering 55% of the territory, woods 39% and sowable land 5.3% (Fig. 5.2). However, considering only the number of land uses named in the cadastre, the highest frequency concerns pastures and sowable land. Trees in the field and in pastures are actually a very important feature of this landscape, confirming the functions and the role described for most traditional rural societies in the world, both in developed and developing countries (Arnold, 1995; Rackham, 1995; Sereni, 1997). The frequency of tree species, in the land uses naming them, presents the following hierarchy: turkey oak 32%, beech 24%, chestnut 23%, white oak 13%, hornbeam 5%, and walnut 2.6%. A few descriptions report shrubs like juniper and heather, as well as some mulberries as trees on boundaries between fields. The prevalence of beech and turkey oak reflect the climatic distribution of these species, while chestnut has mostly been planted. The estate has a nursery where chestnut is the most important species produced, together with maple and poplar used for mixed cultivations, as well as fruit and olive trees.

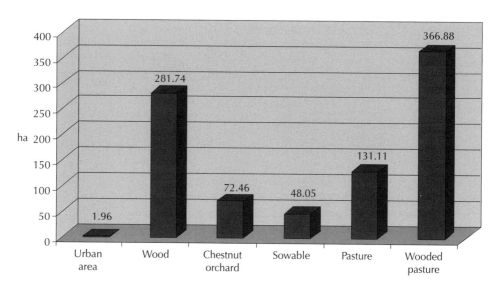

Fig. 5.2. Extent of the main land uses in 1832.

The Landscape between 1832 and 1954

The situation in 1954 (Figs 5.3 and 5.4 and Plate 9) shows a much more simplified structure, with only nine categories of land use remaining of the former 89. This problem is partly due also to the different quality of the source, since the aerial photographs are coming from flights at high altitude in black and white, but they surely show a reduced fragmentation. Woodlands are now covering 64% of the landscapes, with an increase in their extension from 354 ha to 581 ha, pastureland makes up 16% and sowable land 19%. An interesting change is the presence of a greater amount of sowable land in the bottom of the valley, greatly increasing the total amount existing in 1832 and contributing to the total reduction of pastureland (−73%). Particularly strong is the reduction of wood pasture from 366 to 24 ha.

The general dynamics indicated in Fig. 5.4 show the importance of forestation (32%), followed by intensification due to new agricultural areas, indicating a change in farm activities. Most of the new woodlands are clearly growing on former wooded pastures (55%) and only 15% on simple pastures following a successional pattern already observed in other Italian areas after abandonment (Salbitano, 1987; Agnoletti and Paci, 1998). Some new chestnuts are also growing on former pastures, while 42% of the new sowable land is placed on former pastures, 29% on former wood pasture and 11% on woodlands, according to the different degree of difficulty in turning these different land uses into cultivated land (Table 5.1). All these changes have obvious effects on the aesthetic of the local landscape. The mountain slopes are now covered with woodlands, rarely interrupted by pastures, while cultivated areas can be seen in the bottom of the valley (Plate 5). This also confirms a peculiar inverse tendency in the area for its rural population, which has not yet abandoned this part of the Apennine Mountains as is more generally reported in other areas (Agnoletti, 2002b) – in fact 223 people are still resident in the community.

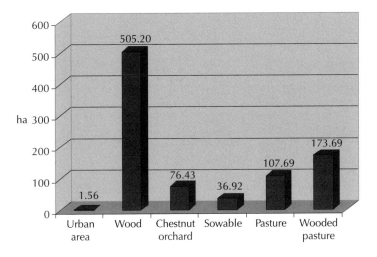

Fig. 5.3. Extent of the main land uses in 1954.

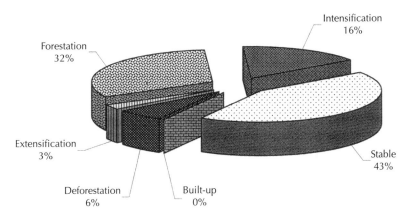

Fig. 5.4. Main landscape dynamics 1832–1954.

Table 5.1. Cross-tabulation 1832–1954.

	Extent in ha	1954						
		Anthropic	High stand	Chestnut orchard	Pasture	Wooded pasture	Sowable	Total
	Anthropic	0.61	0.53		0.57	0.00	0.24	1.96
	Woodland	0.13	226.07	2.67	6.56	27.13	19.18	281.73
	Chestnut orchard	0.28	9.69	58.53	0.95	1.80	1.22	72.46
1832	Sowable	0.13	7.21	2.88	0.78	7.94	29.10	48.05
	Pasture	0.01	34.50	5.09	4.35	13.46	73.70	131.11
	Wooded pasture	0.41	227.84	7.26	23.72	57.41	50.25	366.88
	Total	1.56	505.84	76.43	36.92	107.75	173.69	902.18

The life of the farm described in the account books shows the importance of wood production, demonstrated by 16 different assortments including timber for building, poles, staves, bundles, fuelwood, charcoal and cinder, although all the production decreases towards the Second World War. On the agricultural side there is a strong increase in cereal production, an effect of the new fields in the valley, and a steady decrease of chestnut flour and chestnut production, confirming the decreased importance of chestnut orchards as a food supply for farmers, accompanied by the cutting of many chestnut trees to utilize bark for tannin, turning the chestnut woods into coppice for poles (Agnoletti, 2002a). Although the period after the war saw the sale of the Moscheta farm to the state, the landscape of the 1950s still represented a part of the traditional agrosilvopastoral economy, where the marginalization of unfavourable areas had not yet fully expressed its influence, but the signs of a process affecting the Italian mountains were already there.

The Landscape between 1954 and 2000

The decades after the war confirm the continuous expansion of woodlands, now covering 79% of the territory, followed by 20% of pastures, but sowable land has disappeared from the valley. In terms of land uses, there are now 28 categories, 19 more than in 1954, an effect partly due to the more accurate analysis, allowing also field work to check photo interpretation. However, it is worth noting that landscape diversity in 1954 was given by eight land use types, of which only two were associated with woodlands, while in the year 2000, 21 types out of 28 are woodlands (see Plate 10 and Fig. 5.5). This shows how the local landscape is more and more dominated by the absence of open spaces and characterized by a continuous forest cover. The main dynamics are confirming a large amount of territory remaining unchanged (61%), while 16% of extensification is due to sowable land turned into pastures, and afforestation (8%) is the new element of the

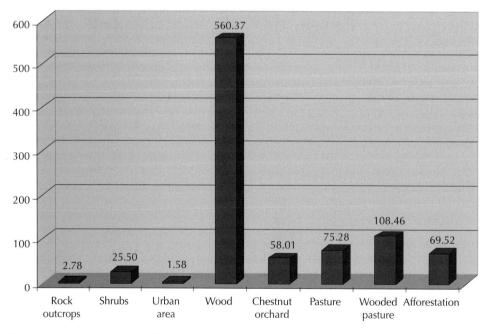

Fig. 5.5. Land use in the year 2000.

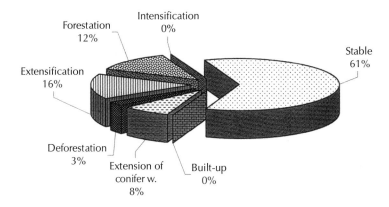

Fig. 5.6. Main landscape dynamics 1954–2000.

landscape (Fig. 5.6). As for the other areas analysed in the project, once again it is confirmed how the first period (1832–1954) is the most important one for landscape changes. The analysis presented in the cross-tabulation of 1954–2000 (Table 5.2) shows that 124 ha of agricultural land is now turned into wood pasture, but another interesting thing is afforestation. As already explained in Chapter 1 about Tuscany, afforestation by the state in Italy started soon after the unification (Agnoletti, 2002b), but a real programme in the study area started only after the sale of the farm to the state and continued after the organization of the Italian state into regions in 1974, when the area passed under the control of the Region of Tuscany (Freschi and Hermanin, 2000).

As clearly shown by Table 5.2, 36% of the plantations occurred on former sowable land, 25% on wooded pastures, 20% on chestnut orchards and 14% on high stands. Looking at the distribution of afforestation on the land-use map, it is clear that there was no real intention to recover degraded areas for soil protection, since very few denuded ridges or high slopes are afforested. The will to extend conifers is instead quite clear, replacing beech and chestnuts that were considered not sufficiently productive. After 50 years the results are insignificant for timber production, dominated in Italy and Tuscany by international markets, but are surely considerable for the landscape, now affected by plantations often having little to do with the cultural landscape of the area, not

Table 5.2. Cross-tabulation 1954–2000.

	Extent in ha	Rock outcrops	Shrubs	Anthropic	Woodland	Chestnut orchard	Pasture	Wooded pasture	Afforestation	Total
						2000				
	Anthropic			0.48	0.57	0.23	0.16	0.01	0.13	1.56
	High stand	0.65	5.26		453.47	12.21	3.57	19.37	10.67	505.20
	Chestnut orchard	0.01			19.61	41.66	0.54	0.43	14.18	76.43
1954	Pasture		4.36	0.22	14.42	0.98	2.41	12.80	1.73	36.92
	Wooded pasture	2.11	12.96	0.20	53.75	1.73	1.01	18.42	17.50	107.69
	Sowable		2.92	0.69	18.54	1.20	67.60	57.43	25.32	173.69
	Total	2.78	25.50	1.58	560.36	58.01	75.28	108.46	69.52	901.49

positively affected from an aesthetic point of view. This at least is the case shown in Plate 7, which needs few words, presenting a problem already posed to Italian foresters in the early 1960s. However, we must also consider that the monks had a fir plantation at the back of the abbey (see Fig. 5.1), presenting a historical and cultural link which accounts for the conservation of a portion of the conifer forests. On the other hand, in the dark internal landscape of the even aged Douglas fir plantation, covered by a layer of dead needles (with no grass or regeneration) one can still see dead stumps, trunks, and the stone huts for drying chestnuts of the pre-existing chestnut grove. The reduction by 23% of chestnut orchards is in line with the pattern observed for Tuscany as a whole, as indicated in Chapter 1, and it is described in terms of dynamics due to natural and human factors by several authors (Cronon, 1983; Vos and Stortelder, 1992; Agnoletti, 2005). This new conifer forest is the clear sign of the absence of farmers and shepherds in the area, replaced by a possible landscape for loggers of lumber companies, but is also a sign of a more widespread European trend (Johann *et al.*, 2004). The population in the year 2000 is now reduced to 15 people in the whole community, but the return of pasture land in the lower part of the valley is due to the activity of the farm 'Le Lame', where horse breeding is now taking place. Pastures in the form of simple pasture, wooded pasture and also pastured woods are still there; in fact a detailed study made on the 'Le Lame' farm shows that 68% of its landscape has not changed since 1832, presenting also a rare example of pastured wood inside a turkey oak stand.

A Synthesis of the Changes 1832–2000

Looking at the changes that took place from 1832, the enormous increase of woodlands is evident, increasing from 353 to 688 ha in the year 2000, a trend already noted in many rural areas submitted to abandonment (Foster *et al.*, 1998). It must also be reported that the period between the 18th and early 19th century is probably the one where we have the lowest extension of forest land, as also noted (although with some local variations) for Europe (Watkins and Kirby, 1998; Agnoletti and Anderson, 2000). A fundamental aspect related to this tendency in Moscheta is the dramatic reduction of landscape diversity, reflecting huge changes from all points of view: social, economic, environmental.

Table 5.3 shows the strong reduction in the number of land uses and patches forming the landscape mosaic, accompanied by the significant increase in their average size and Hill's diversity number, a good example of the transformation from a fine-grained to coarse-grained landscape, an important process also in terms of habitats (McArthur and Levins, 1964; Southwood, 1977). The dominance index is reduced because of the absence of the large patches of pastureland existing in 1832. The diminution of 67% of the land uses and the increase of the average extension of patches by 54% shows the decrease of the complexity of landscape mosaic, since there are no parts of the area presenting fine-grained structure any more. The analysis of the index of Sharpe clearly indicates the existence of two different tendencies for the periods 1832–1954 and 1954–2000 (Fig. 5.7). The first one is characterized by the reduction of wooded pasture

Table 5.3. Indices of landscape ecology for the three years.

	1832	1954	2000
Dominance index	1.03	0.86	0.71
Hill's diversity number	31.85	3.81	14.27
Number of land uses	89	9	29
Average extent of patches	3.39	6.13	5.25
Number of patches	266	147	172

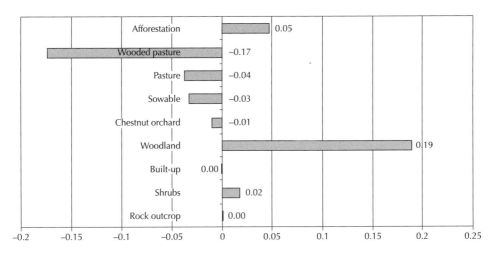

Fig. 5.7. Index of Sharpe. On the left the land uses affected by a reduction, on the right the land uses affected by an increase, the numbers indicating the intensity of the trend.

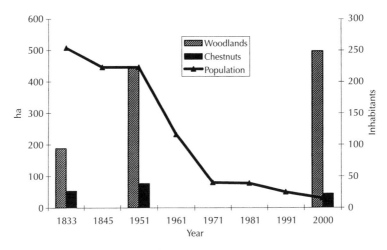

Fig. 5.8. Population and woodlands in Moscheta.

and the increase of woodlands and sowable land, while the second presents the continuous increase of woodlands and afforestation, as well as the decrease of sowable land and wood pasture. It must also be observed that the second period, although not characterized by great changes in terms of surfaces, indicates significant tendencies in a much shorter time period. All these changes are happening because of one single main driving force: the action of man. Figure 5.8 presents the relationship between decrease in population, increase of woodlands and the reduction of chestnut orchards, according to a typical pattern for traditional rural societies where changes in population affect local resources. The map of the landscape dynamics (Fig. 5.9) shows the distribution of changes in the territory, which looks like it is being transformed in most of its parts, with none of the differences due to altitudinal belts.

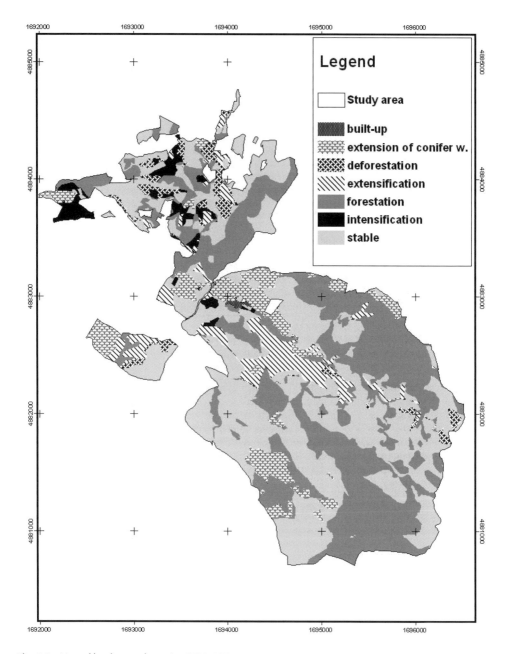

Fig. 5.9. Map of landscape dynamics 1832–2000.

Most of the sampling forays made during the field work were to analyse the dynamics of vegetation in abandoned fields. There is no space here to present the results; however, current successions are dominated by shrubs with *Rosa canina*,

Prunus spinosa and *Juniperus* entering in former pastureland, showing the prevalence of *Juniperus* in most of the situations according to well-known successional patterns (Peterken, 1981). One study area was the historical chestnut orchard existing on the slopes at the back of the abbey, with an extension of 400 m^2, where there are only four ancient trees left with a diameter of 110, 130, 100 and 90 cm, and height of between 15 and 20 m. The bad health of the trees, with many dead branches and stems, will soon lead to the disappearance of this remnant of the historical landscape (Fig. 5.12), a problem quite widespread all over Tuscany and reported for other areas (Agnoletti, 2005). This orchard is related also to a study conducted on the relationships between land uses and soil during the project. The study has collected eight samples of soils, 50 cm deep, located in chestnut orchards (two samples) and pastures (six samples), including different conditions related to areas classified as stable and slide and under erosion, by the geologists carrying out the study. The pastures are located in the bottom of the valley, at the foot of the slope where the chestnut orchards are. Almost all the soils show the prevalence of a clay matrix, with acidity (pH) between 6.90 and 7.42. Considering also the geological and morphological features, the results of the investigation show the absence of any direct relation between soil quality and land use, confirming that the presence of pastures or chestnuts was only due to the choice of farmers and the owner in order to meet the needs of the estate. In other words, no ecological conditions justify the presence of the two land uses in the area where they are today; the chestnuts could instead have been planted in the valley and the pastures have been placed on the slopes.

The Historical Index

In the case of Moscheta, the historical index (Agnoletti and Maggiari, 2004) was applied also to gather indications for the landscape management plan. The index is used to assess the value of a cultural landscape analysing the changes in time and space of any single land use or patch, creating a hierarchy in which every element has a ranking according to the value of the index. The index requires the definition of a *spatial scale* expressed in ha (Sr), and a *temporal scale* expressed in years (Tr) representing the limits in which the index is applied. Other variables are the *historical geographic distribution* (Hgd), that is the past extension of the land uses at the beginning of the period considered, and the *present geographic distribution* (Pgd), that is the present extension of the land uses, both expressed in ha. The other element of the index is the *historical persistence* (Hp), the number of years of presence of a given land use in the Tr considered, and its value will vary from 0 to 1. After choosing the Sr and Tr, the algorithm to calculate the Hi is the following:

$$Hi = Hp \ (Hgd/Pgd)$$

The index attributes a higher value (Hi) to those elements with a long historical persistence (Hp), but a present geographical distribution (Pgd) smaller than the one of the past. Using the database created with GIS, every land use can be analysed, considering its historical persistence (Hp) and the variation in the extension (Hgd–Pgd). The data resulting not only create a hierarchy that can be referred to single or groups of land uses, but can be represented in maps of the area studied. The maps created can refer to the 'general' Hi and the 'topographical' Hi. In the first case the different colours of the map indicate a single land use or a group of land uses that have different values of the index. In the case of Moscheta this means that any land use existing today has a value, according to the result of the calculation, regardless of the fact that its location today is the same as that of 1832. The highest values indicate an emergency, given the fact that land uses with a long persistence have strongly reduced their extension. This means that management should operate to protect or to restore these land uses, particularly important not only for the cultural landscape, but also as habitats that are going to disappear. The topographical Hi map instead indicates and classifies only the land uses that are still present in the same exact location as that of 1832, attributing a greater value to them, as they are not separated from their original topographic position.

The study made for Moscheta shows the values indicated in Fig. 5.10, where the highest level of the index is found for the wooded pastures, followed by pastureland, chestnut orchards and urban areas. The general Hi shows that the land uses with the highest values are concentrated in the bottom of the valley (Fig. 5.11), along the 'Fosso di Moscheta' stream, and in some areas of Mount Acuto, while the topographic Hi shows that land uses with the same locations and high values are also located there, but with a much smaller extension. These areas have the most important meaning for the conservation of the cultural landscape of Moscheta. The present structure of the index is quite simple, but leaves several matters unsolved, a problem currently under investigation through the elaboration of a new index using a 'decision tree' model with disaggregate information, trying to solve the matter of the lack of information in between the Trs chosen. However, the index has already been applied in the guidelines for the environmental impact assessment of wind farms for Tuscany.

The Survey of Social Perception and Economic Value

The creation of a park, as well as any other kind of planning, must include the participation of the public in the process. Therefore, a team interviewed 100 people chosen among residents and tourists, asking their opinion about the landscape and also investigating their willingness to pay a tax for its protection, a very important element to create a credible scenario, supporting the questionnaire with photographic documentation. The methodology used was derived from the open-ended contingent valuation method (Mitchell and Carson, 1989; Loomis, 1990), with a list of questions meant to ascertain the knowledge of the landscape and a maximum or a minimum amount to pay in relation to specific measures to protect the landscape. The list of questions was quite specific regarding the land uses and less focused on more general aspects often investigated in similar studies (Phipps, 1996). Concerning the age classes, 39% of the people interviewed had an age between 31 and 45, a category representative of both the highest productive capacity and income level. The cultural level was another important factor affecting the test: 42% of the people had a university degree, 34% had finished high school, while a smaller percentage (7%) had only attended junior high school and 10% primary school.[1] Concerning origins, 74% came

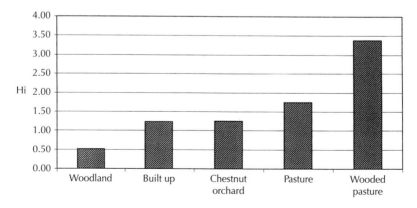

Fig. 5.10. Values of the historical index for Moscheta.

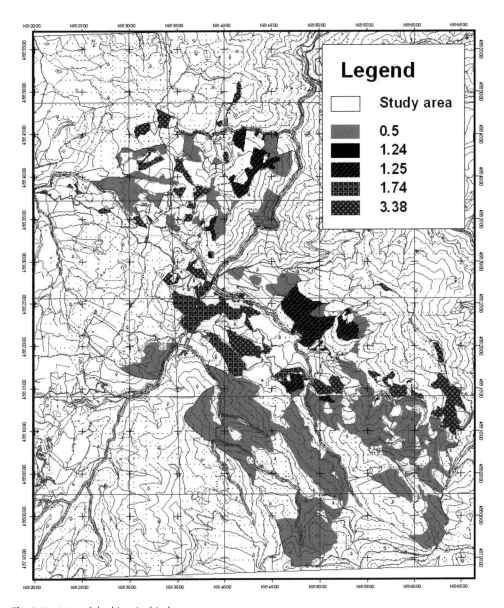

Fig. 5.11. Map of the historical index.

from Tuscany, a common trend since a high percentage of the inhabitants enjoy spending their holidays or free time in the region. Regarding working activities, 35% were private entrepreneurs, 27% were employees, 15% retired people, 6% housewives and 7% students. As for the questions about landscape changes, 55% of those interviewed did not think the landscape had changed in the last decades, while 45% indicated that some changes had occurred, also indicating the reasons.

Most of the people (92%) recognized a positive role of agriculture and silviculture for the landscape, 3% indicated a negative role and 5% did not answer. Abandonment

Table 5.4. Indications coming from interviews. Numbers rating importance are ranked with lowest numbers being most 'important' (i.e. chestnut orchards).

Reason for changes in the landscape (%)		Importance of some elements in the landscape		Indications for public incentives (%)	
Abandonment	43	Chestnut orchards	4	Agriculture	21
Infrastructures	35	Pastures	7	Chestnut orchards	20
Human action	10	Mixed woods	7	Tree rows	11
Lack of mixed cultivations	3	Mixed cultivations	8	Terraces	10
More forests	3	Wooded pastures	8	Coppices	9
More wildlife	3	Fir forests	8	Mixed cultivations	8
Polluted rivers	3	Pine woods	8	Pastures	8
		Tree rows	8	Pine woods	7
		Sowable	9	Wood pastures	6
		Terraces	9		
		Coppice woodland	9		

was clearly indicated as the most important reason for landscape changes (Table 5.4), confirming the result of the historical analysis, while the infrastructures probably refer also to the works for the new railway connecting Bologna and Florence, having a very strong impact on landscape and environment. All the minor aspects indicated, however, represent effective elements of the landscape dynamics occurring in the area. Particularly interesting is the wildlife problem; in fact, the increase of woodlands and also the policy undertaken in recent years of reintroducing animals like wild boar, deer and wolves has favoured a strong increase in their number and in the number of other small mammals. Unfortunately, their high number and the lack of open spaces are contributing to their abnormal increase, allowing them to come closer and closer to the villages and farms, a common problem also in many parks and protected areas in Italy (Filacorda, 1999). The importance given to chestnut orchards also confirms the value given to the role of man and the significance of them as a key element for the cultural identity of the area, although the hierarchy produced by the interviews does not exactly reflect the 'emergencies' revealed by the historical index.

Responding to the need to detect the willingness to pay, the questionnaire proposed some forms of conservation measures, asking which one was considered more important to support. Almost all those interviewed (85%) indicated the need for more than one policy, but the two most important were those to support agriculture and chestnut orchards, while minimum importance was given to the recovery of wooded pastures, which are the element that has nearly disappeared in the landscape. Interviewees were also asked to specify the maximum amount of money acceptable as a form of yearly tax to support these measures. Sixty percent of the people indicated no intention to pay at all, while the rest specified the amount. The resulting average annual amount was 24 euros. Put into context with the social features, it shows that a higher amount was accepted by those having reached only a high-school level education, while those having attended university were prepared to pay 30% less. Concerning working activities, the highest amount was indicated by private entrepreneurs, obviously those with higher availability of income, followed by employees. Another interesting element concerned the reasons for being there, including tourism (25%), recreation (25%), the quality of mountain landscape (11%) and a visit to the small museum in the abbey (12%). In general, the survey showed important attention given to the elements making up the landscape of the area and the role of agriculture and forestry, but also an inverse relationship existing between the cultural level and the will to pay, which was quite surprising.

Back to the Historical Landscape: the Restoration Project

The information collected in the investigation has been elaborated to prepare a plan for the park, which at this stage represents the proposal on which the local authorities will make their final decision concerning the creation of the park. Despite some uncertainties expressed by several studies on what is good or bad concerning grain size in landscape for biodiversity (Angelstam, 1997; Farina, 1998), or the fitness of patch interpretation for the concept of habitat (Mitchell and Powell, 2003), we think that for ecological, cultural and economic reasons it is recommended to recover at least a part of the diversity existing in the past, reducing the process of forestation. A theoretical previous 'non-human-impacted' condition, besides the difficulties in its identification, surely does not correspond to the cultural identity of the area, represented by the period with the deepest links between the people and the land. Nor can one be sure whether this can really be considered the best condition for the health of an ecosystem (Vogt et al., 1997).

It must be pointed out that a project like this, like the forest management plans, must have a long time-period perspective, whose results will only be seen by future generations. In the following paragraphs we are not going to consider other activities related to the inventorying and conservation of a lot of material evidence existing in the territory (dry walls, charcoal places, huts, etc.), as well as the promotion of traditional practices, but rather the intervention needed for the management of the park. In order to create the park there is the need not only to indicate the measures to protect some portions of the territory, but also to 'restore' some of the ancient landscapes. In fact, the simplification of the landscape mosaic occurring in the last two centuries requires a reduction of the amount of certain land-use categories and an increase of others, redistributing them in the territory.

There are several problems connected to this vision. From a technical point of view the actions have been focused on a reduced portion of the territory, in order to find a reasonable size for economic and management reasons. The latter also presents important problems. Official forestry has neglected to incorporate traditional practices in its texts; therefore, even well-known books do not mention them, while others may refer briefly to the existence of traditional silvicultural systems such as wood pastures without detailed information (Perrin, 1954). The other problem is due not only to the lack of any law allowing such a park (in fact only natural parks are foreseen), but also the existence of a regional forest law forbidding the reduction of the extension of woodlands, even in places previously cultivated. This is possible only by doing a compensatory afforestation, or paying an equal amount of money for an afforestation that has to be made somewhere else. Despite special permissions for study and research, this is a problem affecting not only any restoration of former rural mosaics, but even the creation of panoramic sites. Tuscany has hundreds of roads potentially offering outstanding scenery and views of landscapes, but most of them are darkened by trees growing at the sides of these roads, protected by law. Therefore, although this landscape park might find a way to be created, the wider problem of landscape restoration still has a long way to go.

The selection of interventions at landscape scale

The project team prepared a list of potential interventions attributing to each one of them an identification number shown on a map of 'potential interventions', where 193 of them were selected. To each one of them has been allocated an index of 'restoration complexity', classifying them into three categories: (i) low complexity; (ii) medium complexity; and (iii) high complexity.

These categories consider the technical aspects, time and cost of interventions. The main goal of the work has been to recover a fairly good amount of past land uses, considering the historical value, with the aim of increasing the diversity of the landscape and selecting the areas also according to how the future landscape is going to look. The only ecological concern has been the evaluation of the hydro-geological risk concerned with the effects of some of the interventions.

The interventions indicated cover 28% of the total area. The recalculation of the historical index, including the new data resulting from the inclusion of ancient landscape forms, shows the decrease of the value of the index for the categories most threatened and also the inclusion of a category (sowable land) deleted from the area, supporting the effectiveness of the planning (Table 5.5).

Table 5.6 shows the amount of the main land-use categories of 1832 recovered with the percentage of each one of the total area. More details are given in the following paragraph. About 16.2% of the proposed interventions will affect the present surface

area occupied by woodlands, and 5.6% will affect afforestation. The map in Plate 11 shows how the new land uses will be distributed in the land, representing the future landscape of the area.

The interventions

The space available does not allow us to provide details of all the categories recovered and their technical descriptions, as in a real management plan, therefore only some general information will be reported for each main land-use type.

The restoration of *woodlands* in six cases shows the recovery of one unit of pastured woods from a high stand of oak; an oak wood unit from a transitory high stand of oak; a beech wood and pastures unit from a conifer stand; beech woods from mixed conifer stands; and beech woods from a transitory beech stand. The substitution of conifers with beech, which will take some time and will necessitate gradual substitution, constitutes the most difficult problem. An important aspect will be the danger for the young seedlings due to browsing by

Table 5.5. Historical index before and after restoration of the ancient landscape.

Land uses	Hi before	Hi after	Variation
Built up	1.24	1.24	0.00
Woodland	0.50	0.61	0.11
Chestnut orchard	1.25	0.95	−0.27
Sowable land		3.91	3.91
Pasture	1.74	1.21	−0.38
Wooded pasture	3.38	1.87	−1.51

Table 5.6. List of the main land-use categories recovered from the landscape of 1832 and their extent.

Main land uses of 1832 recovered	Area (ha)	% of the total area
Woodlands	50.52	5.60
Chestnut orchards	41.65	4.61
Sowable land	12.30	1.36
Pastures	33.50	3.71
Wooded pastures	114.02	12.63
Total	252.02	27.91

Fig. 5.12. The ancient chestnut orchard near the abbey.

ungulates, today also affecting regular regeneration in beech forests.

The restoration of *cultivated* patches concerns 13 interventions creating very interesting units, from a cultural and natural point of view, bringing back into this landscape elements that no longer exist. Sowable land with chestnuts, but also with hazelnuts, mulberry trees, walnuts, oaks, as well as kitchen gardens, all represent an opportunity to increase biodiversity, recovering fruit trees usually not considered in the list of endangered species and therefore not protected, but disappearing from this cultural landscape. Cultivated areas were often alternated with pasture land and put close to farmhouses and the abbey, where still today they could present an interesting option for small-scale landscape units, providing typical food for the little local restaurant. All the patches possible to recover are presently occupied by forest trees, such as Douglas fir, oak, beech and some chestnuts. For all of them a gradual removal has been planned.

The restoration of *chestnut* typologies is referred to in seven interventions to recreate ancient patches. The most extended pattern suggested for restoration is the 'mixed chestnut and pasture', presently occupied by partly abandoned chestnut orchards (20 ha), and a Douglas fir stand (12.4 ha) (see Fig. 5.12 for the ancient

chestnut orchard and Fig. 5.13 as an illustration of use of the whole tree). For the chestnuts, it is simply a matter of reintroducing grazing, as farmers did to keep the soil clean, while the Douglas firs must be removed and replaced by new chestnut trees. The easiest intervention among those listed is the one related to the conversion of chestnut coppice to 'high stand of chestnut with pastures', although big trees will take several decades to grow, while a moderate

Fig. 5.13. The traditional use of forest trees concerned the whole tree. In this photograph of the Garfagnana valley (1923), the chestnut leaves are harvested and transported with the 'cavagnada' and then put on the floor of the stable (Regione Toscana, 1997).

complexity is given to the realization of mixed chestnut/beech stand.

The restoration of *pastureland* represents a much more complex problem considering how much of this land use has been removed from this landscape and the 62 different interventions foreseen. The possibility to recover pastureland is also linked to the possibility of having animals graze in the area, an opportunity already given by horse breeding and included in the future economic plan for the area, although this has not always worked as expected (Laycock, 1991). It is also relatively more difficult to go back to a pasture compared to developing a wood pasture (Filacorda, 1999); however, these formations can be submitted to a management plan as still occurs in Spain, in order to replace trees. The easiest situations are those related to the transformation of pastures with shrubs into simple pastureland, while the most difficult situations are those considering the restoration of pastures from dense woodlands. In this case, after the removal of trees and their root systems, it is important to facilitate the fast growth of a grass layer to protect the soil from erosion. This intervention has been excluded from steep slopes in order to reduce the danger of erosion.

Conclusions

The research in Moscheta has clearly shown that traditional knowledge was able to create very complex landscape mosaics, richer in biodiversity and cultural values than those existing today after the abandonment of traditional farming activities. From this point of view this study can be considered a sort of regressive long-term experiment on biodiversity, as suggested many years ago (Platt, 1964), but more recently considered impossible (Christiansen *et al*,. 1996). Today the local ecosystem incorporates history, with all the effects of economic, social and environmental factors that acted over this time period that must be considered in future management, as ecologists have also acknowledged (Allen and Hoekstra, 1994; Kay and Schnei-

der, 1994). The real emergency, even at an ecosystem level, seems to be the dramatic reduction of the diversity of spaces, considering also that a healthy ecosystem is not necessarily one with a specific composition and functional organization that is comparable to the natural ecosystems in the same geographic region (Karr and Dudley, 1981). This clearly suggests the introduction of a landscape approach, which the Regional Government of Tuscany has promoted also for the management of protected areas (Agnoletti, 2005), considering that the values reflected by the protected areas, now covering almost 18% of the territory, are basically better expressed by the concept of landscape.[2]

Although the area has been submitted to huge changes, it still shows some evidence of the ancient landscape, in the form of cultural and ecological values expressed by the landscape mosaic, as for many cultural landscapes (Mitchell and Buggey, 2000), presenting also outstanding beauty and aesthetic values. The landscape of Moscheta can be restored, protected and improved. This can be done not by excluding it from present social and economic conditions, but by integrating it into a rural development plan, utilizing both internal and external resources as for other marginal areas (Bartos *et al.*, 1999), taking advantage of the growing importance of services in rural development (Cox *et al.*, 1994), especially those related to landscape (as with agritourism) (Chang *et al.*, 1997–1998; Casini, 2000). The project is surely an ambitious proposal, especially considering the way forestry, environmental protection and rural development is today conceived and applied. Besides the inevitable difficulties there is also the need for such projects in order to generate a debate, hoping that new ideas and visions can develop; but surely something must be changed in the way we are interpreting both environmental changes and sustainable development. However, although the 'context' of the project can be quite different from other situations, we believe that in many other countries, also in developing ones, there is the need to develop such an approach, especially keeping a close link between local people, traditional knowledge and the landscape they have created (Sardjono and Samsoedin, 2001). The recognition, conservation and sustainable management of landscape systems facing economic and cultural globalization seems to be a better way of interpreting sustainable development, rather than the application of paradigmatic views on nature conservation. The degradation of cultural landscapes is not inevitable, as is often suggested. They are a significant resource that needs to be understood, preserved and allowed to evolve.

Notes

1. In the Italian school system children attend 5 years of primary school starting at the age of 5, then 3 years of middle school, 5 years of high school and 5 years of university. The new system has changed university studies according to the Bologna process.
2. Mauro Agnoletti is the author of the project for the guidelines on the conservation and development of landscape resources in the network of protected areas in Tuscany, not yet published.

References

Agnoletti, M. (ed.) (2002a) Il Paesaggio Agro-forestale Toscano, Strumenti per l'Analisi la Gestione e la Conservazione. ARSIA, Firenze, Italy.

Agnoletti, M. (2002b) Le sistemazioni idraulico-forestali dei bacini montani dall'unità d'Italia alla metà del XX secolo. In: Lazzarini, A. (ed.) Processi di Diboscamento Montano e Politiche Territoriali. Alpi e Appennini dal Settecento al Duemila. Franco Angeli, Milano, pp. 389–416.

Agnoletti, M. (2005) Landscape Changes, Biodiversity and Hydrogeological Risk in the Area of Cardoso between 1832 and 2002. Regione Toscana, Tipografia Regionale, Firenze, Italy.

Agnoletti, M. and Anderson, S. (eds) (2000) Forest History: International Studies on Socioeconomic and Forest Ecosystem Change. CAB International, Wallingford, UK.

Agnoletti, M. and Maggiari, G. (2004) La valutazione dell'impatto sul paesaggio e sul patrimonio storico, architettonico e archeologico (The evaluation of the impact on landscape and cultural heritage of windmill farms). In: Regione Toscana, Linee Guida per la Valutazione dell'Impatto Ambientale degli Impianti Eolici, Firenze, Italy.

Agnoletti, M. and Paci, M. (1998) Landscape evolution on a central Tuscan estate between the eighteenth and twentieth centuries. In: Kirby, K.J. and Watkins, C. (eds) The Ecological History of European Forests. CAB International, Wallingford, UK, pp. 117–127.

Agnoletti, M. and Paci, M. (2001) Monks, foresters and ecology: Silver fir in Tuscany from XIV to XX century. In: Corvol, A. (ed.) Le Sapin. L'Harmattan, Paris, pp. 173–194.

Allen, T.F.H. and Hoekstra, T.W. (1994) Toward a definition of sustainability. In: Covington, W.W. and De Bano, F.L. (eds) Sustainable Ecological Systems: implementing and ecological approach to land management. US Department of Agriculture, Technical Report RM-247, Fort Collins, pp. 98–107.

Angelstam, P. (1997) Landscape analysis as a tool for the scientific management of biodiversity. Ecological Bulletins 46, 140–170.

Arnold, J.E.M. (1995) Framing the issues. In: Arnold, J.E.M. and Dewees, P.A. (eds) Tree Management in Farmer Strategies. Oxford University Press, Oxford, pp. 3–17.

Bartos, M., Tesitel, J. and Kosova, D. (1999) Marginal areas – historical development, people and land-use. In: Kovar, P. (ed.) Nature and Culture in Landscape Ecology. The Karolinium Press, Prague, pp. 109–113.

Baudry, J. and Baudry-Burel, F. (1982) La mesure de la diversité spatiale. Relation avec la diversité spécifique. Utilisation dans les évaluations d'Impact. Acta Ecologica, Oecol. Applic. 3, 177–190.

Casini, L. (2000) Nuove prospettive per uno sviluppo sostenibile del territorio. Studio Editoriale Fiorentino, Firenze.

Casini, L. and Ferrini, S. (2002) Le indagini economiche. La valutazione economica del paesaggio toscano. In: Agnoletti, M. (ed.) Il Paesaggio Agro-forestale Toscano, Strumenti per l'Analisi la Gestione e la Conservazione. ARSIA, Firenze, pp. 49–68.

Chang, Ting Fa M., Piccinini, L.C. and Taverna, M. (1997–1998) Agricoltura futuribile: primario o terziario? *Agribusiness Paesaggio e Ambiente* 4, 237–255.

Christiansen, N.L., Bartuska, A., Am., Brown, J.H. *et al.* (1996) The report of the ecological society of America committee on the scientific basis for ecosystem management. *Ecol. Appl.* 6, 665–691.

Cox, L.J., Hollyer, J.R. and Leones, J. (1994) Landscape services: an urban agricultural sector. *Agribusiness* 10, 13–26.

Cronon, W. (1983) *Changes in the Land.* Hill and Wang, New York.

Farina, A. (1998) *Principles and Methods in Landscape Ecology.* Chapman and Hall, London.

Filacorda, S. (1999) L'animale come elemento mobile e modificabile del paesaggio. *Agribusiness Paesaggio e Ambiente* 3, 172–183.

Foster, D.R, Motzkin, G. and Slater, B. (1998) Land-use history as long-term broad scale disturbance: regional forest dynamics in central New England. *Ecosystems* 1, 96–119.

Fowler, P.J. (2001) *Cultural Landscapes: great concept, pity about the phrase.* ICOMOS-UK, London.

Fowler, P.J. (2003) *World Heritage Cultural Landscapes 1992–2002.* UNESCO, Paris.

Freschi, M.L. and Hermanin, L. (2000) A brief history of Italian forest policy. In: Agnoletti, M. and Anderson, S. (eds) *Forest History: International Studies on Socioeconomic and Forest Ecosystem Change.* CAB International, Wallingford, UK, pp. 351–362.

Fuentes Sanchez, C. (1994) *La Encina en el Centro y Suroeste de Espana.* Servantes, Salamanca.

Johann, E., Agnoletti, M., Axelsson, A.L., Burghi, M., Ostlund, L., Rochel, X., Schmidt, U.E., Schuler, A., Skovsgaard, J.P. and Winiwarter, V. (2004) History of secondary spruce forests in Europe. In: Spiecker, H., Hansen, J., Klimo, E., Skovsgaard, J.P, Sterba, H. and von Teuffel, K. (eds) *Norway Spruce Conversion. Option and Consequences.* EFI research report 18. Brill, Leiden, pp. 25–62.

Karr, J.R. and Dudley, D.R. (1981) Ecological perspective on water quality goals. *Environmental Management* 5, 55–68.

Kay, J.J. and Schneider, E. (1994) Embracing complexity: the challenge of the ecosystem approach. *Alternatives* 20, 33–39.

Laycock, W.A. (1991) Stable states and thresholds of range conditions on North American rangelands: a viewpoint. *Journal of Range Management* 44(5), 427–433.

Levins, R. (1968) *Evolution in Changing Environments.* Princeton University Press, Princeton.

Loomis, J.B. (1990) Comparative reliability of dichotomous choice and open-ended contingent valuation techniques. *Journal of Environmental Economics and Management* 18, 78–85.

McArthur, R.H. and Levins, R. (1964) Competition, habitat selection and character displacement in a patchy environment. *Proc. Natl Acad. Sci. USA* 51, 1207–1210.

Mitchell, N. and Buggey, S. (2000) Protected landscapes and cultural landscapes: taking advantage of diverse approaches. *The George Wright Forum* 17(1), 35–46.

Mitchell, R.C. and Carson, R.T. (1989) *Using Surveys to Value Public Goods: The Contingent Valuation Method.* Resource for the Future, Washington, DC.

Mitchell, M.S. and Powell, R.A. (2003) Linking fitness landscapes with the behaviour and distribution of animals. In: Bissonette, J.A. and Storch, I. (eds) *Landscape Ecology and Resource Management.* Island Press, Washington, pp. 93–125.

Ozenda, P. (1975) Sur les etages de vegetation dans les montagnes du Bassin Méditerranéen. *Doc. Cart. Ecol.* 16, 1–32.

Pavari, A. (1916) Studio preliminare sulla coltura di specie forestali esotiche in Italia. *Ann. R. Ist. Sup. For. Naz.* I, 159–379.

Perrin, H. (1952–1954) *Sylviculture (Tome II).* Ecole National des Eaux et Foréts, Nancy, France.

Peterken, G. (1981) *Woodland Conservation and Management.* Chapman and Hall, London.

Phipps, S. (1996) Perceptions of landscape and protection of the wider countryside. In: Jones, M. and Rotherham, I.D. (eds) *Landscapes – Perceptions, Recognition and Management: reconciling the impossible?* Conference Proceedings, Wildtrack Publishing, Sheffield, UK, pp. 25–30.

Platt, J.R. (1964) Strong inference. *Science* 146, 347–353.

Rackham, O. (1995) *Trees and Woodlands in the British Landscape.* Weidenfeld and Nicholson, London.

Regione Toscana (1987) *L'Uomo e la Terra, Campagne e Paesaggi Toscani.* Italia Grafiche, Firenze, Italy.

Rotherham, I.D. and Jones, M. (2000) The impact of economic, social and political factors on the ecology of small English woods in South Yorkshire, England. In: Agnoletti, M. and Anderson, S. (eds) *Forest History: International Studies on Socioeconomic and Forest Ecosystem Change.* CAB International, Wallingford, UK, pp. 397–410.

Salbitano, F. (1987) Vegetazione forestale e insediamento del bosco in campi abbandonati in un settore delle Prealpi Giulie (Taipana – Udine). *Gortani, Atti del Museo Friulano di Storia Naturale*, 83–143.

Sardjono, M.A. and Samsoedin, I. (2001) Traditional knowledge and practice of biodiversity conservation. In: Pierce, C.J. and Byron, Y. (eds) *People Managing Forests*. Resources for the Future Press, Washington, pp. 116–134.

Sereni, E. (1997) *History of the Italian Agricultural Landscape* (reprint, first edition 1961). Princeton University Press, Princeton.

Southwood, T.R.E. (1977) Habitat, the templet for ecological strategies? *Journal of Animal Ecology* 46, 337–365.

Vogt, K.A., Gordon, J.C., Wargo, J.P. and Vogt, D.J. (1997) *Ecosystems: Balancing Science with Management*. Springer-Verlag, New York.

Vos, W. and Stortelder, A. (1992) *Vanishing Tuscan Landscapes*. Pudoc Scientific Publisher, Wageningen, Regione Toscana, Firenze, Italy.

Watkins, C. and Kirby, K.J. (1998) Introduction – historical ecology and European woodland. In: Kirby, K.J. and Watkins, C. (eds) *The Ecological History of European Forests*. CAB International, Wallingford, UK, pp. ix–xv.

6 Long-term Vegetation Dynamics in Southern Scandinavia and Their Use in Managing Landscapes for Biodiversity

R.H.W. Bradshaw[1,3] and G.E. Hannon[2,3]

[1]Environmental History Research Group, Geocenter, Copenhagen, Denmark
[2]Southern Swedish Forest Research Centre, SLV, Alnarp, Sweden
[3]Present address: University of Liverpool, UK

Introduction

Sweden and Finland are sparsely populated and have the highest proportion of forest cover within Europe. About 36,000 km^2 of Sweden (approximately 8% of the total area) is protected within national parks and nature reserves with relatively virgin mountain environments comprising almost 90% of the land protected within national parks. One might imagine that the approaches to conservation of biodiversity would not give a particularly high priority to cultural landscapes in a country where the hunter-gatherer lifestyle persisted long after sophisticated civilizations in southern Europe had brought about major landscape change. This would be a misleading impression. Research has shown that cultural impact in Sweden not only has a longer and geographically more extensive history than was once believed, but also that many of the prized biodiversity values within Sweden owe their existence to a complex inter-relationship between cultural and natural processes. The increasing knowledge that is developing about earlier cultural impact on the Swedish landscape is now becoming incorporated into practical management for biodiversity. In this chapter we review the long-term history of a Swedish forest meadow or wood pasture (Swedish: lövāng)

and an open meadow (Swedish: slåtterāng), which illustrate the importance of former cultural activity for current biodiversity values. Building on these examples, we analyse the longer term development of biodiversity values in southern Scandinavia and finally we discuss approaches to landscape management that aim to maintain some of these values.

Swedish Forest Meadows

Forest meadows or wood pastures are among the most cherished and visually attractive cultural landscapes that exist in southern Scandinavia and this landscape element has counterparts in central and southern Europe. They were developed from forested landscapes close to farms and formed part of the in-field system. They are maintained by selective thinning of trees, mowing and controlled burning to create a mosaic of vegetation types, comprising sparse deciduous trees and shrubs and herb-rich grassy meadows. Many of the trees are pollarded and in the past were shredded for leaf fodder. Characteristic tree species include *Tilia cordata, Fraxinus excelsior, Ulmus glabra, Corylus avellana, Quercus robur* with *Malus sylvestris* and *Crataegus monogyna*. The herb layer is often species-rich and typically includes *Anemone nemorosa, Geranium sylvaticum, Filipendula vulgaris, Leucanthemum vulgare, Galium verum* and many orchid species (Fig. 6.1).

Forest meadows are known to be of cultural origin and most have been abandoned during the last century. Swedish forest meadows are presumed to be the oldest type of grassland community with possible origins in the Neolithic period up to 6000 years ago (Lagerås, 1996), although their detailed history is poorly known. Recent speculation suggests that they have a close relationship with pre-existing 'natural' wood pastures that were maintained by populations of aurochs and tarpan (Vera, 2000), although other explanations are possible (Svenning, 2002; Bradshaw and Hannon, 2004). Their chief purpose was for grazing animals and provision of winter fodder. Berlin (Berlin, 1998) reviewed the status of north European semi-natural

Fig. 6.1. View of the forest meadow at Råshult, Sweden with *Tilia cordata* in the foreground to the left. (Photo: Torbjorn Larsson)

grasslands including forest meadows, which on calcareous soils are among the most species-rich habitats for vascular plants with up to 63 species/m^2 (Kull and Zobel, 1991). There were an estimated 2,000,000 ha of grasslands in Sweden at the beginning of the 1800s with only about 200,000 ha surviving today, of which only 3300 ha are forest meadows (Bernes, 1994). The parish of Mara in the province of Blekinge comprised just five farms within an area of about 10 km^2 in 1769, yet included 130 separate forest meadows (Thomas Persson, personal communication). This illustrates the former importance of this form of land-use in southern Sweden.

Råshult, Småland

One such forest meadow is found at Råshult estate, south-west Sweden (56° 37′N, 14° 12′E), which was the birthplace of Carl Linnaeus and was declared as one of only ten Swedish cultural reserves in 2003. Lindbladh and Bradshaw (Lindbladh and Bradshaw, 1995) used high-resolution palaeoecological methods to investigate the origin, development and subsequent abandonment of the forest meadow at Råshult. They collected 1 m of peat from a small, wet depression located in the Råshult infields and analysed the pollen and charcoal content (Fig. 6.2).

The record covered the last 4500 years and showed that the meadow system had its origins about 1000 years ago. Prior to the development of the forest meadow the data recorded forested conditions with the most important trees being *Quercus, Tilia, Ulmus* and *Corylus*, with *Alnus* and *Betula* growing on wetter sites. Charcoal fragments indicated occasional burning, at least of forest floor communities. Pollen from *Plantago lanceolata* and other indicators of anthropogenic activity indicate some human impact, but the dominance of tree and shrub pollen recorded in this small forest hollow until about 1200 AD strongly suggests that the local vegetation was forest, whose composition had been relatively

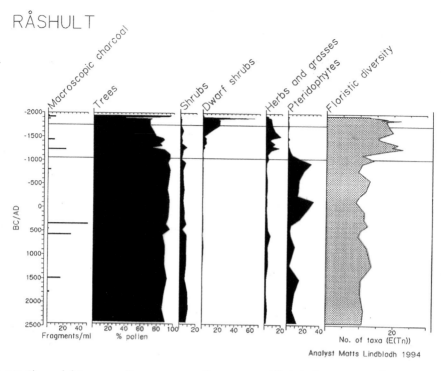

Fig. 6.2 Charcoal data, composite percentage pollen curves and floristic diversity curve for profile from Råshult. A total pollen calculation sum is used excluding pteridophytes; pteridophyte representation is based on the pollen sum plus pteridophytes (after Lindbladh and Bradshaw, 1995).

stable for several millennia. The relative abundance of *Quercus* and *Corylus* and a low but significant presence of a suite of herbaceous pollen types show that the forest had an open structure, creating habitat for light-demanding species (Vera, 2000). This open structure may have been maintained by occasional burning, perhaps in conjunction with light grazing by domestic animals. Aurochs and tarpan have been extinct in Sweden for many millennia (Bradshaw and Hannon, 2004).

The pollen diagram recorded abrupt changes in vegetation and increased floristic diversity beginning about AD 1100 that are clearly associated with cultural impact and the establishment of the forest meadow system (Fig. 6.2). Pteridophytes characteristic of forest conditions almost totally disappeared and the pollen percentages of herbs and grasses increased markedly. The opening of the vegetation is also reflected by the

decrease in percentages of tree pollen to about 80% at this time. Cereal pollen grains, including *Secale cereale*, became more frequent. Grasses and several herbaceous taxa that are strongly associated with agriculture (Behre, 1981; Gaillard *et al.*, 1992) occurred for the first time, or became abundant during this period. Examples of this group are Poaceae (<40 µm; mowing and grazing), *Plantago lanceolata* (mowing and grazing), *Potentilla* (grazing) and *Rumex acetosa/acetosella* (grazing). *Calluna*, which is an indicator of strong grazing pressure (Gaillard *et al.*, 1992), also became more abundant.

The pollen data show a striking correspondence with the archaeological and historical data. Larsson (Larsson, 1975) identified the early Middle Ages as the time of colonization of the forested and remote parts of southern Småland. He further emphasized that livestock-raising was the main activity because of the natural conditions of the area.

The forest meadow period (around AD 1100–1900) was not static, but the pollen assemblages changed through time in response to socio-economic developments. The pronounced dip in floristic diversity around AD 1400 at the in-field site coincides with a critical period in agricultural history. A period of 150 years of economic and population decline began in Sweden following the Black Death of the mid-14th century; the so-called 'agrarian crisis' (Gissel *et al.*, 1981). The initiation and duration of the dip in floristic diversity of the Råshult forest meadow shows a striking similarity with that of the agrarian crisis, and the dating of this part of the profile is securely based. We propose a causal connection between the historical events and the decreasing number of species that would be expected if the meadows and arable fields were abandoned and overgrown. Furthermore, the event is also noticeable in the pollen diagram as a decrease in the pollen percentage values of herbs and grasses and an increase in that of trees, particularly *Betula* – expected observations if cultivation were to cease. The decline in cultivation also coincided with the beginning of a colder period in Europe, namely 'the little ice age' (Bradley and Jones, 1992), and it is probable that the agricultural regression was partly caused, or at least amplified, by the changing climatic conditions. Whatever the ultimate cause of the regression may have been, it is striking that it can be detected in both historical and biological archives. It emphasizes the detailed temporal and spatial resolution that can be obtained from the pollen analysis of small hollows (Bradshaw, 1988).

A dramatic increase of ericaceous dwarf shrubs occurred around AD 1750 and *Calluna* pollen reached values of more than 40% by 1900. *Calluna* is generally considered to be a good indicator that grazing and livestock-raising had been the dominant land-use in this part of Scandinavia for a long time (Larsson, 1975; Emanuelsson, 1987). During the 18th and 19th centuries

the local parish experienced a sharp population increase, from 206 inhabitants in 1703 to 1913 inhabitants by 1850 (Nilsson *et al.*, 1994). As this increase paralleled that of Sweden as a whole, it is likely that the substantial peak in *Calluna* pollen was associated with an increasing number of animals managed by the growing human population (Myrdal and Söderberg, 1991). Another grazing indicator, *Rumex aceotsa/ acetosella*, showed a minor peak at the same time as *Calluna*. This ericaceous phase ended abruptly around 1900 and was replaced by a secondary forest succession and tree pollen percentages returned to high levels. The use of fertilizers was introduced more generally to Scandinavia at the beginning of the 20th century and the pressure on available land was considerably reduced, resulting in reforestation of marginal areas such as Råshult (Emanuelsson, 1987).

The development of floristic diversity, as detected by pollen analysis of this small hollow, reflects the development and demise of the forest meadow system. The rapid loss of diversity during the last century, coupled with the historical interest of the estate being the site where Linnaeus was born and raised, has led to the recent restoration of the forest meadow system.

Slottet, Bjärehalvön

The forest meadow at Råshult developed in a landscape that was probably only permanently settled rather late with place names dating from the Mediaeval Period. The age of the forest meadow may simply reflect the late exploitation of this landscape. The situation around the coasts of southern Sweden is rather different with extensive evidence for human impact dating back many millennia (Berglund, 1991). The hay meadow 'Slottet' on the Bjäre peninsula lies in a landscape that has been influenced by human activity for many millennia and the main landscape structure was established during the Bronze Age. The peninsula contains one of the greatest concentrations of Bronze Age burial mounds in Scandinavia (Gustafsson, 1996).

The meadow has been recently restored and comprises about 2 ha of species-rich open grassland that is occasionally flooded by a small stream. The meadow is first mentioned in historical archives dating from 1596 (Hannon and Gustafsson, 2004). The area was almost totally deforested by 1670 and a map from 1841 showed the present meadow to be part of a larger complex of meadows and arable fields. By 1928 some woody successions were reclaiming open areas and the restored meadow is today surrounded by *Alnus glutinosa* stands with groves of deciduous trees.

A 1-m sediment core was collected for analysis of plant macrofossils and charcoal to investigate the timing and development of the meadow. The sediment had accumulated close to the stream in the meadow, where the permanently waterlogged conditions had preserved plant remains. Plant macrofossils are plant remains visible to the naked eye. They are not so abundantly preserved as pollen but travel less far from their point of production and can usually yield a greater taxonomic resolution. The core covered the period from about 200 BC until present (Fig. 6.3).

Quercus acorns were the earliest macrofossil evidence, dating from just before AD 0. Subsequent records of *Stachys sylvatica* and *Carex* suggest that the earliest recorded vegetation was wet, deciduous woodland. Large fragments of charcoal began to appear around AD 350 and persisted in varying quantities until the 1200s. Natural fire was very rare in the western coastal area of Sweden, so the charcoal is likely evidence for human impact in the form of tree clearance, slash-and-burn agriculture, coppicing and forest grazing. The dominating macrofossil finds are from *Alnus glutinosa*, *Betula pubescens*, *Frangula alnus*, *Polygonum* spp. and *Rubus fruticosus* suggesting that these first cultural activities took place in an open forest setting. The first appearance of *Fagus sylvatica* during the AD 500s occurs in association with both fire and cultural activity which is characteristic of its history in southern

Scandinavia (Bradshaw and Lindbladh, 2005).

The abundance of charcoal reduces after AD 1000–1100 and there is a corresponding increase in the diversity of meadow and fen plants such as *Viola palustris*, *Rumex acetosa*, *Stellaria alsine*, *Stellaria media*, *Cardamine pratensis*, *Glyceria fluitans*, *Chenopodium rubrum*, *Urtica dioica*, *Linum catharticum* and Poaceae. Burning had been replaced by hay-making and coppice as the main cultural activities and a true meadow was created. It is interesting to note that this development was roughly synchronous with the creation of the forest meadow at Råshult, indicating the importance of the early Mediaeval Period in the development of the south Swedish cultural landscape.

During the 1500–1600s there is a slight resurgence of forest species including *Betula*, *Alnus* and conifers, followed by their disappearance again during the 1700–1800s. These trends are in agreement with the fragmentary documentary evidence available and reflect varying pressures of land-use and particularly timber exploitation on this area during the 1700–1800s. From the late 1800s onwards, both the macrofossil and documentary records indicate a gradual fall into disuse of the meadow system until the partial restoration during the 1990s.

These two examples share some common features that give insight into the role of cultural activity in the development of plant diversity in southern Sweden. In both cases a 'natural' disturbance regime involving fire and grazing developed into a more managed regime that stabilized around 1200 AD. The landscape that developed then was not stable, but reflected economic developments and social change. There was heavy exploitation during the 1600–1800s followed by a major change in land use that led to abandonment of parts of the landscape and intensification on other parts. This major change has led to the current crisis in management of biodiversity. We have summarized this development in a conceptual model with three distinctive periods of land use (Fig. 6.4).

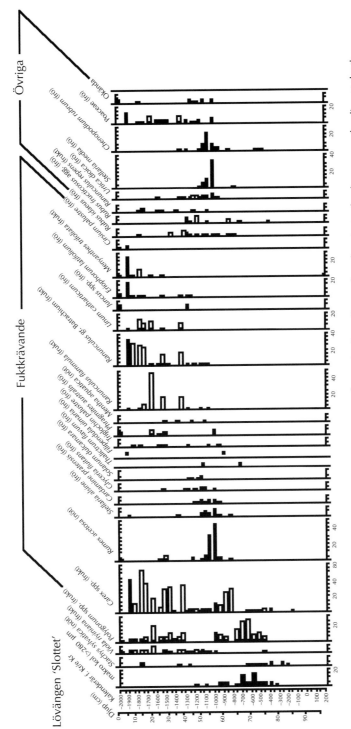

Fig. 6.3. Number of plant macrofossils and charcoal fragments in 20 cm³ samples from Slottet plotted against calendar years and sediment depth. Fuktkrävande – moisture-demanding; övriga – others (after Hannon and Gustafsson, 2004).

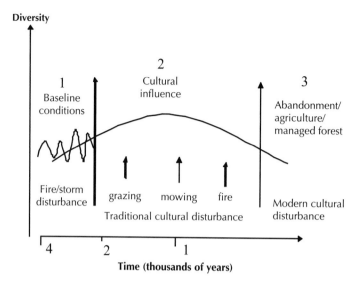

Fig. 6.4. Model of changes in vascular plant biodiversity associated with the development and abandonment of traditional agricultural methods.

1. Baseline conditions

The concept of baseline conditions with minimal anthropogenic impact and maximal 'naturalness' is superficially attractive, but hard to characterize in Europe and is more of academic than practical interest. The concept is more widely used in North America, Australia and New Zealand and other regions where European settlement is relatively recent. Pre-settlement vegetation and fire regimes are often used as a reference for nature conservation in North America, although there is now increasing recognition of the impact of aboriginal cultures, particularly on fire regimes. Important aspects of biodiversity are strongly affected by the disturbance regime that has been systematically modified by cultural activity, so it is of some relevance to reconstruct disturbance regimes prior to major human intervention as a biodiversity baseline.

The most widespread baseline in Northern Europe is various types of forest. Natural forest dynamics and diversity are driven by rare but significant disturbance events such as wind, fire, drought, flooding, land-slippage and disease (Pickett and White, 1985; Peterken, 1996; Gardiner and Quine, 2000). These driving forces interact

with the continuous disturbance of browsing and grazing animals and the small-scale processes of single tree replacement by seedling establishment, growth, competitive interactions and senescence (Falinski, 1986; Woods, 2000). The disturbance regime creates the conditions necessary for regeneration, generates dead wood, influences the size and age structure of a stand and controls continuity of local niches and habitats and hence biodiversity.

Fire

The pivotal role of fire in many natural boreal ecosystems has now been widely acknowledged, reviewed and even incorporated into forest management (Johnson, 1992; Goldammer and Furyaev, 1996; Bradshaw *et al.*, 1997; Angelstam, 1998). Fires have always been of importance in boreal regions, although on longer time-scales this importance has varied dependent upon prevailing climatic conditions. Individual forest fires have tended to be very large in Canada and Siberia, but were much smaller in Scandinavia, at least during the last few centuries where dendrochronological methods can be used to reconstruct fire size (Niklasson and Granström, 2000). It is still an open question whether this difference in scale can be attributed to differences in landscape form and the distribution of natural fire-breaks, or whether cultural factors play a role. Road-building and fire suppression activities have impacted Fennoscandian fire regimes during the last 100 years, but the effect of these cultural factors may have been over-estimated (Carcaillet *et al.*, 2002).

The historical legacy and role of fire in temperate regions and at the boreal-temperate ecotone has been less discussed (Bradshaw *et al.*, 1997), but raises issues of importance for forest management in these regions. Sediment records from small forest hollows throughout southern Scandinavia contained abundant charcoal remains in the past, but charcoal is less frequent in recent centuries and decades. Typical records from south-east Sweden record fires of probable natural origin between 4000 and 500 BC. The charcoal records often change form after 500 BC and become more uniform. This can be interpreted as the onset of management of the natural fire regime, although there was also a widespread change in climate around 500 BC that would also have impacted the fire regime (Lindbladh and Bradshaw, 1998). *Quercus* populations, which regenerate under rather open conditions, were important forest components throughout the last 6000 years but declined in abundance during recent centuries. The decline of *Quercus* is associated with a corresponding increase in local importance of *Picea* and the cessation of charcoal in the sediments. The 'natural' fire regime changes through time but has certainly been greatly modified within the temperate forest zone of Europe during the last few centuries.

Fire is one of the main disturbance factors in natural forest, but because of its long association with human societies and the ease with which a fire regime can be influenced by cultural activity it is difficult to describe a 'natural' fire regime for all but the most remote boreal forest regions of the world. Information about former fire regimes is valuable as an indication of the disturbance intensity and frequency to which the present flora and fauna are adapted. This information is valuable in conservation and the assessment of likely future fire hazard.

Browsing animals

The role of browsing and grazing mammals in natural forest has been largely ignored by forest researchers, partly because of the difficulty in obtaining relevant data. Again, the palaeoecological record documents large changes in the fauna of boreal and temperate forests and a long history of anthropogenic modification of faunas through hunting, herding and later domestication of large mammals (Bradshaw and Mitchell, 1999; Bradshaw *et al.*, 2003).

By the early Holocene, 9000 years ago, the forests of north-west Europe had been re-colonized not only by species still surviving in these countries, such as red and roe deer, but also by elk, aurochs, bear, wolf, lynx, beaver and others (Aaris-Sorensen, 1998; Yalden, 1999). However, this fauna was seriously impoverished compared to previous interglacials, specifically in large herbivores and carnivores. Both natural and anthropogenic factors have affected the Holocene faunas of north-west Europe. Some species left the region long ago, due to natural changes in Holocene climate and vegetation, such as the horse *Equus ferus*; others are now totally extinct, such as the aurochs *Bos primigenius*. Most of the remaining species, however, disappeared after the Neolithic agricultural revolution or even in very recent centuries, as a result of habitat clearance and hunting. In theory, these species still 'belong' to the region. Models for their persistence in a reasonably aboriginal ecosystem exist in Poland and Russia. However, crucial to the current survival of these communities is that they are embedded in very large areas of habitat (particularly, the Russia taiga), so the mammals form a semi-natural metapopulation, which can survive as a whole even if local areas become unsuitable or unavailable for various reasons (Bradshaw *et al.*, 2003).

Liljegren and Lagerås (1993) summarized the Holocene records for larger mammals recorded from Sweden. The range of sources and the quantity of data available permit the development of models describing changing grazing regimes during the Holocene. The changing balance between grazers and browsers (*sensu*) (Hofmann and Stewart, 1972) in southern Swedish forests suggests that the present large populations of roe deer and moose are a recent development. There have been continual fluctuations in the balance between browsers and grazers chiefly due to the early Holocene local extinction of bison (*Bison bonasus*) and aurochs (*Bos primigenius*) (Ekström, 1993) and the subsequent introduction of domestic cattle nearly 3000 years later. The removal of domestic cattle from south Swedish forests during the last 100 years was a significant event in long-term grazing–vegetation interactions and has had a major influence on forest composition and structure (Andersson and Appelqvist, 1990).

The large-scale, anthropogenic alteration of landscape structure in north-west Europe makes the present-day situation significantly different from previous interglacials or from the early- to mid-Holocene. As with fire, we may gain new insight into processes of ungulate–vegetation interactions from palaeoecological data, but the current situation is unique when viewed from the long-term perspective. Studies of the past reconstruct the trajectory of events that created the present condition and give insight into processes, but they also show that no single set of equilibrium 'baseline' conditions can be recognized in the recent geological past, so there is no secure reference from the past to use as a model for future management. The much higher ungulate diversity (and presumably total biomass) of past interglacials strongly suggests a diverse vegetational environment that varied between interglacials, with elephants, rhinos and many species of deer that must have been niche-separated. It is hard to conceive how the ungulate communities could have caused the apparent differences in habitat diversity between past interglacials themselves (Vera, 2000; Svenning, 2002). Grazing-adapted species, by

definition, are unlikely to have been responsible for the clearing of forest in the first place. One might also ask why they did not appear and open out the forest in previous interglacials. The conclusion is that other, probably physical, factors were responsible for the primary habitat structure (Bradshaw *et al.*, 2003).

Storm

In Europe there are only a few, isolated studies of storm damage from near-natural forests (Falinski, 1978; Pontailler *et al.*, 1997; Wolf *et al.*, 2004) and most information is derived from coniferous forest or plantations. Tree size and species are important factors influencing the extent of storm damage.

In December 1999 the most powerful hurricane ever recorded in Denmark caused the greatest destruction of forest volume in Denmark during the 1900s. In Draved Forest, a semi-natural deciduous forest in southern Jutland, long-term observations of tree growth and mortality were available. Analyses of mortality characteristics through time showed that storm was the major mortality factor affecting large trees in this forest. For smaller trees, competition was an important cause of death, as trees that were found standing dead had a slower growth rate than survivors. Individual species showed different mortality patterns. *Betula*, *Fagus* and *Tilia* were mainly windthrown, whereas for *Alnus* and *Fraxinus* 50% of the mortality was observed as standing dead trees. So this study concluded that both wind and competition are important mortality factors in Draved Forest (Wolf *et al.*, 2004). An important finding from the Draved study was that about 4% of all trees larger than 10 cm diameter at breast height (dbh) were damaged by the storm, while in neighbouring commercial coniferous plantations, damage was almost total. Natural forest mortality rates range between 0.5% and 3% per year in the boreal and temperate zone (Runkle, 1985; Peterken, 1996) and natural forests are clearly better adapted than managed forests to survive severe storm damage.

Pests and disease, flooding and land-slippage are other natural disturbance factors that influence forest structure and the biodiversity of natural forest, but the greatest disturbances that have affected forests in the boreal and temperate zones are those brought about by human activities, both agricultural and silvicultural. Forest has been cleared for agricultural use for over 5000 years in Europe, and even if agricultural activity has subsequently been abandoned, the break in forest continuity can affect the species composition for a long time.

Commercial forestry has altered the size structure of forests throughout boreal Scandinavia by first the removal of large trees and subsequent short rotation periods. Linder and Östlund (1992) documented the enormous modification of forest structure that has typically occurred in northern Sweden during the last 125 years. Densities and basal areas of large trees in surviving old-growth North American and European temperate forests were found to be comparable (Nilsson *et al.*, 2002). The adequate description of natural disturbance regimes and full recognition of how they have been modified, both directly and indirectly by anthropogenic activities, is one of the major challenges in the identification and study of natural forest ecosystems.

Other important characteristics of natural forest systems

The hydrological regime of natural forests has not been studied in detail, but regional and local drainage of forest and agricultural areas has been relentlessly pursued in the temperate and boreal zones for many decades without proper economic assessment of the consequences (Møller, 2000). Regional pollen diagrams record a general reduction in amounts of *Alnus* and *Salix* on the landscape and the field experience of all European forest ecologists is of a systematic alteration of micro-habitats whose effects on biodiversity are only now being recognized.

The role of dead wood and its importance for forest biodiversity has also been stressed by many researchers (Bobiec, 2002;

Krankina *et al.*, 2002; Mountford, 2002). The natural level and dynamics of dead wood varies greatly in space and time and is generated as a consequence of the disturbance regimes discussed above.

2. Cultural influence

The history of the last 5000 years of northern European forest cover is primarily one of deforestation and the spreading of diverse cultural habitats with open, species-rich vegetation structure. The two Swedish case studies presented earlier are detailed examples of this development. Reliable quantitative data are only available for the last few decades, but generalized deforestation trajectories have been estimated for several countries based on palaeoecological and historical sources and some general features emerge. Typically, deforestation trajectories fall to a minimum value of national forest cover and then reafforestation takes place either due to socio-economic collapse of the traditional agricultural systems and natural woody succession, or to active tree planting. The timing and minimum value of forest cover reached varies somewhat between regions. For Sweden as a whole the minimum value was probably reached between 1850 and 1900, and whereas the minimum percentage cover was considerably greater than in other European countries, most of this cover lay in the relatively inaccessible and sparsely populated western and northern regions. The trajectory for Denmark is rather typical for north-western Europe with a minimum forest area of around 3% reached around 1800. A detailed survey was made of the Danish landscape between 1780 and 1820 that generated a map of non-forest area.

The deforestation trajectory in Denmark typically began around 5500 years ago during the Neolithic Period (Odgaard and Rasmussen, 2000). The onset of widespread deforestation can be detected from pollen diagrams and is typically associated with the time-transgressive spread of settled agriculture across Europe from south-east to north-west (Roberts, 1998). All pollen diagrams show a significant increase in landscape-scale species diversity as relatively homogeneous forest communities were replaced by a mosaic of cultural habitats (Fig. 6.5).

Arable crops were cultivated on land cleared of forest, often by burning. However, grazing of livestock was probably a greater driving force for reduction of forest area through restriction of regeneration and winter browsing of young growth (Bradshaw and Mitchell, 1999). The effects of Neolithic agriculture on general forest structure are also actively debated, but it is likely that many European forests began to develop more open structures from this time onwards (Vera, 2000; Svenning, 2002; Bradshaw *et al.*, 2003).

Thus, the phase of cultural influence is of variable length and intensities throughout Europe, with central and northern Scandinavia showing the least impact. However, in all areas the phase of 'traditional' cultural impact generated the biodiversity legacy that survives today and is now under pressure from recent, rapid changes in land use. The phase of traditional cultural impact lasted many millennia in southern Europe, and at least the last 1000 years in southern Sweden. As the

examples above have illustrated, important new vascular plant biodiversity value was added during this period. It also acted as a bridge to the less-modified 'baseline' conditions which generated more 'natural' biodiversity. Both these biodiversity elements have been severely threatened by recent developments in land use.

3. Abandonment of traditional land use and the biodiversity crisis

Much has been written about rural depopulation, urbanization, the collapse of traditional systems of land use and their replacement by intensive agriculture and forestry. The abandonment of extensive agricultural use of marginal land such as the forest meadow at Råshult, but also summer grazing pastures in upland areas and hay meadows has placed particular pressure on floristic diversity in Scandinavia. Changes in land use have also been rapid, giving little opportunity for species to relocate on the landscape. Pollen analytical data can record the rate of change of vegetation by numerical comparison of consecutive samples. Analysis of 13 small hollow sites from southern Sweden showed a rapid increase in the rate of vegetational change since AD 1500 (Fig. 6.6).

These changes in land use and new pressures have contributed to a general decline in vascular plant diversity in southern Scandinavia that has become the focus of conservation efforts.

Conclusions

Landscape management is a new concept in Sweden that is being practised both by the forest authorities and by government organizations with responsibility for nature protection. Halting biodiversity loss is one of the major aims of this type of management. Important information needed for management includes inventories of existing species and habitats of high biodiversity (e.g. the key biotope inventory and red lists

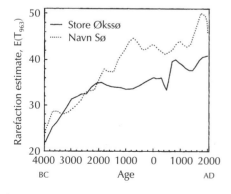

Fig. 6.5. The changing number of plant species detected in the pollen records from two lakes in Jutland, Denmark (after Odgaard, 1999).

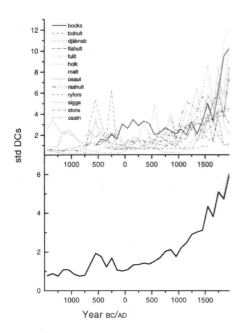

Fig. 6.6. 'Rate of change' analysis for 13 stand-scale sites from southern Sweden displayed for individual sites and as a mean. The analysis included all pollen types that exceeded 2% of the pollen sum in at least one sample. std DCs: chord distance per 100 years (after Lindbladh *et al.*, 2000).

of threatened species) and information about the past.

The characterization of past disturbance regimes is proving to be important within the forestry sector. The basic principle is to mimic these former regimes in current timber exploitation and fire frequency is one popular model (Rulcker *et al.*, 1994). Former cultural diversity, as represented by the forest meadows in this chapter, is managed within systems of nature reserves. In the province of Halland in south-west Sweden, almost half of the land protected in nature reserves is valued for communities of cultural origin that require active management for their maintenance. Increasingly small reserves are being merged into larger complexes where 'natural' and cultural values are combined. This formal recognition of the biodiversity value of cultural landscapes in southern Scandinavia, with its relatively modest scale of anthropogenic impact compared with the Mediterranean region indicates the importance of the cultural heritage to European nature conservation.

References

Aaris-Sorensen, K. (1998) *Danmarks forhistoriske dyreverden*. Gyldendal, Copenhagen.

Andersson, L. and Appelqvist, T. (1990) Istidens stora växtätare utformade de nemorala och boreonemorala ekosystemen. *Svensk Botanisk Tidskrift* 84, 355–368.

Angelstam, P. (1998) Maintaining and restoring biodiversity in European boreal forests by developing natural disturbance regimes. *Journal of Vegetation Science* 9, 593.

Behre, K.E. (1981) The interpretation of anthropogenic indicators in pollen diagrams. *Pollen et Spores* 23, 225–245.

Berglund, B.E. (ed.) (1991) The cultural landscape during 6000 years in southern Sweden – the Ystad project. *Ecological Bulletins* 41. Munksgaard, Copenhagen.

Berlin, G. (1998) *Semi-Natural Meadows in Southern Sweden*. Lund University, Lund, Sweden.

Bernes, C. (ed.) (1994) Biologisk mångfald i Sverige - en landstudie. *Monitor* 14.

Bobiec, A. (2002) Living stands and dead wood in the Bialowieza forest: suggestions for restoration management. *Forest Ecology and Management* 165, 125–140.

Bradley, R.S. and Jones, P.D. (eds) (1992) *Climate Since A.D. 1500*. Routledge, London.

Bradshaw, R.H.W. (1988) Spatially-precise studies of forest dynamics. In: Huntley, B. and Webb, T.I. (eds) *Vegetation History. Handbook of Vegetation Science*. Kluwer Academic Publishers, Dordrecht, pp. 725–751.

Bradshaw, R.H.W. and Hannon, G.E. (2004) The Holocene structure of north-west European temperate forest induced from palaeoecological data. In: Honnay, O., Verheyen, K., Bossuyt, B. and Hermy, M. (eds) *Forest biodiversity: lessons from history for conservation*. CAB International, Wallingford, UK, pp. 11–25.

Bradshaw, R.H.W. and Lindbladh, M. (2005) Regional spread and stand-scale establishment of *Fagus sylvatica* and *Picea abies* in Scandinavia. *Ecology*, 86, 1679–1686.

Bradshaw, R. and Mitchell, F.J.G. (1999) The palaeoecological approach to reconstructing former grazing-vegetation interactions. *Forest Ecology and Management.* 120(1–3), 3–12.

Bradshaw, R.H.W., Tolonen, K. and Tolonen, M. (1997) Holocene records of fire from the boreal and temperate zones of Europe. In: Clark, J.S., Cachier, H., Goldammer, J.G. and Stocks, B.J. (eds) *Sediment Records of Biomass Burning and Global Change.* Springer-Verlag, pp. 347–365.

Bradshaw, R.H.W., Hannon, G.E. and Lister, A.M. (2003) A long-term perspective on ungulate-vegetation interactions. *Forest Ecology and Management* 181(1–2), 267–280.

Carcaillet, C. *et al.* (2002) Holocene biomass burning and global dynamics of the carbon cycle. *Chemosphere* 49(8), 845–863.

Ekström, J. (1993) The late Quaternary history of the urus (*Bos primigenius* Bojanus 1827) in Sweden. LUNDQUA Thesis, 29. Lund University, Lund, Sweden.

Emanuelsson, U. (1987) Overview of the Scandinavian cultural landscape. In: Emanuelsson, U. and Johansson, C.E. (eds) *Biotoper i det Nordiska Kulturlandskapet.* Naturvårdsverkets rapport, pp. 13–52.

Falinski, J.B. (1978) Uprooted trees, their distribution and influence in the primeval forest biotype. *Vegetatio* 38, 175–183.

Falinski, J.B. (1986) *Vegetation Dynamics in Temperate Lowland Primeval Forests.* Kluwer Academic Publishers, Dordrecht.

Gaillard, M.-J., Birks, H.J.B., Emanuelsson, U. and Berglund, B.E. (1992) Modern pollen/land-use relationships as an aid in the reconstruction of past land-uses and cultural landscapes: an example from south Sweden. *Vegetation History and Archaeobotany* 1, 3–17.

Gardiner, B.A. and Quine, C.P. (2000) Management of forests to reduce the risk of abiotic damage: a review with particular reference to the effects of strong winds. *Forest Ecology and Management* 135, 261–277.

Gissel, S., Jutikkala, E., Osterberg, E., Sandnes, J. and Teitsson, B. (1981) *Desertion and Land Colonization in the Nordic Countries c. 1300–1600.* Almqvist and Wiksell International, Stockholm.

Goldammer, J.G. and Furyaev, V.V. (eds) (1996) *Fire in Ecosystems of Boreal Eurasia.* Kluwer, Dordrecht.

Gustafsson, M. (1996) *Kulturlandskap och flora på Bjarehalvon.* Lunds Botaniska Förening, Lund, Sweden.

Hannon, G.E. and Gustafsson, M. (2004) Slottet – historien om en slåtteräng. *Svensk Botanisk Tidskrift* 98(3–4), 177–187.

Hofmann, R.R. and Stewart, D.R.M. (1972) Grazer or browser: a classification based on the stomach-structure and feeding habits of east African ruminants. *Mammalia* 36, 226–240.

Johnson, E.A. (1992) *Fire and vegetation dynamics. Studies from the North American boreal forest.* Cambridge University Press, New York.

Krankina, O.N., Harmon, M.E., Kukuev, V.A., Treyfeld, R.F., Kashpor, N.N., Kresnov, V.G., Skudin, V.M., Protasov, N.A., Yatskov, M., Spycher, G. and Povarov, E.D. (2002) Coarse woody debris in forest regions of Russia. *Canadian Journal of Forest Research* 32, 768–778.

Kull, K. and Zobel, M. (1991) High species richness in an Estonian wooded meadow. *Journal of Vegetation Science* 2, 711–714.

Lagerås, P. (1996) Vegetation and land-use in the Småland Uplands, southern Sweden during the last 6000 years. LUNDQUA Thesis, 36. Lund University, Lund.

Larsson, L.O. (1975) *Det medeltida Värend. Kronobergsboken, 1974-75.* Kronobergs läns hembygdsförbund, Växjö.

Liljegren, R. and Lagerås, P. (1993) *Från mammutstäpp till kohage. Djurens historia i Sverige.* Wallin and Dalholm, Lund, p. 48.

Lindbladh, M. and Bradshaw, R. (1995) The development and demise of a Medieval Forest-Meadow system at Linnaeus birthplace in southern Sweden – Implications for conservation and forest history. *Vegetation History and Archaeobotany* 4(3), 153–160.

Lindbladh, M. and Bradshaw, R. (1998) The origin of present forest composition and pattern in southern Sweden. *Journal of Biogeography* 25(3), 463–477.

Lindbladh, M., Bradshaw, R. and Holmqvist, B.H. (2000) Pattern and process in south Swedish forests during the last 3000 years, sensed at stand and regional scales. *Journal of Ecology* 88(1), 113–128.

Linder, P. and Östlund, L. (1992) *Förändringar i Sveriges boreala skogar 1870-1991.* 1, Department of Forest Ecology, Swedish University of Agricultural Sciences, Umeå.

Møller, P.F. (2000) *Vandet i skoven – hvordan får vi vandet tilbage til skoven?* GEUS, Copenhagen.

Mountford, E.P. (2002) Fallen deadwood levels in near-natural beech forest at La Tillaie reserve, Fontainebleau, France. *Forestry* 75, 203.

Myrdal, J. and Söderberg, J. (1991) Kontinuitetens dynamik. *Acta Univ Stockholmensis* 15.

Niklasson, M. and Granström, A. (2000) Numbers and sizes of fires: long-term spatially explicit fire history in a Swedish boreal landscape. *Ecology* 67, 1254.

Nilsson, S.G., Arup, U., Baranowski, R. and Ekman, S. (1994) Trädbunda lavar och skalbaggar i ålderdomliga kulturlandskap. *Svensk Botanisk Tidskrift* 88, 1–12.

Nilsson, S.G., Niklasson, M., Hedin, J., Aronsson, G., Gutowski, J.M., Linder, P., Ljungberg, H., Mikusinski, G. and Ranius, T. (2002) Densities of large living and dead trees in old-growth temperate and boreal forests. *Forest Ecology and Management* 161, 189–204.

Odgaard, B.V. (1999) Fossil pollen as a record of past biodiversity. *Journal of Biogeography* 26(1), 7–17.

Odgaard, B.V. and Rasmussen, P. (2000) Origin and temporal development of macro-scale vegetation patterns in the cultural landscape of Denmark. *Journal of Ecology* 88(5), 733–748.

Peterken, G.F. (1996) *Natural woodlands*. Cambridge University Press, Cambridge

Pickett, S.T.A. and White, P.S. (1985) *The Ecology of Natural Disturbance and Patch Dynamics*. Academic Press, Orlando, Florida.

Pontailler, J.-Y., Faille, A. and Lemee, G. (1997) Storms drive successional dynamics in natural forests: a case study in Fontainebleau forest (France). *Forest Ecology and Management* 98, 1–15.

Roberts, N. (1998) *The Holocene*. Blackwell, Oxford.

Rulcker, C., Angelstam, P. and Rosenberg, P. (1994) Naturlig branddynamik kan styra naturvård och skogsskötsel i boreal skog. *Skogforsk Resultat* 8, 1–4.

Runkle, J.R. (1985) Disturbance regimes in temperate forests. In: Pickett, S.T.A. and White, P.S. (eds) *The Ecology of Natural Disturbance and Patch Dynamics*. Academic Press, Orlando, Florida, pp. 17–3.

Svenning, J.-C. (2002) A review of natural openness in north-western Europe. *Biological Conservation* 104, 133–148.

Vera, F.W.M. (2000) *Grazing ecology and forest history*. CAB International, Wallingford, UK.

Wolf, A., Moller, P.F., Bradshaw, R.H.W. and Bigler, J. (2004) Storm damage and long-term mortality in a semi-natural, temperate deciduous forest. *Forest Ecology and Management* 188(1–3), 197–210.

Woods, K.D. (2000) Dynamics in late-successional hemlock-hardwood forests over three decades. *Ecology* 81, 110–126.

Yalden, D. (1999) *The History of British Mammals*. Poyser, London.

7 Mountain Landscape, Pastoral Management and Traditional Practices in the Northern Pyrenees (France)

J.-P. Métailié

*GEODE – UMR 5602 CNRS, Maison de la Recherche,
Université Toulouse–le Mirail, Toulouse, France*

Introduction

Mountain pastoral landscapes represent in all of southern Europe an immense heritage, shaped by successive phases since the Neolithic era. They are at the same time an economic resource, which is difficult to restore when it is degraded, a factor of biodiversity (as much botanical as faunistic), and also a cultural heritage. These landscapes were built and maintained for centuries or millennia by an agrosilvo pastoral system integrating all the functions of production and activation of the resources in the same places: agricultural production, pasture, litter and manure, gathering, charcoal, firewood, framework and craft industry. Today, in many regions, the crisis of livestock farming and land abandonment involve dramatic environmental changes, which are often very fast and threaten these landscapes. Spread of fallow land and afforestation are the main effects, but pastoral degradation includes phenomena of decline, in particular in heathlands.

The contemporary management of these extensive lands threatened by abandonment is involved today in a net of contradictions between, on the one hand, ideology of nature and conservation (some seeing the pastures as a pristine landscape), and on the other hand the willingness for management and economic development of the mountain.[1] In the French Pyrenees as in other mountains, the agropastoral landscapes evolved during the last decades, but they are still relatively well preserved, especially in high mountain areas, because of the maintenance of livestock and recent reorganization of modes of exploitation (Eychenne-Niggel, 2003). The recovery of the agropastoral landscapes is today a local consensus, and the support for some traditional practices is recognized as one of the essential tools for this purpose. We will present here the past and contemporary role of two great 'groups of practices' (the management of the pastures by fire, and the management of the *bocages* and wooded pastures), and the prospects for their future use (Fig. 7.1).

From the Historical Crisis to Precarious Revival: the Present Condition of Pastoralism in the French Pyrenees

The disappearance of the practices and traditional knowledge is closely related to the economic and demographic crisis that the Pyrenees underwent more than a century

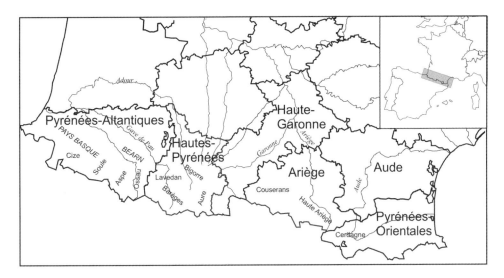

Fig. 7.1. Locality of the places discussed in the text.

ago. Rural emigration began in the 1830s in the eastern part of the chain (as in all the French Mediterranean mountains), but it only spread from the years 1880–1890, provoked by agricultural crises of the end of the 19th century. Until the 1930s, rural emigration mainly affected the poorest classes of the population: peasants without land, small artisans, and the workmen in metallurgical industries in decline. This emigration and the losses of World War I had paradoxically little consequence on the landscape dynamics because the main farms were maintained. Most of the great geographic and ethnographic investigations on the pastoral society were carried out during this time (Cavaillès, 1930; Lefebvre, 1933; Chevalier, 1952), and aerial photographs of 1942–1948 offer the vision of an agrosilvopastoral landscape still well maintained.

The great change occurred after World War II, in the context of economic growth and development of industrial agriculture between 1945 and 1975. During this period, 75% of the exploitations disappeared, involving a landscape upheaval without equivalent since the Middle Ages. However, this evolution was far from being uni-

form, and each Pyrenean department, even each valley, presented specific characteristics. Depopulation, land abandonment and spread of fallow were more accentuated towards the east of the chain, and less dramatic and even unknown in the west. In Pyrénées-Orientales or Ariège there were many villages with only one farmer, and many others without any, whereas in Pays Basque there were still several tens of stockbreeders without estate, obliged to rent meadows on the plain in winter and mountain pastures in summer. The context of the maintenance of practices was thus very different according to the valleys and agropastoral systems.

From 1972 a new period started in France marked by the promulgation of the law on the 'pastoral development of the mountain', in particular creating Pastoral Land Associations (*Associations Foncières pastorales*) and Pastoral Groups (*Groupement pastoraux*), which were in the following decades the principal tools of the stock farming reorganization. The 'policy of the mountain' continued to develop at the end of 1970s, supported by the beginnings of the European policy of assistance for less-favoured areas. In 1985, the new law for the development of the mountain led to an increased engagement of the local communities, even when the Common Agricultural Policy set up agro-environmental measures, which were very important for environment and land management. In the prolongation of these various policies the agencies of pastoral development were set up in each Pyrenean department. These dynamics lead today to a relative stabilization, even to a revival of farming, with maintenance of pastoral landscapes and with renewed attention to some traditional practices. Although the maintenance of mountain farming, essential for the management of the landscape, has consensus in France, that evolution remains fragile because of the repeated crises of the meat market (especially ovine) and of the uncertainty related to the perspectives of the Common Agricultural Policy. The number of pastoral exploitations continues to decrease regularly.

The last general census of agriculture (SUAIA Pyrénées, 2002) gives an overview of the current situation of the Pyrenees and evolutions foreseeable in the years to come (Table 7.1). It shows that the Pyrenees are divided today into three zones of pastoral economy:

- The eastern part (Aude, Pyrénées-Orientales) where the agri-businesses are very few, but with rather young farmers and a high level of qualification. The greater landscape changes took place in the past, and dynamics of recovery are beginning. The productions are primarily of bovine and ovine meat. Traditional practices and knowledge can be con-

sidered to have practically disappeared in these areas because of the strong decrease and renewal of the farming population, and consequent cultural change.

- The central part (Ariège, Haute-Garonne, Hautes-Pyrénées), with comparatively numerous agri-businesses but of the old, little-qualified farmers and a low number of young successors. Ovine and bovine herds for meat production largely use the collective pastures. Due to the ageing and progressive disappearance of the stock-breeders, there are clear prospects for abandonment and fallow land spreading in decades to come. Practices and knowledge are in the process of degradation and unequally maintained according to valleys.

- The western part (Pyrénées-Atlantiques) has the most intensive and dynamic farming, with a great number of businesses (half of the pastoral businesses of the whole mountain zone) and a high proportion of young and qualified farmers. Moor clearings and intensification have been carried out since the 1960s. Production is diversified, mainly based on sheep milk and cheese, but also on bovines for meat and milk. The good social and cultural continuity (especially in Basque Country) has allowed traditional practices and knowledge to be maintained.

What are the effects of these developments on the mountain pastoral landscape? Their dynamics are a complex phenomenon and difficult to evaluate at the Pyrenean scale, even for recent decades. The inventories

Table 7.1. The pastoral exploitations of the 'Massif zone' of the Pyrénées (SUAIA, 2002).

Departments (from east to west)	Number of pastoral exploitations	Number of cattle units (UGB)	Farmers of more than 50 years
Aude	268 (4.5%)	12,069 (4.3%)	31%
Pyrénées-Orientales	326 (5.5%)	16,995 (6.1%)	38%
Ariège	924 (15.4%)	40,452 (14.5%)	42%
Haute-Garonne	303 (5%)	10,384 (3.7%)	53%
Hautes-Pyrénées	1449 (24%)	45,352 (16.3%)	50%
Pyrénées-Atlantiques	2748 (45.6%)	153,243 (55.1%)	35%
Total	6018	278,495	2466

available (land registers, forest inventories, general census of agriculture) use different typologies and methodologies for all that concerns 'moors', 'fallows' and 'waste lands', and cannot give reliable information on dynamics nor even real surface areas. To get more details, one can try an approach using two sources:

1. The National Forest Inventory (IFN), which was repeated at least three times with intervals of 10 years in all the Pyrenean departments since the end of the 1960s with a uniform typology.[2] It offers a negative of the pastures, by quantifying the growth of the woodlands in each 'forest area' defined by the inventory. In the Pyrenees, one can note that the growth of the surface area of woodlands reaches on average 1–2% per decade. In some areas at the edge of the central Pyrenees, which are very wet and where the forest has dominated the landscape for a long time, the growth reaches up to 5% per decade.[3] Tree colonization in the upper valleys is much slower; the only dynamics are perceptible in the moors and pastures, for which the inventory is not very reliable.[4] In the western Pyrenees, the dynamics are still very weak, both in the upper valleys and the low mountains.
2. The systematic studies which began from the pastoral survey of 1999, to inventory the pastures, and various research by scientific teams. These works give detailed information on the pastoral environment, but on the other hand the problem of contemporary and historical dynamics of the pastures remains very little investigated. In Ariège, a first assessment of spatial dynamics was made on 13 pastoral units; one could note that, according to the units, 20–50% of the surface of the mountain pastures suffered strong dynamics since the 1950s, either by spontaneous afforestation or by scrub colonization of the pastures. Numerous observations in the central and eastern Pyrenees (in particular by repeated photographs) (Métailié and Paegelow, 2004) confirm this dramatic evolution, which is much less pronounced or absent in the western Pyrenees.

In synthesis, one can say that the evolution of the pastoral landscape was dramatic in the Pyrénées-Orientales and Ariège, in particular in low and medium mountains (below 1600 m); the high mountain pastures (above 1600–1700 m) are more stable, but do not escape either from important changes during the last decades. In the central Pyrenees (Haute-Garonne, Hautes-Pyrénées), significant changes touched low and medium-sized mountains, while the high pastures remained rather stable until now because of the maintenance of a good pastoral pressure. In Béarn (Pyrénées-Atlantiques), the good quality of the soil (many pastures are on limey ground) and a heavy pastoral pressure, related to the cheese economy, made it possible to preserve the pastoral landscapes until the 1980s; but a tendency for scrub and broom colonization has definitely been perceptible since 10 years ago. In the Basque mountains, on the contrary, stability seems the rule in the upper pastures, because of the maintenance of a very heavy pastoral pressure, both ovine and bovine (it is the only part of the Pyrenees where the ovine livestock increased because of the strength of the dairy economy). On the other hand, the hillsides of medium-sized mountains saw a contrasting development: new clearings and cultivations (meadows and maize) of the flatter areas, resulting in abandonment of the steep slopes, invaded by dense stands of bracken and *Ulex*.

The Historical Role of Traditional Practices in the Construction of the Pyrenean Landscape

Fire, the main tool for construction and management of agro-pastoral landscapes

Research in environmental history makes it possible today to highlight the multiform and generalized role of fire in the construction of the Pyrenean mountain landscapes (Fig. 7.2). The first evidence of clearings by fire in upper mountains is found in

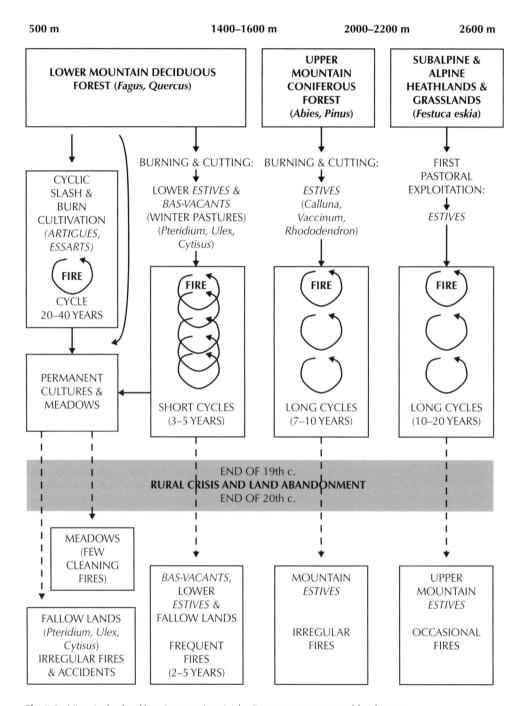

Fig. 7.2. Historical role of burning practices in the Pyrenean agro-pastoral landscapes.

Catalonia (Cerdagne) as of the Neolithic era, towards 4000 BC (Vannière *et al.*, 2001); at the same time, slash and burn cultivations appear on the northern central piedmont (Galop *et al.*, 2002). They intensify considerably towards the end of the Neolithic and the beginning of the Bronze Age, which represents the first great phase of creation of the pastures in all the Pyrenees. As at this time, several fire practices will coexist: the *fire of pastoral clearing*, by burning the edges or the whole woodlands; the *running fire* for maintenance of the pastures (which can easily lead to the clearing fire); *slash and burn cultivation* based upon a long cycle of coppice burning, every 15 to 30 years. It should be noted that maintenance by fire of silvo-pastoral forests, in particular the oak and beech stands of sunny slopes, had to start at that time. The impact of fire intensifies from the early Middle Ages up to the 14th century, a period of generalization of the agricultural and pastoral clearings, which will set up the Pyrenean soils. In the 14th century, censuses of population show that practically all the current Pyrenean villages exist (a small number will disappear during the 15th century, while some others will appear during the 18th–19th centuries), with already important populations, equivalent to those of the middle of the 20th century. After the demographic crisis of the end of the Middle Ages, the last phase of construction of the soils corresponds to the 16th–18th centuries. Many overexploited forests disappear during this period; the pastures, systematically managed by fire, reach their greatest extent, and the woodlands evolve into coppices or grazing forests managed by current fires, except some high stands protected in the state or common forests (Métailié, 1996; Bonhôte *et al.*, 2000).

The stabilization of the soils makes agricultural fire disappear gradually and the last clearings occur at the beginning of the 19th century.[5] After a long continuity of construction of the soils and generalized practice of fire, a short phase of stability succeeds, which remains today in the local memory as the apogee of the 'traditional' landscape. Fire is then primarily a manage-ment tool of pastoral lands, well under control on account of the abundance of labour and the intensity of the agro-pastoral exploitation.

All this was over as from the 1950s: the acceleration of rural emigration, the abandonment of the land and the fast reduction in the herds cause a rapid spread of fallow lands which standardize landscapes, increase the combustible biomass and make the natural firebreaks disappear. The practice of fire then changes completely: aerial photographs and statistics show the change from frequent burnings on small surface areas distributed over the whole of the mountain, to fires at wide intervals, very vast and concentrated on the sunny slopes which are easiest to burn (Métailié, 1981; Faerber, 1996, 2000). In Ariège, for example, the average surface area burnt by fire increased between the 1940s and the 1980s from 5–8 ha to 50 ha, while the number of fires fell by half. The reduction in the number of stock-breeders also resulted in a disappearance of the collective practices and knowledge, worsened by the social changes affecting the shepherds, with fewer and fewer coming from the local population.

Former practice was regular, 'when necessary', and was made possible by a numerous labour force and a constant presence in the pastures; today fire is used 'when one can', according to the availability of the farmers and their friends. This evolution of the practice was paralleled by a major cultural change (Table 7.2); until the 18th century, one can consider that fire profited from a general social consensus: knowledge and standards were shared by all the mountain dwellers, and lords and peasants had the same culture.

From the 19th century, new standards and new scientific knowledge were emerging, carried out by the engineers of the state administrations (*Eaux et Forêts, Ponts et Chaussées*) and by the first agronomists or phytogeographers. They are contradictory to those of the Pyrenean peasants and are opposed in particular to the traditional practices, like fire. This

Table 7.2. Historical periods of fire practices and socio-cultural background.

Time	Practices	Knowledge	Rules	Social actors
From Neolithic to 18th century	Building of landscape: slash and burn cultivation (*essarts*, *artigues*), clearing fire in forests, pastoral fire in pastures and moors	Vernacular	Local	Peasant society
19th century	Stabilization of landscape, disappearance of agricultural fire, pastoral fire exclusive	Opposition between local knowledge and scientific knowledge	Local rules vs. national rules	State engineers and administrations
20th century	Agro-pastoral crisis Fallow lands spreading Degradation of common fire practices	Crisis of local knowledge, increasing power of technical and scientific knowledge	National and European	Multiple protagonists
21st century	Use of fire for global land management Potential difficulties due to changes in local society, climate and vegetation changes?	Adaptation of local knowledge? Generalization of prescribed burning?	European?	Multiple protagonists on European scale?

situation evolves quickly to a prohibition, in fact, of the burning practices (although burning remains legal and simply regulated), which will bring the stockbreeders to clandestine practices and uncontrolled fire. This tendency to prohibition and conflict is accentuated during the 20th century, in a context of disintegration of the mountain society, increasing power of the administration and multiplication of the social or institutional groups intervening in mountains. The practice of fire is bound more and more to national or European standards, in spite of the scientific rehabilitation in progress since the 1980s. It is clear that there is an evolution towards an increased technical and legal framing of the practice, in spite of the efforts of the agents of development to manage it at the local level, on the basis of adapted traditional knowledge. One cannot exclude either a return to strong regulations or prohibition.

The peasant forest: bocage and wood pastures, a pastoral design of the forests

The use of the forest as an essential element of the agro-pastoral system probably began at the same time as pastoralism (Sigaut, 1987). The leaves of many trees provide good fodder and are convenient to use: before the scythe, cutting branches was easier than mowing grass, storage was possible without barns, and many trees have persistent foliage by winter. The beechmasts and oak acorns are also very important resources, mainly for pigs but also for sheep (even for man). Recent archaeopastoral research shows that the farming and transhumance of pigs in the oak and beech forests were already widespread in the Bronze Age in the Basque mountains (massif of Irati and Arbailles, for example[6]); this tradition remained until the 19th century, as in many other Pyrenean valleys (Bigorre, Aure).[7]

Fraxinus excelsior is the best known forage tree, but the list of the taxa used in the Pyrenees is very long: *Populus nigra*,[8] *P. italica*,[9] *Quercus petraea*, *Q. pyrenaica*, *Q. pedunculata*,[10] *Q. ilex* (for goats), *Prunus avium*, *Tilia cordata*, *Alnus glutinosa*, *Salix, Acer; Abies alba* was also bred for bovines and sheep,[11] like *Fagus sylvatica*.[12] It is necessary, of course, to add *Castanea sativa*, which was not a fodder tree, but whose plantations constituted important grazing forests on lower slopes. Many types of foliage are used only after drying, because they are considered too 'strong' in a green state for the animals (*Populus, Fraxinus*). There is, unfortunately, little historical documentation on these practices during the centuries when they were most widespread (the 17th to 19th century), when stock farming was under the permanent threat of the winter forage shortage, because of the extreme extension of the cultivations to the detriment of the meadows. Mention appears only indirectly in the archives, when shortage crises led the stock-breeders to make requests for cutting branches in the forests ruled by the Forest Administration. [13] Foresters and agronomists regarded them as aberrations, caused by the insufficiency of the meadows, and did not give any attention to them; only some authors gave an account of it. First is Louis de Froidour, *Réformateur* and *Grand Maître des forêts* in 1667, who visited all the Pyrenees and, parallel to his administrative and legal action, noted the agricultural as well as pastoral uses. Thus, he describes the pollard forests of the Basque Country, 'true orchards of oaks', used for intensive breeding of pigs; generally, he protests against the practice of pruning which he considers too frequent, without care and leading to the exhaustion of the trees. At the end of the 19th century, in 1889, Henri de Lapparent, agronomist and inspector of agriculture, made a study trip to the Pyrenean pastures; being at variance with the opinions of this time, he justified the pastoral use of the forest and recommended even the transformation of the less productive stands into 'fodder coppices' (Lapparent, 1890). One can find a similar

analysis in the works of Felix Briot on the pastoral economy of the Alps (Briot, 1907);[14] he describes many silvopastoral woodlands like fodder coppices of beech or alders, or pruned fir mixed with coppices, and analyses their profitability.

The country forests had many other well-known uses, like firewood, and the production of charcoal, tools, wooden shoes and frame timber. The demographic growth and the intensity of the exploitation during the 17th–19th centuries led to the disappearance of many forests, especially in the areas with strong metallurgical activity (Ariège and Catalonia). In the valleys with predominating agropastoral economy (Bigorre, Béarn), the destruction was of less importance but the forest landscapes were deeply transformed: generalization of coppices, of sparse forests with herbaceous undergrowth and of pollard trees.[15] The practices of maintenance of sylvo-pastoral woodlands disappeared during the 20th century, but much of the grazed forests last to this day because of the maintenance of a heavy pastoral pressure and use of fire: in the Hautes-Pyrénées, the Forest Inventory individualizes a category 'sparse forests', which accounts for 8.8% of total forest area.[16] Many wood pastures are still managed by fire (Fig. 7.3). In the Basque Country, the wood pastures and oak-pollard landscapes are still very widespread in all the hills and lower mountains (but pollarding is no longer practised).

Probably a long time after the construction of the agrosilvopastoral forests, a second type of peasant forest was created: the Pyrenean bocage. Its forms are rather different from those of western France: it is discontinuous and limited to soils with irrigated meadows; barns are very numerous, up to one per plot. Generally low fences (1–1.20 m) prevent the passage of the animals, sometimes stone-built but mainly vegetal (*Corylus, Buxus, Cratægus, Prunus, Cornus*), associated with fodder trees. The branches of hazel are cut every year in winter and are used for basket-making and firewood. Ashes surmount the hazel fence at intervals of a few metres, formerly pruned every 4–5 years in

Fig. 7.3. Winter burning of bracken and *Brachypodium* in a pastoral oakland, Lavedan (Hautes-Pyrénées), 1994. (J.P. Métailé).

August–September, but many sources relate the collection of the leaves every year. In general, people brought together branches and foliage to make faggots and stored them in the barns or in the plots; sometimes it was even stored in the tree, beyond the reach of the cattle (in Occitan: *la fullera,* Fig. 7.4) and given to the animals throughout January and February. The large branches of ash were used to make firewood, fences, plates and tools (cattle collars, rakes, forks, etc.). In the cultivated soil (cereals, potatoes, etc.) the obligation of common grazing prevented the construction of fences to allow free grazing after harvesting. The meadows were, on the other hand, for private use, and it was important to protect them from the bordering herds from spring onwards and to maintain their owners' own cattle in winter.[17]

The chronology of the construction of the Pyrenean bocage is not clear, and its origin, like all the other bocages, is difficult to know. Anthracological studies on the Gallo-Roman forges of Riverenert (Couserans, Ariège) showed that, during the 1st century AD, charcoal was made using taxa such as *Corylus, Prunus, Acer* and *Juglans.* This does not prove the existence of a bocage, but reveals a landscape with hedges and sparse woods. *Fraxinus*, which is the present key element, is present in the Pyrenean pollen diagrams as early as the Neolithic period (Galop *et al.*, 2002), but the real importance of ash in the landscape is difficult to reconstruct, because of the pruning, which prevents flowering and pollination. In the 17th century, some forest maps of the *Reformation* reveal a net of hedges and fences[18] and at the end of the 18th century there are dendrochronological evidences of ash hedges in the Spanish Pyrénées (Gomez and Fillat, 1984). The most credible hypothesis is that the construction of bocage was probably parallel with the creation and private appropriation of the irrigated meadows during the Middle Ages and modern times, and with the rarefaction of forest resources. This evolution is highlighted in Champsaur (Southern Alps) (Court-Picon, 2003): the beginning of the ash bocage is noted during the Middle

Fig. 7.4. 'La fullera': leaves of *Fraxinus excelsior* stored in the tree after pruning (Ariège, 1952). (coll. Office National des Forêts).

Ages, and the greatest development takes place during the 17th and 18th centuries. The latest increase of ash pollens in the diagrams corresponds to the cessation of pruning and the growth of trees during the 20th century. This type of evolution can be seen in present times in Nepal, where recent deforestation and forest regulations have limited access to woodlands and induced the plantation of hedges for leaf fodder and firewood (Bruslé *et al.*, 1997); in this subtropical environment, 56 taxa are used.

Between Heritage and Development: the Precarious Survival of Traditional Practices in the Pyrenees

From peasant fire to prescribed burning

The question of the management of pastoral fires in the Pyrenees showed a complete inversion during the last 15 years. For a long time, fire was blamed for the degradation of pastures and forests; the agents of the administration and the foresters tried to remove this 'archaic' practice, or at least to regulate it strictly. However, according to the scientific research of the last 20 years, which highlighted the logic and the role of pastoral fire, it is now recognized as a tool and not a constraint or a threat. The rehabilitation even reached a point that was unbelievable 20 years ago: prescribed burning was officially included in the forest law of 9 July 2001 as a technique for the prevention of forest fires.

The logic of the practice

In the Pyrenees, the majority of the pastures are made up of heathland, fern, broom and gorses, on poor and acid substrates, unfavourable for herbaceous vegetation. To have a fodder resource, the stock-breeders need periodic fire, burning scrub and dry biomass and regenerating (cleaning) the grassland. The idea of *cleanliness* is closely related to the image of fire in the mountain society: when the mountain is *dirty*, you have to *clean* it by fire (Ribet, 1996); grass is clean and scrub is dirty. This cultural logic of fire is based on a precise knowledge of the dynamics of the vegetation and its rhythms: running fires are practised during the period of rest for the plants – in the autumn or at the beginning of spring; the duration of cycles is linked to the speed of growth and to pastoral pressure. The planning and technical management of burning is based on an intimate knowledge of the mountain, of local climate and slope microclimates, of fire behaviour in general and in precise places. In each village there were skilled men, 'fire leaders' who organized the burning when it could be carried out officially or, in contrary cases, did it clandestinely (Métailié, 1981, 1998; Faerber, 2000).

Contemporary decline of the practice and knowledge

Over the last 50 years, the consequences of rural emigration and pastoral abandonment have resulted in a major change in the nature of Pyrenean fire: the pastures are fallowing, in particular on the lower slopes close to the villages, the combustible biomass increases, the former limits and landmarks are disappearing. We can say that the territory escapes to the stock-breeders, and by consequence fire itself: it no longer has the same behaviour, no longer stops at the usual places and spreads unusually far. The multiplication of afforestations, of equipment (power lines, telephones, etc.) and of tourists' homes constitute increasing fire risks.

In parallel, the mountain society changed: the number of stock-breeders unceasingly decreases, which makes the collective control of burning more difficult, even improbable. The know-how disappears at the same time as the old fire leaders, or it does not evolve and becomes unsuited to the new environment. Lastly, the power of the administration and other social groups increases: foresters, of course, but also firemen, hunters, agricultural agencies, police, ecological associations, tourists and 'neo-rural' residents, etc. They constitute special interest groups and it becomes difficult for stock-breeders to maintain a less and less controlled practice, causing increasing damage; they are becoming a minority in the country and they have to take that into account.[19]

The development of the local practices takes, today, approximately three forms in the Pyrenees. In some valleys, mainly in the west, in Basque Country, the stock-breeders are in great number, the pastoral pressure is heavy, and the traditional practices are maintained, which allows a local, traditional management of burning, which is relatively well controlled. In other valleys, the absence of the control of pastures and fallows is accompanied by dangerous individual practices, heirs of a long history of conflicts and clandestineness. This is the case, for example, in some valleys of Ariège, where individual stock-breeders and pensioners are desperately burning 'to clean the mountain' in abandoned zones, without concern for the consequences. In the majority of other cases, the stock-

breeders who need the fire to manage the pastures are receptive to a transformation of its organization. It is in this general context that the experiments of management of burning have developed over the last 10 years, in all the Pyrenean departments, in the form of technical groups or of local committees. They represent a true rupture in the history of the practice, a test to modify at the same time the social and technical patterns.

From burning to land management

The forms of dialogue and management of burning are different according to departmental contexts in the chain, and are mainly of two types: on the one hand, dialogue and realization of burning by a professional team; on the other hand dialogue and organization of fires by the local people. The key word remains always that of dialogue, a departure from the preceding repressive attitudes of administrations (Métailié and Faerber, 2003).

In the Pyrénées-Orientales, the local pastoral agency set up a specialized team in 1987, in order to deal with burnings necessary for stock-breeders (Lambert and Parmain, 1990). It was considered that the level of risk was too high in the Mediterranean environment, and local technical capabilities too limited, to leave the management of fires to the stock-breeders alone. Here the traditional practice was banished. The team, made up of pastoralists and foresters, collects the requests for burning, discusses them with the local partners, prepares burnings and carries them out, in collaboration with units of firemen. This organization took advantage of the Mediterranean context and the policy of forest-fire prevention. Such fire management is today developing in all the French Mediterranean zone where, according to departments, foresters or firemen organize burning campaigns, either with a pastoral objective or for prevention of forest fires (PASTUM, 1998).

In the remainder of the Pyrenees, to take technical responsibility for all the burnings would be unrealistic: on the one hand, the stock-breeders are still numerous and would not agree to be excluded from the management of their lands; on the other hand, the very great number of fires necessary each year in the various valleys would make impossible their realization by a professional team. In the Hautes-Pyrénées, since 1989, the choice was that of local committees (at the level of the canton or the valley), allowing first the dialogue between all the partners. In these committees, the various local protagonists present their points of view and discuss the benefits or risks of burnings, setting up zones of land and fire zones. One of the essential actions of the committees remains provision of local information on regulation and techniques. The objective is the end of clandestine burnings and the revitalization of collective fire practices, using traditional knowledge and modern tools (drip torch, water, fire-beater). Accordingly, fire must become again an agricultural tool like any other. The progressive installation of the committees in the mountain cantons allows a slow but relatively sure diffusion of this new collective management of fire.

In Ariège, a burning committee was set up at the level of the whole department. Parallel to the usual work of dialogue and information, the committee led to the constitution of a semi-professional burning team, made up of shepherds. The team was created based on the evidence that, in many cases, the local protagonists needed help because they were not numerous enough or no longer had the know-how to control their burnings. After 2 years, the team had to stop, through lack of funds, and the firemen took over the programme again; but, unfortunately, they do not have appropriate relations with stock-breeders. At the present time, the problem of fire management and evolution of practices has still not been worked out. On the one hand, the stock-breeders organized within Pastoral Land Associations or Pastoral Groupings generally changed their practices by integrating fire in the current and lawful management. However, on the other hand, the clandestine practices were not controlled and continue to cause important scrub and forest fires.

The Pyrénées-Atlantiques represents a particular problem because the department includes as many stock-breeders as the remainder of the Pyrenees. That has two important consequences. First, the number of farmers in only one commune can be equivalent to that of a whole valley in another department; the problems of dialogue and the potential conflicts are thus multiplied. Secondly, the requests for burnings have also multiplied – in some communes the mayors receive 60 to 80 declarations of burning per annum, which is equivalent to several cantons in the neighbouring departments. Furthermore, traditional practices are still very much alive, and the farmers are not generally disposed to accepting criticism.

A first attempt at installation of a burning committee was made in 1989, in the valley of Soule (Basque Country), based on a similar model to the cantonal committee of the Hautes-Pyrénées. However, the experiment failed because of the lack of funds and continuity in the animation of the committee, but also of contradictions between the 43 communes of a very large and populated valley. The committee also encountered a strong inertia of the local farmers, who think they have nothing to learn about fire from technicians. Other actions were then started in Béarn in 1994–1995 (Aspe valley), where a local request for management of fires was better expressed. After 2 years, the first assessment highlighted good participation of shepherds and stock-breeders, and of the mayors, a change in the practices and good control of burnings. The second time, this commission profited from a local agro-environmental measure ('Patrimonial management of pastoral landscapes and protection of the bear in the valleys of Béarn') including the management of natural risk, within which prescribed burning was financed (1998–2002).

In spite of that, no dialogue was really developed at the departmental level during all these years. It will be started again only after the shock of the dramatic accidents in 2000 and 2002. On 10 February, 2000, eight hikers on the GR10 trail were trapped by a fire in the valley of Estérençuby in Basque Country – five of them died and two others were seriously wounded. The site of burning was perfectly banal for the Basque mountains: a steep slope with dense grassland of *Brachypodium pinnatum*, frequently burned, where the fire reached a mortal velocity for tourists without knowledge of fire behaviour. It should be noted that the stock-breeders regard themselves as the only inhabitants of the mountains (especially in winter) and burnings are usually started without any specific precaution or information; the stock-breeders of Estérençuby are indifferent to the welfare of passing tourists. The gravity of the accident and the questioning of the practices which underlay it caused a shock in the department, restarting the discussions on the management of burnings and the creation of a departmental committee.

In this context another catastrophe occurred in February 2002. After several weeks of winter dryness and snow melting, there was a spell of strong, hot southerly wind over the whole of the Pyrenees. On 3 February 2002 the maximum temperature in the lower valleys reached 28 °C, while on the high slopes the winds reached 80–100 km/h. Logically, no burnings should have been carried out under these conditions – any control is impossible with winds exceeding 20–30 km/h, and the extreme dryness of the vegetation meant that the moisture which normally stops the fires in the thalwegs or on the northern slopes was absent. In spite of that, many stock-breeders started burnings in situations where they were difficult or impossible to control. In the Pyrénées-Atlantiques, especially in Basque Country, usually the wettest area of the chain, the damage was the most severe: more than 5000 ha of forests were burned, in sectors never touched before (high beech stands of northern slopes). The impact on the forests was thus considerable, even if in the majority of the cases the fire was an under-wood fire; burned woodlands lose any commercial value, and the losses were estimated at 1.8 million euros. An old farmer was trapped by a fire and died. The gravity of the consequences again re-ignited

the debate in favour of departmental input and the creation of local committees in 2001–2003, under the care of the Association of Mountain Mayors and of the Chamber of Agriculture.

The principal effect of these repeated serious accidents was to highlight the deficiencies of traditional burning practices in Basque Country. Paradoxically, the risks are increasing in a context of intensive pastoralism, which is confronted to changing a social and biological environment. For the years to come, the challenge of the committees is to convince the stock-breeders to adapt their practices and organize burnings better.

The rehabilitation of fire during the last 20 years was particularly spectacular in the Mediterranean area, where it led to the integration of prescribed burning into fire prevention as a clearance technique associated with stock-breeding. From 1989, the creation of the Prescribed Burning Network, an association of agronomists, firemen, foresters and pastoral agents, led to a technical framework of burning which made it acceptable for the forest services (PASTUM, 1998). The Forest Law of 9 July 2001 confirmed this development by recognizing the prescribed burning as a technique for fire prevention. However, this 'cultural revolution' at the level of the Forest Administration does not resolve the problem of the traditional fire. The 'prescribed burning' is becoming an administrative standard, with legal codification corresponding to a very technical organization with the objective of forest-fire prevention. These constraining rules are suitable for the burning teams of the Mediterranean area, made up of foresters and firemen who are heavily equipped, and who carry out a small number of burnings every year in high-risk zones and use the technical model of American prescribed burning. Unfortunately, it is impossible to apply this model to the practice of the shepherds and stock-breeders in the Pyrenees. Hundreds or thousands of burnings are made every year, and in this case the main objective is responsibility, observance of the legal rules and good technical management of the fire.

There is thus a risk of the practice of fire again being limited by new technical standards, even prosecutions if new serious accidents happen. Now the conditions for such accidents are becoming more frequent, especially if the current climatic evolution causes more frequent winter dryness. The practice of pastoral burning thus arrives at a historic turning-point, similar to the disappearance of the agricultural uses of fire 200 years ago. The change of the second half of the 20th century seems irredeemable and will oblige the stock-breeders and shepherds to adapt their ancestral practice and integrate new technical and social arrangements. The pastoral logic of fire must be integrated today in a global mountain land-management policy.

The precarious future of the bocage

What is the current evolution of bocage management in the French Pyrenees? The pruning of the ashes continues in some valleys (Ariège, in some communes of Couserans; Hautes-Pyrénées, in Lavedan, valley of Barèges; Pyrénées-Atlantiques, in the valleys of Aspe, Ossau, Soule and Cize), where the collection of leaf fodder is relict. In most of the valleys, it is extinct, except during droughts, and pruning is practised mainly for the firewood[20] and a few crafts. The chronology of disappearance of the practices is clearly shown by photographs and ethnographic investigation: during the years 1950–1960, the bocage was still busily used and the landscape well maintained in all the valleys. The decline began during the 1970s, but at the time of the drought of 1976, the ashes were pruned again for forage in all the Pyrenean valleys, even where the practice was already extinct. At the time of the drought of 2003, in spite of a great forage shortage, the farmers preferred to buy costly hay, and the ashes were pruned only in the few valleys where the practice is still alive. In some places, farmers simply cut whole trees into the meadows to allow breeding. The change of working conditions (increase in the size of the farms and of the herds, lack of time)

and the increased integration of intensive agriculture explain this disaffection.[21]

The decline of the practice of pruning occurred especially in the 1980s, but recovery attempts took place as of the 1990s. At that time, management and restoration of rural landscapes became stakes for the local elect and the agencies of development. They were based on a succession of procedures and financings, national or European: European agro-environmental measures, from 1985, to maintain practices compatible with good environmental management (pasture and mowing, but also burning in some cases); the 'Grass Grant' in 1993; Rural Areas Management Funds in 1995; and Territorial Exploitation Contracts in 1998, replaced in 2002 by the Sustainable Agriculture Contracts.[22] The conservation of the bocage remains one of the principal objectives in the majority of the projects. Generally, these measures are used to compensate for the lack of local labour by financing the realization of work by specialized companies. This is what occurred in the hamlet of Laspe, in the commune of Sentein (Couserans, Ariège) – until about 1990–1995, the bocage was still particularly well managed and all its uses maintained by two families, especially the old people. This landscape was considered as an important heritage in a valley where fallows were quickly spreading. From 1995, a programme of fence building, maintenance of the byways and ash pruning was undertaken by local pastoral groups, and mainly carried out by one stock-breeder, helped by a company (Fig. 7.5). The principal problem on this level remains to obtain regular financing to care for the bocage at least every 4–5 years.

Conclusion

The traditional practices in the Pyrénées are today in a paradoxical situation. For the majority they are clearly disappearing, but they are also perceived as a cultural heritage indispensable for the maintenance of the landscape of pastures, bocage and terraces. In the case of the bocage, the

Fig. 7.5. Landscape of Pyrenean bocage in winter: hamlet and barns of Laspe, in the Biros valley (Ariège), 2002. (J.P. Métailié).

practices are disappearing, but the social consensus on the landscape moves towards an assumption of management by the local communities, in forms related to agro-environmental policies. In the case of fire, it is the reverse: the practice is considered as 'too alive' by certain people, its social acceptance is not guaranteed and it is necessary to adapt the burning to a new social and biological environment.

Notes

1. The current conflict over the reintroduction of large predators (bear and wolf) characterizes this opposition between the perceptions of the mountain; even the Natura 2000 project is a case of conflict, because the local elect perceive it as a loss of control of the territory.
2. National Forest Inventory (see www.ifn.fr). Ariège: 1968, 1978, 1990; Hautes-Pyrénées: 1974, 1986, 1997; Haute-Garonne: 1972, 1987, 2000; Pyrénées-Atlantiques: 1971, 1984, 1995.
3. For example, in the Couserans (Ariège) forests increased between 1978 and 1990 from 36,290 ha to 40,597 ha. On the northern edge of Hautes-Pyrénées, between 1986 and 1997, forests increased from 53,502 ha to 55,314 ha.
4. The typology of woodlands is homogeneous from one inventory to another, but this is not the case for moors, pastures, wastelands, etc., and such data are not useful for our purpose.
5. There is evidence of clearings at the beginning of the 20th century in forest archives and ancient photographs, but they are relict practices.
6. Collective research project 'Palaeo-environment and anthropisation dynamics in Basque mountain', 2000–2003, coordinated by D. Galop.
7. At the end of the 19th century, in the valley of Bareilles (Aure), herds of pigs were assembled for breeding in the oak lands of the state forest (Métailié, 1986).

8. The bocage with *P. nigra* still exists in Cerdagne and the Spanish valleys, but is not maintained.
9. *Populus italica* was the last bocage tree to arrive in the Pyrenees. Coming from central Asia, the first clones were imported by the Royal Garden at the end of the 18th century. Its success was very quick, because of its multiples uses, and it was one of the major elements of the Pyrenean landscape during the 19th century. The poplar bocage declined in France after the 1950s, but it is still rather important in the Spanish Pyrenees.
10. These three taxa make up the Basque pollard landscape.
11. Young firs still constitute an under-wood pasture for bovines.
12. The beech seedlings are bred during the spring. In the Ligurian Apennines, pollard beeches were used for fodder until the 1950s (Moreno and Poggi, 1998).
13. Archives and other sources are numerous in Italy and the comparison can help us (Moreno and Poggi, 1996, 1998).
14. Briot, an atypical 'pastoralist forester', was one of the founder members of the Society of Alpine Economy (1913), which has evolved into the current French Federation of Mountain Economy.
15. This woodland type, much extended throughout all of southern Europe (see the dehesa and montado) was conceptualized as 'savannas' by Rackham (1996).
16. IFN, Hautes-Pyrénées, 1996.
17. In some valleys, common grazing was practised in the private meadows from October to March (Chevalier, 1952).
18. Bethmale Valley: Arch. Départ. Haute-Garonne, série B, Réformation de Comminges, 0–3 (1668).
19. Some mayors of mountain communes, who are townsmen in the countryside, simply forbade the burnings.
20. It is interesting to note that the bocage exploitation produces two-thirds of the firewood of the Midi-Pyrenees region.
21. The ash fodder is still well used in the Spanish valleys and it is possible to find some practices, like the 'fullera', which have disappeared in France.
22. In French: *Prime à l'herbe, Fonds de Gestion de l'Espace Rural, Contrat Territorial d'Exploitation, Contrat d'Agriculture Durable.*

References

Bonhôte, J., Davasse, B., Dubois, C., Galop, D., Izard, V. and Métailié, J.P. (2000) Histoire de l'Environnement et cartographie du temps dans la moitié est des Pyrénées: pour une "chrono-chorologie". In: Barrué-Pastor, M. and Bertrand, G. (eds) *Les Temps de l'Environnement.* Presses Universitaires du Mirail, Toulouse, pp. 501–515.

Briot, F. (1907) *Nouvelles Études sur l'Économie Alpestre.* Berger-Levrault, Paris.

Bruslé, T., Fort, M. and Smadja, J. (1997) Un paysage de bocage. Masyam et le hameau de Kolang. In: Smadja, J. (ed.) *Histoire et Devenir des Paysages en Himalaya.* Paris, CNRS Editions, pp. 485–527.

Cavaillès, H. (1930) *La Vie Pastorale et Agricole dans les Pyrénées des Gaves, de l'Adour et des Nestes.* Colin, Paris.

Chevalier, M. (1952) *La Vie Humaine dans les Pyrénées Ariégeoises.* Génin, Paris.

Court-Picon, M. (2003) Approche palynologique et dendrochronologique de la mise en place du paysage dans le Champsaur (Hautes-Alpes, France) à l'interface des dynamiques naturelles et des dynamiques sociales. Thématique, méthodologie et premiers résultats. *Archéologie du Midi Médiéval* 21, 211–224.

Eychenne-Niggel, C. (2003) Trente ans de relance pastorale pastorale en Ariège: le temps de la maturité. Les enseignements de l'enquête pastorale de 1999 et du recensement agricole de 2000. *Sud-Ouest Européen* 16, 5–14.

Faerber, J. (1996) Gestion par le feu et impact sur la diversité : le cas des friches sur les anciennes terrasses de culture dans les Pyrénées centrales. *JATBA (Journal d'Agriculture Topicale et de Botanique Appliquée),* 273–293.

Faerber, J. (2000) De l'incendie destructeur à une gestion raisonnée de l'environnement: le rôle du feu dans les dynamiques paysagères dans les Pyrénées centrales françaises. *Sud-Ouest Européen* 7, 69–80.

Galop, D., Vanniere, B. and Fontugne, M. (2002) Human activities and fire history since 4500 BC on the northern slope of the Pyrenees: a record from Cuguron (central Pyrenees, France). In: Thiébault, S. (ed.) *Charcoal Analysis. Methodological Approaches, Palaeoecological Results and Wood Uses. Proceedings of*

II Intern. Meeting of Anthracology, Paris, 2000. *BAR International Series* 1063, 43–51.

Gomez, D. and Fillat, F. (1984) Utilisation du frêne comme arbre fourrager dans les Pyrénées de Huesca. In: Lazare, J.J., Marty, R. and Dajoz, R. (eds) Ecologie et biogéographie des milieux montagnards et de haute altitude. *Documents d'Écologie Pyrénéenne* III–IV, 481–489.

Grove, A.T. and Rackham, O. (2001) *The Nature of Mediterranean Europe. An Ecological History.* Yale University Press, New Haven, p. 384.

Lambert, B. and Parmain, V. (1990) Les brûlages dirigés dans les Pyrénées-Orientale. De la régénération des pâturages d'altitude à la protection des forêts. *Revue Forestière Française,* T.42, pp. 140–155.

Lapparent, H. de (1890) Voyage d'étude dans les hauts pâturages de la chaîne des Pyrénées. *L'Avenir, Journal de l'Ariège,* 27 April to 11 September 1890.

Lefebvre, Th. (1933) *Les Modes de Vie dans les Pyrénées Atlantiques Orientales.* Colin, Paris.

Métailié, J.P. (1981) *Le Feu Pastoral dans les Pyrénées Centrales.* Toulouse, CNRS Editions.

Métailié, J.P. (1986) Les chênaies des montagnes pyrénéo-cantabriques, un élément forestier du système agro-pastoral. *Revue Géog Pyrénées Sud-Ouest,* 57(3), 313–324.

Métailié, J.P. (1996) La forêt du village et la forêt charbonnée. La mise en place des paysages forestiers dans la chaîne pyrénéenne. In: Cavaciocchi, S. (ed.) *L'uomo e la foresta, secc. XIII–XVIII.* Istituto Datini, Atti della 20° settimana di studi, Prato, 8–13 mai 1995. Le Monnier, Florence, pp. 397–422.

Métailié, J.P. (1998) Le savoir-brûler dans les Pyrénées. De "l'écobuage" au "brûlage dirigé", la transformation d'une pratique traditionnelle en outil de gestion de l'espace. In: Rousselle, A. (ed.) *Monde Rural et Histoire des Sciences en Méditerranée. Du bons sens à la logique.* CRHISM, PUP, pp. 165–179.

Métailié, J.P. and Faerber, J. (2003). Quinze années de gestion des feux pastoraux dans les Pyrénées: du blocage à la concertation. *Sud-Ouest Européen* 16, 37–52.

Métailié, J.P. and Paegelow, M. (2004) Land abandonment and the spreading of the forest in the Eastern French Pyrénées in the 19th to 20th centuries. In: Mazzoleni, S. *et al.* (eds) *Recent Dynamics of the Mediterranean Vegetation and Landscape.* John Wiley & Sons, Chichester, UK, pp. 219–236.

Moreno, D. and Poggi, G. (1996) Storia delle risorse boschive nelle montagne mediterranee: modeli di interpretazione per le produzioni foraggiere in regime consuetudinario. In: Cavaciocchi, S. (ed.) *L'uomo e la Foresta, secc. XIII–XVIII.* Istituto Datini, Atti della 20° settimana di studi, Prato, 8–13 mai 1995. Le Monnier, Florence, pp. 635–653.

Moreno, D. and Poggi, G. (1998) Identification des pratiques agro-sylvo-pastorales et des savoirs naturalistes locaux : mise à contribution de l'écologie historique des sites. In: A. Rousselle (ed.) *Monde Rural et Histoire des Sciences en Méditerranée. Du bons sens à la logique.* Centre de Recherches Historiques sur les Sociétés Méridionales, Presses Universitaires de Perpignon, PUP, pp. 151–163.

Moreno, D. and Raggio, O. (1990) The making and fall of an intensive pastoral land use system in the eastern Liguria (XVI–XIXth c.). In: Maggi, R., Nisbet, R. and Barker, G. (eds) *The Archeology of Pastoralism in Southern Europe.* Rivista di studi liguri, LVI, 1990, pp. 193–217.

PASTUM (1998) *Numéro spécial "Brûlages dirigés".* AFP, 51–52, 120.

Rackham, O. (1996) Forest history of countries without much forest: questions of conservation and savanna. In: Cavaciocchi, S. (ed.) *L'uomo e la Foresta, secc. XIII–XVIII.* Istituto Datini, Atti della 20° settimana di studi, Prato, 8–13 mai 1995. Le Monnier, Florence, pp. 297–326.

Ribet, N. (1996) Le feu pastoral dans le Parc Naturel des Volcans d'Auvergne. *Revue d'Auvergne,* 539, 103–119.

Sigaut, F. (1987) L'arbre fourrager en Europe: rôle et évolution des techniques. In: *La Forêt et l'Élevage en Région Méditerranéenne Française. Fourrages,* no. hors série, 45–54.

SUAIA Pyrénées (2002) Les exploitations pastorales pyrénéennes, entre résistance et dynamisme. *Agreste,* Données no. 9, p. 4.

Vannière, B., Galop, D., Rendu, Ch. and Davasse, B. (2001) Feu et pratiques agro-pastorales dans les Pyrénées-Orientales: le cas de la montagne d'Enveitg (Cerdagne, Pyrénées-Orientales, France). *Sud-Ouest Européen* 11, 29–42.

8 Maintaining Cultural and Natural Biodiversity in Europe's Economic Centre and Periphery

P. Angelstam

Faculty of Forest Sciences, School for Forest Engineers, Swedish University of Agricultural Sciences, Skinnskatteberg, Sweden

Introduction

Europe is located in the so-called Old World. This means that the history of economic use of renewable natural resources is very long. In principle, the natural potential vegetation of terrestrial ecosystems in Europe is forest and woodland (e.g., Mayer, 1984; Vera, 2000). However, in most of this continent humans have been virtually everywhere at all times since the last glaciation. Neolithic agriculture entered southeast Europe about 8000 years ago, and covered all of central Europe 3000 years later (Vos and Meekes, 1999). This means that most landscapes can be viewed as a total phenomenon where man and the physical landscape are integrated based on the use of arable land, grasslands and trees (Angelstam, 1997; Antrop, 1997, 2005; Jongman, 2002).

Broadly speaking, the maintenance of biodiversity encompasses two main visions, depending on the history of the actual landscape. The first vision involves biodiversity in forest ecosystems and with explicit reference to the concept of naturalness (Peterken, 1996; Rametsteiner and Mayer, 2004) including natural disturbance regimes at the scales of stands and landscapes (Angelstam and Kuuluvainen, 2004).

In spite of the ambiguity of the naturalness concept (Egan and Howell, 2001), it is obvious that compositional, structural and functional forest biodiversity indicators should represent elements found in naturally dynamic forests (Noss, 1990). This vision dominates the boreal forest management, is widespread in mountain forests and to some extent implicit in near-to-nature silviculture and even plantation forestry (Grabherr *et al.*, 1998; Mason, 2003; Angelstam and Dönz-Breuss, 2004).

The second vision is that of the pre-industrial agricultural landscape, which is an important aspect of Europe's cultural heritage (Rackham and Moody, 1996; Fraser Hart, 1998; Agnoletti, 2000; Jongman, 2002; Antrop, 2005; Sauberer *et al.*, 2004). This traditional cultural landscape includes a wide range of more or less wooded vegetation types ranging from wooded grasslands with natural forest elements to ancient forms of agroforestry, including agriculture, animal husbandry, and with tree management by pollarding, lopping and coppicing. Hence, although influenced by human land use for a very long time, the pre-industrial cultural landscape contained structural elements such as dead wood and large old trees that are typically found in naturally dynamic forests. As a consequence, remnants of pre-industrial cultural landscape provide a refuge for many species adapted to a pristine or near-natural forest environment. However, very rapid changes in land-use patterns due to socio-economic changes, especially since World War II (e.g., Angelstam *et al.*, 2003a; Mikusiński *et al.*, 2003; Bender *et al.*, 2005), means that the maintenance of biodiversity and other values are no longer automatically provided as a by-product of traditional land use (von Haaren, 2002).

Thus, most of the European continent and especially the Atlantic and lowland broad-leaved deciduous forest regions has been severely altered throughout history (Heckscher, 1941; Darby, 1956; Mayer, 1984). Today, reference landscapes for the visions of natural forests and pre-industrial cultural landscapes are usually confined to regions in the periphery of economic development. The Bialowieza forest in north-east Poland is one of the few remaining reference areas for lowland temperate forests (Vera, 2000). Even if this transformation of landscapes started a long time ago, the rate of change increased with the advent of the industrial revolution (Darby, 1956; Good, 1994). Already von Thünen (1875) noted that the type and intensity of land use was related to the distance from the market. During about 200 years of gradually intensified land use, the accumulated human footprint of a growing industrialized population of Europe has resulted in replicated gradients in landscape alteration from the centres of economic development into more remote regions (Mikusiński and Angelstam, 1998, 2004). The demand for timber, grain and other primary products was satisfied by import from the periphery of the spreading industrial revolution (Gunst, 1989). The first region to be exploited was the North Sea coast, in the 16th to 17th centuries. Eventually the exploitation reached Romania and Ukraine for grain in the 18th to 19th centuries (Powelson, 1994), and Russia for timber as a gradually spreading frontier developed from the 18th–20th centuries (Björklund, 1984; Redko and Babich, 1993). The exploitation of these resources was depend-ent on the development of facilities for transportation of bulky products such as railways and roads (Turnock, 2001; Angelstam *et al.*, 2004a). As an example, Hungarian exports were initially restricted to live cattle, herded to the destination countries until the mid-19th century when the development of railways reached Hungary and grain replaced cattle for export (Gunst, 1989).

In general, such gradual landscape changes have eventually negatively affected the viability of populations of specialized species (e.g. Tucker and Heath, 1994; Mikusiński and Angelstam, 1998). Moreover, the associated intensification of human land use resulted in an increase in hunting and persecution of larger vertebrates (Breitenmoser, 1998). This means that landscapes having virtually the same stand-scale local forest structures could be affected differently at the regional scale regarding the loss of species with large area requirements, depending on the landscape's location in relation to the centre and periphery of human economic activity (Mikusiński and Angelstam, 2004).

In his review of rural Europe since the 16th century, Whyte (1998) concluded that areas of retardation and tradition are concentrated in northern Europe, the Atlantic periphery and mountain areas in central Europe and the Mediterranean. There is also a clear west–east trend with increasing ecological diversity of species (Angelstam *et al.*, 2004b). Broadly speaking, economic remoteness in Europe has thus both a west–east dimension and a lowland–mountain dimension. In the Carpathian Mountains the two dimensions co-occur, which explains why the region is still a hotspot for natural and cultural biodiversity, and cultural heritage (Miya, 2000; Turnock, 2002; Angelstam *et al.*, 2003a; Opelz, 2004; Oszlányi *et al.*, 2004).

In this chapter I first briefly introduce the Carpathian Mountain ecoregion, having a range of economically peripheral landscapes, in some of which biodiversity and cultural heritage are still relatively intact. I then propose how this can be used as a facility and a 'labscape' approach (Kohler,

2002) for landscape-scale research and development based on an improved understanding of the mechanisms behind the different trajectories of development towards maintenance, or degradation, of natural and cultural biodiversity. I then outline how studies designed to quantify how much habitat is needed to maintain biodiversity can be carried out. Finally, I discuss the need for establishing arenas for practical management of relationships among elements of biodiversity such as species, habitats and ecosystem processes on the one hand and the interaction with nature and cultural heritage on the other.

The Carpathian Mountains as a Landscape Laboratory

Geography and economic history

The Carpathian Mountains form an arc-shaped mountain range, about 1400 km long, from north-eastern Austria, via the Czech Republic, Slovakia, northern Hungary, southern Poland and south-west Ukraine, into Romania (Turnock, 2002; Opelz, 2004). During the times of the Hapsburg Empire and until World War I most of the Carpathians were one geo-political unit. Today the area is divided among seven different countries. Already in the 19th century there was a clear gradient in the economic history from the centre to the periphery of the former Hapsburg Empire and the mountain regions were characterized as traditional cultural landscapes based on animal husbandry (Good, 1994). Later, the development of the railway network stimulated the demand for natural resources, which was increasingly met by the growing economy at the end of the 19th century (Turnock, 2001). The trends in agricultural and even tourism development correlated strongly with the expansion of the transport structure.

Because of its location at the periphery of economic development, there are in the Carpathian Mountains still villages where the pre-industrial traditional cultural landscape, once widespread in Europe, remains

(Miya, 2000). At the local scale, usually centred on the village itself, there is a characteristic zonation from the centre to the periphery with different kinds of economic use of renewable natural resources ranging from gardens and fields to pastures and forest (Angelstam et al., 2003a; Mikusiński et al., 2003; Bender et al., 2005). Open field systems and old forms of cultural heritage give characteristic expressions to the landscape, and several species that are declining or even endangered in intensively managed regions are thriving (Tucker and Heath, 1994). The landscape configuration and farming models that most Europeans understand as a part of their history is the everyday reality for the people in this part of the world. This landscape is characteristic not only in parts of the Carpathian Mountains, and further to the south-east in the Balkan and Rodopi Mountains, but was common in many other European regions in the past (Sporrong, 1998; Vos and Meekes, 1999).

Since the break-up of the Austro-Hungarian Empire the geo-political history has produced different trajectories of development in the central Carpathian Mountains within a very small area where Poland, Slovakia, Ukraine and Romania meet. With the pre-industrial cultural landscape as a common ancestor, different parts of the Carpathian Mountains have radiated in different directions. Using a combination of historical maps and satellite data, Angelstam et al. (2003a) found that in south-east Poland villages had developed along three different trajectories: (i) remained traditional; (ii) intensified agriculture; and (iii) become abandoned with encroaching forest as a consequence.

Within the first trajectory different consequences for biodiversity are likely to be dependent not only on whether the traditional structure is found at the scale of the individual village or not, but also whether or not most villages in the local landscape have the traditional structure and function. This is linked to differences in the area requirements of different species, and thus the area of habitat required to maintain populations in the long term (Angelstam et *al.*, 2003b, 2004c). For example, a focus on vascular plants in the short term versus specialized and area-demanding species in the long term yields quite different conclusions regarding the size of the management unit (from hectares to thousands of square kilometres) for the maintenance of biodiversity.

The second trajectory is the characteristic gradual degradation of traditional cultural landscape due to intensification of the agricultural practices, which is characteristic of most agricultural landscapes today (Larsson, 2004). In the north-east part of the Carpathian Mountains the past collectivization of traditional agriculture in Ukraine is a good example.

The third trajectory reflects abandonment and subsequent encroachment of natural vegetation for socio-economic reasons. As a consequence, cultural landscapes may return to near-natural forest conditions, and with few or no people. For example, the abandonment of the forest pastures and wooded meadows in the traditional cultural landscape now is the main habitat for deciduous forest birds in Swedish boreal forest (Enoksson *et al.*, 1995; Jansson and Angelstam, 1999; Mikusiński *et al.*, 2003). Similarly, a very specialized and area-demanding natural deciduous forest species like the white-backed woodpecker (*Dendrocopus leucotos*) is increasing in numbers in Poland as the same process takes place there, but 50 years later than in Sweden. The Bieczszady national park established in 1973 in south-east Poland is a good example of this. Here thousands of villagers were expelled after World War II, and the landscape returned to a near-natural 'wilderness' condition. The area is now a hotspot for a wide variety of species including area-demanding and specialized birds and mammals (Glowaciński, 2000), and is a major tourist attraction.

As a consequence of increased inaccessibility, urbanization, globalization and even calamities such as war and rapid socio-economic changes, traditional cultural landscapes have been altered in three clear steps. Initially, cultural biodiversity in traditional pre-industrial landscapes was an unintentional by-product of extensive agro-forestry. To encourage economically efficient growth, active measures were then taken to make both forestry and agriculture more economically viable by separating them in space. A diverse landscape with many shades of grey became essentially black and white. Finally, there is now a desire to maintain the cultural landscape for anthropocentric reasons (Vos and Meekes, 1999; Jongman, 2002; Antrop, 2005; Table 8.1).

Natural and cultural biodiversity and its management

The Carpathian Mountain range is a European hotspot for both cultural and natural biodiversity. Of the European continent's different eco-regions the Carpathian Mountains form one of the most valuable areas for biodiversity conservation in temperate coniferous and broadleaved forest (Schnitzler and Borlea, 1998; Opelz, 2004; Oszlányi *et al.*, 2004), as well as for conservation of cultural heritage (Turnock, 2002). The region hosts populations of large carnivores and herbivores that have become extirpated elsewhere (Perzanowski *et al.*, 2004), area-demanding specialist vertebrates (Mikusiński and Angelstam, 2001, 2004), as well as many endemic species (Oszlányi *et al.*, 2004). This has long made the Carpathians a focal area for multiple-functional landscape management (Oszlányi *et al.*, 2004; S. Stoyko and M. Elbakidze, personal communication) and eco-regional planning aiming at functional networks of habitats for the maintenance of viable populations of a range of specialized and/or area-demanding focal species (Turnock, 2002). In addition, ancient wooden churches and village building traditions have been maintained and are still practised (Miya, 2000).

The mixture of natural and cultural biodiversity, however, makes it necessary to formulate strategies for how conservation tools such as habitat protection, management, and, if necessary, restoration can be integrated. A common view is that satisfying the natural biodiversity vision is

Table 8.1. The changing role of cultural landscapes over three different periods of economic development.

Level of organization (see Fig. 8.3)	Traditional cultural landscape	Modification of landscapes during the agricultural and industrial revolutions	Post-industrial society
Policy-level desires	Merge local regions into states	Encouraging intensified use	Maintain biodiversity and cultural heritage
Administrative activities	Collecting tax	Land reforms to increase the intensity of use	Attempts to mitigate negative consequences of intensive production by subsidies and restoration, rural development
Role of people in villages	Maintain local sustainable landscapes	Urbanization, intensive production	Recreation, tourism
The landscape	Semi-natural, diverse	More and more industrial	Strong contrast between intensively managed matrix and semi-natural remnants
Role of biodiversity and cultural heritage	Biodiversity and cultural heritage a by-product	An obstacle for economic development; degradation of natural and cultural values	The role of non-tangible values has increased in post-industrial society

associated with non-intervention, and that the cultural biodiversity vision is associated with management. In this chapter it will be argued that this is too simplified a view, and that there is sometimes a need for management to reach the 'naturalness' vision, and non-intervention for that of the cultural landscape (Fig. 8.1).

There is scientific agreement that the conservation of biodiversity requires a range of disturbance regimes (Table 8.2) that have resulted in forest and woodland environments to which species have adapted. To maintain biodiversity this ecological dimension must be understood and a sufficiently wide range of different land-management

Table 8.2. List of the different kinds of disturbances that are important for the maintenance of natural and cultural biodiversity.

	Natural biodiversity	Cultural biodiversity
Wind	Uprooting creates dead wood, bare soil and special microhabitats	Dead wood is often removed and used as fuel
Flooding	Natural stream dynamics create important aquatic and riparian habitat	Irrigation and draining often occur, as well as active flooding to benefit meadows and pastures
Fire	Larger patches, lower frequency	Smaller patches, higher frequency
Large herbivores	Domination of browsers	Domination of grazers
Human	Not important, unless restoration measures are needed	Vital, and includes mowing, pollarding, coppicing, shredding, etc.

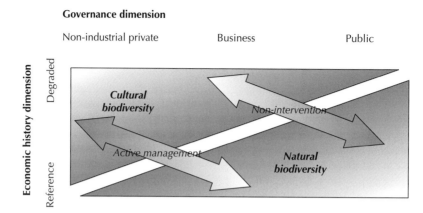

Fig. 8.1. Based on the two complementary main visions of natural and cultural landscapes for maintaining biodiversity in Europe's landscapes, this figure illustrates the need to balance active management and non-interventional strategies. The shading on the arrows indicates a gradient in the need for active management (lower left, darker) and non-intervention (upper right, darker) to maintain the different elements of biodiversity. Additionally, the vision of cultural landscapes prevails where the economic history is long and land is privately owned, whereas the vision of natural landscapes prevails where the economic history is short and land is publicly owned (Angelstam and Törnblom, 2004).

regimes must then be at hand. As advocated within the natural disturbance regime paradigm for near-to-nature forest management (Hunter, 1999), the management regimes chosen for different forest environments must harmonize with the ecological past of different forest types (Angelstam, 2003, 2005). The wide range of different even-aged, multi-aged and uneven-aged silvicultural systems available (Matthews, 1989) means that there is in principle good

potential for emulating natural disturbance regimes by combining protection and management for both maintenance and restoration of biodiversity (Table 8.3). The Fennoscandian boreal forest is a well-studied example, where management alternatives have been developed (Fries *et al.*, 1997). Three main disturbance regimes are characteristic (Angelstam and Kuuluvainen, 2004): (i) succession from young to old growth with shade-intolerant deciduous

Table 8.3. Table illustrating that to maintain natural biodiversity, depending on the vegetation type and its authentic dynamics, both active management and non-intervention may be needed.

		Landscape-scale management	
		Non-intervention	Management
Patch-/stand-scale management	Non-intervention	Ravine with old-growth forest, natural spring	To link remnant patches of old-growth forest on wet sites along a stream, management of hydrology is restored
	Management	Fire-dependent plants (small area requirements)	Fire-dependent insects and species using different stages in the succession require habitat connectivity at the landscape scale

species in the beginning and coniferous shade-tolerant species later on; (ii) cohort dynamics in dry Scots pine, *Pinus sylvestris,* forest; and (iii) gap dynamics in moist and wet Norway spruce, *Picea abies,* forest. In addition, approaches for landscape-scale planning have been developed (Fries *et al.,* 1998), and are partly also applied (Angelstam and Bergman, 2004).

Similarly, to maintain cultural biodiversity the methods employed in the pre-industrial cultural landscape need to be considered (Table 8.4). Due to the occurrence of species found in elements of naturally dynamic forests such as large old trees, both cultural and natural disturbances need to be considered. Agri-environmental schemes do consider the maintenance of hedgerows, pollarded trees and processes including mowing and grazing. However, there are few efforts towards encouraging the maintenance of functional connectivity. This also applies to reforestation programmes, which need to employ a spatially explicit approach (Møller Madsen, 2002).

However, because of the small amount remaining of traditional cultural landscapes and temperate deciduous forests, quantitative knowledge about their authentic composition, structure and dynamics is limited. Regarding the evolutionary background of the temperate deciduous forest, the ideas revolve around both abiotic disturbances, such as wind, and the interaction between herbivores and vegetation (Bengtsson *et al.,* 2003). The latter may also have included megaherbivores such as the forest elephant

present during most of the evolutionary history of the present fauna and flora, as well as wild and domestic large herbivores including deer (Cervidae), cattle (Bovidae) and horses (Equidae) (Vera, 2000).

Discussion

Implementing ecological sustainability and the governance of landscapes

The inhabitants of economically remote areas desire the same development and improvement of economic standards as the rest of Europe has experienced in the last decades. It looks like a paradox, but it is reality: at the same time that the EU distributes more and more of its budget to improve biodiversity and rural sustainability, we can find an increasing modernization in economically remote regions, which is promoted with increasing financial support, sometimes from similar sources. The extensive plan to develop the transport infrastructure within the expanding European Union is a good example. Unless effective mitigation measures are implemented, this will subsequently result in a decrease in the functionality of existing habitat networks, and threaten the last remaining reference landscapes for both natural and cultural biodiversity (e.g., Antrop, 2005; Bender *et al.,* 2005).

The development of the cultural landscape as a social-ecological system has gone through three main phases (see Table 8.1). In the past, traditional low-intensity

Table 8.4. Table illustrating that to maintain cultural biodiversity, depending on the vegetation type and its authentic dynamics, both active management and non-intervention 'laissez-faire' are needed.

		Landscape-scale management	
		Non-intervention	Management
Patch/stand-scale management	Non-intervention	Not relevant	Temporarily relaxed management intensity should be encouraged to allow regeneration of trees
	Management	Grazing and tree management favour a wide range of plant and animal species	Species using different stages in the succession require habitat connectivity at the landscape scale

agriculture was the norm (Vos and Meekes, 1999), and the maintenance of biodiversity was an unplanned by-product (von Haaren, 2002). Then followed a transition from primary to secondary economics, and industrialized agriculture led to loss of biodiversity and other values (Höll and Nilsson, 1999). Finally, the present post-industrial society represents a third phase – a desire to maintain the 'grandparents' landscape'. While it is certainly possible to satisfy some cultural heritage values and elements of biodiversity, the maintenance of sustainable rural landscapes and ecosystem integrity is a major challenge (Anon., 2004). Because there is a risk that traditional economic development will eventually result in a decrease in landscape diversity regarding both biodiversity and cultural heritage, remedial measures are needed.

An example of a large-scale attempt to mitigate such problems is the development of agri-environmental schemes, for example, within the framework of the Common Agricultural Policy (CAP) of the EU (Arler, 2000; Buller et al., 2000; Oñate et al., 2000; Jordan, 2002; Valve, 2002; Primdahl et al., 2003). However, according to Larsson (2004) the system of subsidies in place in the European Union to support the maintenance of biodiversity linked to agricultural landscapes seems not to be a remedy; the EU has bureaucracy and considers non-modernized agriculture as a problem to handle, and as a hindrance to EU enlargement. In this process, it is obvious that different interests and ideals collide. For example, stereotypical advice and lack of landscape-scale incentives in CAP agri-environmental schemes (Larsson, 2004) hamper planning and management for functional connectivity of habitat patches.

How can the maintenance and, if necessary, restoration of the natural and cultural biodiversity and cultural heritage of pre-industrial cultural landscapes really be planned? To avoid the gradual degradation of cultural landscapes iterated use of both top-down and bottom-up approaches for governance are needed (Anon., 2003; Sayer and Campbell, 2003, 2004). As discussed in detail by Angelstam (1997), Vos and

Meekes (1999), Jongman (2002) and Antrop (2005), an inclusive holistic approach using the landscape concept is needed. To understand landscapes in this way requires interaction among different actors in society. This applies to policymakers, institutions and the actual actors within one sector affecting landscapes on the one hand, and among the different sectors acting at all levels with the chosen landscape on the other. Within a given sector or policy area there are several levels (Primdahl and Brandt, 1997; Larsson, 2004; see Fig. 8.2).

First, at the international policy level, the Convention on Biological Diversity's 'ecosystem approach' can be used as one starting point. The ecosystem approach is a strategy for the integrated management of land, water and living resources that promotes conservation and sustainable use in an equitable way. Application of the ecosystem approach will help to reach a balance of the ecological, economic and socio-cultural objectives of the Convention. The approach should be based on the application of appropriate scientific methodologies focused on levels of biological organization, which encompass the essential processes, functions and interactions among organisms and their environment. It recognizes that humans, with their cultural diversity, are an integral component of ecosystems (see http://www.biodiv.org/programmes/cross-cutting/ecosystem/). For forests, sustainable forest management as defined by Rametsteiner and Mayer (2004) can be interpreted as an example of an ecosystem approach (Angelstam et al., 2004d, 2005).

Second, at the national level, policy instruments are then gradually developed, and may include legislation, information, subsidies, monitoring, vocational training, etc. However, the maintenance of natural and cultural biodiversity is usually not maintained by institutions, but rather by local people acting in different formal and informal governance systems. Consequently, several policy areas with their respective planning traditions coincide: forestry, agriculture, transport infrastructure and the energy sector, as well as regional and urban planning.

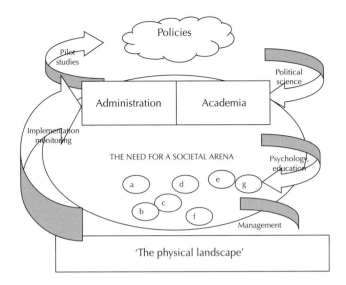

Fig. 8.2. Sustainable landscapes – from policy to practice, and back again – a schematic illustration of the interactions between four levels of the socio-ecological system from the policy level, the institutions and the actors, (a to g) to actual landscapes.

Third, because different landscapes have different governance systems, it is important to understand the actors' knowledge, attitudes and willingness to act in line with the policy (Clark, 2002). The suite of policy instruments used should ideally be adapted to the composition and structure of the actors in the actual landscape in focus. The effects of policies on actual landscapes are thus indirect, and therefore subject to several potential barriers (Clark, 2002).

Fourth, the effectiveness of the whole policy implementation process cannot be evaluated unless the development of different indicators is monitored in actual landscapes (Busch and Trexler, 2003). Additionally, results from monitoring should be compared with quantitative performance targets (Angelstam *et al.*, 2003b, 2004e, f).

Using the compositional dimension of biodiversity (i.e., with species as an exam-ple), this means that planners and managers need to understand that different species have different habitat affinities. Moreover, species also have different quantitative requirements, and viable populations need more habitats than are found in one patch of habitat (Angelstam *et al.*, 2004c). Hence, the successful maintenance of all representative land-cover types in a landscape can be viewed as a series of partly overlapping and complementary 'green infrastructures', each of which has different properties to which species are adapted. The required quality and extent of such habitat networks depends on the requirements of the species. For example, an old-growth forest-specialized species with large area requirements will need more habitat area than one with small area requirements. To steer towards agreed policy goals, there is a need for hierarchical planning with increasing resolution from broader to finer spatial and temporal scales (Angelstam *et al.*, 2005). In addition, there is a need for bottom-up approaches to engage the range of owners, managers and workers in local landscapes (Angelstam et al., 2003c; Sayer and Campbell, 2004). There is also often a need for international cooperation between adjacent regions in different countries (Opelz, 2004).

For large-scale forestry, hierarchical planning from the top to the bottom within a forest management unit (FMU) is well developed. In European boreal forests the size of a FMU for the ecological landscape plans in Fennoscandia ranges from 10^3 ha in the south to 10^4 ha in the north. FMUs can thus be viewed as replicates of landscape units on coarser scales such as ecoregions. The planning problem is usually divided into three sub-processes. The first level is strategic planning to decide on long-term goals covering an entire rotation. The second level is tactical planning to select from different alternatives based on the strategic goals, but on a shorter time scale and for a smaller area. Finally, operational planning is needed to administer the actual operations within a year (annual plan of operations).

The same logic can be used to build a toolbox of analytical tools for the evaluation of the structural elements of forest biodiversity, such as trees and riparian corridors left during logging, tree species composition and the age-class and patch-size distribution in the landscape, all of which managers affect by planning and operational management (Angelstam et al., 2005). At the strategic level, gap analysis (Scott et al., 1996) is a way to estimate the relative loss of different vegetation cover types. Gaps can also be quantified by comparing knowledge about how much habitat is needed to maintain viable populations on the landscape scale with the actual amount (Angelstam and Andersson, 2001). Next, at the tactical level, habitat suitability index modelling (Scott et al., 2002; Angelstam et al., 2004c) can be used for selecting patches or clusters of patches to be protected, managed or restored to maintain functional connectivity.

The same approach can be employed in urban landscapes. In Sweden local munici-

pal governments are responsible for water supply, refuse disposal, social welfare and education. The local governments also have a comprehensive planning monopoly (Alfredsson and Wiman, 1997). I argue that in principle this should alleviate systematic efforts towards the maintenance of biodiversity in urban environments. Municipal governments thus have a key role in explaining to the public what biodiversity maintenance requires in practice (Nilon et al., 2003). Young citizens of today are future decision-makers and need knowledge about the consequences for biodiversity of exploitation of land, which is closest to the people. Consequently, municipal governments are crucial in realizing sustainable development in practice, both directly in the urban setting and indirectly by taking decisions about the whole landscape (Sandström et al., 2006). An urban landscape usually contains parks and other near-natural environments. However, these constitute a green infrastructure only if they are organized with an overriding strategy; for example, with identified valuable green core areas with connecting greenways both among core areas and between core areas and the surrounding land (Dramstad et al., 1996). By and large, agriculture appears to lack landscape planning, even if there are attempts at the local level (Smeding and Joenje, 1999). Similarly Møller Madsen (2002) did not find landscape-level planning efforts in afforestation planning. Nevertheless, two critically important additional prerequisites to be satisfied for the maintenance of biodiversity are:

1. Landscape-scale performance targets for the amount and configuration of habitats needed to maintain species.
2. Arenas for governance where owners, managers and stakeholders can resolve conflicts within a landscape.

Deriving performance targets using gradients in economic development

To maintain, and if necessary restore, different elements of biodiversity such as species' populations, habitats or processes, reference areas are essential as benchmarks for both natural and cultural biodiversity (Egan and Howell, 2001). Further, both visions need to be complemented by knowledge about the quantity and spatial configuration of different habitat elements – and their dynamics over time – that is required to maintain viable populations of different species (Angelstam *et al.*, 2004c). Theoretical and empirical studies clearly indicate that there are thresholds for how much habitat can be lost without extirpation of specialized species (Fahrig, 2002), and that the absolute value varies among species (Angelstam *et al.*, 2004f). As an analogy to the concept of critical load of airborne pollution the forest ecosystems can stand without lost ecological integrity (Nilsson and Grennfelt, 1988), one could speak about critical loss of habitat leading to non-functional habitat networks.

If and when such critical values for the acceptable loss of habitat are available for individuals and populations for a suite of focal species, there is an empirical base for assessment and planning for the maintenance of sufficiently connected areas with habitat that make up functional habitat networks. Gap analysis is a strategic and habitat modelling tactical tool for such analyses. While this is a well-established procedure for the natural biodiversity vision (Scott *et al.*, 2002), it is apparently not discussed for the cultural biodiversity vision.

Even if knowledge about ecological thresholds is of paramount importance for the maintenance of biological diversity in the long term, the present empirical knowledge is limited (Fahrig, 2002; Angelstam *et al.*, 2004f). One important reason is that establishing studies of dose-response, where the amount of habitat is the dose given to a population and the response can be measured as the occupancy, or preferably fitness, of a species requires systematic macroecological studies that are costly and complicated to design (e.g., Bütler *et al.*, 2004; Mikusiński and Angelstam, 2004). Angelstam *et al.* (2004c) presented a gen-

eral procedure for identifying multiple thresholds to be used in the determination of conservation targets in forests in six steps. These were:

1. Stratify the forests into broad cover types as a function of their natural disturbance regimes.
2. Describe the historical spread of different anthropogenic impacts in the boreal forest that moved the system away from naturalness.
3. Identify appropriate response variables (e.g. focal species, functional groups or ecosystem processes) that are affected by habitat loss and fragmentation.
4. For each forest type identified in step 1, combine steps 2 and 3 to look for the presence of non-linear responses and to identify zones of risk and uncertainty.
5. Identify the 'currencies' (i.e. species, habitats, and processes) which are both relevant and possible to communicate to stakeholders.
6. Combine information from a suite of different indicators selected.

This procedure can be employed using at least three approaches. The first involves geographical comparisons with gradient in the variable of interest (Kohler, 2002; Angelstam and Dönz-Breuss, 2004). The plot size must then be scaled to the response variable. Using plot sizes of 1, 3, 16 and 2500 km^2, Bütler *et al.* (2004), Jansson and Angelstam (1999), Angelstam (2004) and Mikusiński and Angelstam (2004), respectively, found that the local occurrence and viability of different bird species responded in a non-linear fashion. Regions with the different economic histories needed to achieve a sample with sufficient variation in the dose of a certain habitat element, such as dead wood or old trees, are usually geographically far apart. There is thus a risk that the requirements of species are not the same. A second quasi-experimental approach can be used in the Carpathian Mountain landscapes, which once had a common culture, but has radiated in different directions for socio-political reasons. By revisiting replicates of the different treatments it is possible to study the consequence for biodiversity (Angelstam *et al.*, 2003a). The break-up of the former Austro-Hungarian Empire can thus be viewed as a landscape laboratory with borders between new countries that have created stark contrasts regarding biodiversity and cultural heritage, with clearly visible state borders in satellite images. Finally, historical ecology methods can be used (Egan and Howell, 2001). This could for example involve retrospective studies of how the species composition has changed over time in relation to land-cover changes as revealed by analyses of historical maps (Vourela *et al.*, 2002). Naturally, there will be a trade-off between the resolution and sample size of the different approaches (Table 8.5).

The need for arenas encouraging transdisciplinary applied approaches

Managers in different sectors and researchers in different disciplines traditionally accomplish most of their work in isolation. The outcomes of landscape changes become gradually evident by monitoring, but are evaluated at the policy level with a considerable time lag. Moreover, planners have a top-down approach that often ignores the 'real' world with the people in it (von Haaren, 2002). What should ideally form an iterated policy cycle based on effectiveness monitoring in local actual landscapes is, in fact, a rather slow process of policy implementation with inefficient feed-back loops from landscapes and

actors to policymakers. There are hence a number of barriers, in particular when attempting to apply a landscape approach for the conservation of natural and cultural biodiversity by a wise combination of management and non-intervention (Holling, 1995; Gutzwiller, 2002; see Fig. 8.1).

Successful planning for the maintenance of functional habitat networks for natural and cultural biodiversity requires simultaneous consideration of ecological, cultural and economic dimensions of landscapes. To handle such a diversity of information, modern techniques such as geographical information systems (GIS) are very useful. Sandström *et al.* (2006) evaluated urban planners' use of this tool in six Swedish cities. GIS was available in five of the six cities, but it was not developed or used to its full capacity because of insufficient knowledge about spatial data, ecological variables and planning algorithms. The planners did not realize the value of this tool. Cooperation between the city and a university may be one fruitful way to deal with obsolete education and conservative thinking, and to develop planning tools based on GIS. This offers advantages for both parties. For example, in a specific planning project the planners can make use of relevant research results and receive education at the same time as their ordinary work is being done. Such a system would open up possibilities for further education for planners, save time for the department, and develop the urban green planning as well as the local planning department. A

Table 8.5. Pros and cons of different approaches to landscape-scale dose–response studies.

	Geographical comparisons	Quasi-experiments using borders between countries	Historical ecology
Landscape data	Good	Good	Variable
Species data	Good	Good	Limited
Sample size	Large	Very limited	Limited
Other aspects	Different species–habitat relationship in different regions may preclude valid comparisons	Same ecoregion	Hard to find relevant data about the occurrence of different species

prerequisite is an attitude of a 'learning organization': in other words, the organization must be flexible and allow personnel to work and learn on a project at the same time (Lee, 1993; Chapin *et al.*, 2004; Sayer and Campbell, 2004).

There are several approaches to establish arenas for integration towards building bridges among stakeholders and different kinds of land use in a landscape. The international model forest network, which forms a partnership between individuals and organizations sharing the common goal of sustainable forest management, is one example (Besseau *et al.*, 2002), and UNESCO's biosphere reserve concept is another (UNESCO, 2002). Both approaches imply that a management unit consisting of an actual landscape with its characteristic ecosystems, actors and economic activities is used as a site for syntheses, innovation, development and education. Ideally, adaptive management teams (Boutin *et al.*, 2002) should be formed whereby researchers, land managers and policymakers share decisions and responsibilities toward the success or failure of the strategy they jointly adopted. I thus argue in favour of a novel win–win oriented approach to research and development, which is based on exchanging knowledge and experience gathered over a long time in different countries and regions. This will be of mutual benefit for both science and practice as a whole, and thus for continued sustainable use of natural resources providing a basis for human welfare in a changing world. Ultimately, acknowledging and adopting this perspective requires the gradual development of a new transdisciplinary profession able to facilitate ecosystem management at the landscape level. This requires an improved mutual feedback between the science, engineering and art of integrated natural resource management.

Reference areas for cultural and natural biodiversity, but for how long?

Depending on the level of ambition regarding maintenance of biodiversity one can formulate at least four different performance target levels. A first level is that the compositional elements of biodiversity are maintained. This is represented by occupancy of one of several species within a stand or landscape in the short term. Occupancy is often the only information available for conservation areas. However, policies are usually explicit about the fact that occupancy is insufficient, and include statements such as 'all naturally occurring species should maintain viable populations'. A second target level is therefore to ensure population viability over a long period of time. Increases in the area of habitat needed for probability of occupancy vs. probability of breeding (Angelstam, 2004) suggest considerably higher conservation costs of this increased level of ambition. The explicit reference to area-demanding specialized species, such as those listed in the EU Birds and Habitats Directives, makes it imperative to define thresholds (Mönkkönen and Reunanen, 1999) for a suite of relevant focal species (Roberge and Angelstam, 2004). As ecosystems are open and dynamic, the area needed to ensure the long-term persistence of interacting species, such as herbivores and predators, increases further. A third level of ambition is consequently to ensure ecosystem integrity (Pimentel *et al.*, 2000). To achieve this, minimum dynamic areas are needed that continuously provide habitat for representative suites of viable populations, as well as for the interactions among them (Bengtsson *et al.*, 2003). Finally, a fourth target level is to ensure ecological resilience measured as the magnitude of disturbance that can be absorbed before the system is unable to recover to its previous state (Gunderson and Pritchard, 2002). These four target levels for the maintenance of biodiversity illustrate that there is a continuous gradient with increasing spatial dimensions required for maintaining biodiversity. There is thus a suite of targets that can be specified for the maintenance of biodiversity in an area, each target representing an increasing probability of maintaining a functional ecosystem (Fig. 8.3). To illustrate this, examples are given for three structural

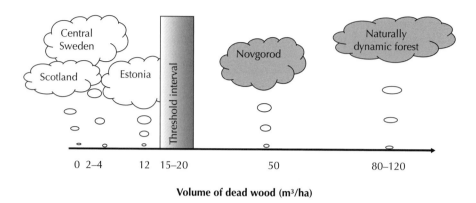

Fig. 8.3. Using the gradient from Scotland (very long history) to Novgorod region (short history) and naturally dynamic forest landscapes as examples, the effects of the history of forest use on the amount of dead wood are illustrated (Angelstam and Dönz-Breuss, 2004). Empirical studies of vertebrates, invertebrates and fungi clearly indicate that there are thresholds in the amount needed to maintain specialized species dependent on dead wood (Angelstam et al., 2002; Humphrey et al., 2004).

elements: dead wood, sufficiently connected patches of semi-natural woodland and large intact landscapes.

Dying and dead trees have been recognized as being of prime importance for numerous animals, fungi and lichens (Jonsson and Kruys, 2001). The amount of dead wood has also been accepted as a new indicator of forest biodiversity by the Ministerial Conference on the Protection of Forests in Europe (Rametsteiner and Mayer, 2004). For spruce-dominated forests, Bütler et al. (2004) used empirical thresholds to develop performance targets for standing dying and dead trees at the level of forest habitat patches. This was based on the quantitative habitat requirements of the three-toed woodpecker, Picoides tridactylus, a focal species of naturally dynamic spruce-dominated forests (Mikusiński et al., 2001). Both a theoretical model and empirical data resulted in similar estimates (15–20 m^3 of standing dead wood per ha over a 100 ha area) of the minimum snag quantities for woodpecker occurrence. In hemiboreal forest, Angelstam et al. (2002) studied the relationships between dead-wood variables and the presence of different woodpecker species in five different coarse landscape types in north-eastern Poland. The mean number of woodpecker species per km^2 varied from 0.6 (plantations) to 4.8 (Białowieża National Park) and was positively correlated with the amount of dead wood. The required volume of deciduous dead wood was estimated at about 20 m^3 of dead wood per ha over a 100 ha area (Angelstam et al., 2002). Comparisons of the species-richness of saproxylic beetles and the amount of dead wood in Norwegian and Finnish forests also indicate a threshold at about 20 m^3/ha (Martikainen et al., 2000; Humphrey et al., 2004). The threshold values for both deciduous and coniferous snags at the level of habitat patches are thus 5–10 times higher than the volume found in a managed forest in Fennoscandia, and about a fifth of what is found in naturally dynamic forest (see Siitonen, 2001; Fig. 8.3).

With a broad perspective, fragmentation is a unifying theme of the history of the European forests as well as an explanation for these local and regional extinctions of forest species. Fragmentation occurs when a continuous habitat is transformed into a number of smaller patches of decreasing area, isolated from each other by a matrix of habitats unlike the original. Both of these two components may cause extinctions; reduction in total habitat area affects population size and thus extinction rates, and redistribution of the remaining habitat into more or less iso-lated patches affects dispersal, and thus immigration rates. This is evident from the historic loss of species in more central regions of economic development, for exam-ple in southern Sweden (Osbeck, 1996).

On the scale of entire ecoregions, the maintenance of large intact forest areas is necessary for the maintenance of wide-ranging species such as large carnivores and herbivores (Breitenmoser, 1998; Mikusiński and Angelstam, 2004). The European bison is a good example. In a review of the constraints for re-establishing a meta-population of this large herbivore, Perzanowski *et al.* (2004) identified several barriers, and concluded that due to fragmentation and loss of large areas of natural habitats this species would have no chance of natural exchange of genes. The population viability is thus dependent on active conservation management.

Conclusions

Biodiversity conservation in Europe's land-scapes is based on both natural and cultural dimensions. The introduction of sustained-yield forest management and intensive agri-culture generally leads to a reduction of the amount of dead wood, functional connectiv-ity and intact areas of natural woodland and cultural landscapes. In a globalized world, the major challenge is to use ecological tar-gets rather than the state of the environment in already managed and altered landscapes as guidelines for management. It is thus crit-ically important that land management becomes spatially explicit on several spatial scales ranging from trees and stands to land-scapes and regions. To develop societal arenas for combining top-down planning with bottom-up implementation is a major challenge.

References

Agnoletti, M. (2000) Introduction: the development of forest history research. In: Agnoletti, M. and Anderson, S. (eds) *Methods and Approaches in Forest History*. CABI, Wallingford, UK, pp. 1–20.

Alfredsson, B. and Wiman, J. (1997) Planning in Sweden. In: Guinchard, D.G. (ed.) *Swedish Planning. Towards Sustainable Development*. The Swedish Society for Town and Country Planning, pp. 11–8.

Angelstam, P. (1997) Landscape analysis as a tool for the scientific management of biodiversity. *Ecological Bulletins* 46, 140–170.

Angelstam, P. (2003) Reconciling the linkages of land management with natural disturbance regimes to main-tain forest biodiversity in Europe. In: Bissonette, J.A. and Storch, I. (eds) *Landscape Ecology and Resource Management: Linking Theory with Practice*. Island Press, Covelo, CA and Washington, DC, pp. 193–226.

Angelstam, P. (2004) Habitat thresholds and effects of forest landscape change on the distribution and abun-dance of black grouse and capercaillie. *Ecological Bulletins* 5, 173–187.

Angelstam, P. (2005) Sustainability of forests and woodlands: the need to match ecological and management dimensions at the landscape level. *Visnyk Lviv University, Ser. Geogr.* 32, 5–21.

Angelstam, P. and Andersson, L. (2001) Estimates of the needs for forest reserves in Sweden. *Scandinavian Journal of Forest Research Supplement* 3, 38–51.

Angelstam, P. and Bergman, P. (2004) Assessing actual landscapes for the maintenance of forest biodiversity – a pilot study using forest management data. *Ecological Bulletins* 51, 413–425.

Angelstam, P. and Dönz-Breuss, M. (2004) Measuring forest biodiversity at the stand scale – an evaluation of indicators in European forest history gradients. *Ecological Bulletins* 51, 305–332.

Angelstam, P. and Kuuluvainen, T. (2004) Boreal forest disturbance regimes, successional dynamics and land-scape structures – a European perspective. *Ecological Bulletins* 51, 117–136.

Angelstam, P. and Törnblom, J. (2004) Maintaining forest biodiversity in actual landscapes – European gradi-ents in history and governance systems as a "landscape lab". In: Marchetti, M. (ed.) *Monitoring and Indi-cators of Forest Biodiversity in Europe – From Ideas to Operationality*. EFI Symposium No. 51, pp. 299–313.

Angelstam, P., Breuss, M., Mikusiński, G., Stenström, M., Stighäll, K. and Thorell, D (2002) Effects of forest structure on the presence of woodpeckers with different specialisation in a landscape history gradient in NE Poland. In: Chamberlain, D. and Wilson, A. (eds) *Proceedings of the 2002 Annual IALE(UK) Meeting held at the University of East Anglia*, pp. 25–38.

Angelstam, P., Boresjö-Bronge, L., Mikusiński, G., Sporrong, U. and Wästfelt, A. (2003a) Assessing village authenticity with satellite images – a method to identify intact cultural landscapes in Europe. *Ambio* 33(8), 594-604.

Angelstam, P., Bütler, R., Lazdinis, M., Mikusiński, G. and Roberge, J.M. (2003b) Habitat thresholds for focal species at multiple scales and forest biodiversity conservation – dead wood as an example. *Annales Zoologici Fennici* 40, 473–482.

Angelstam, P., Mikusiński, G., Rönnbäck, B.-I., Östman, A., Lazdinis, M., Roberge, J.-M., Arnberg, W. and Olsson, J. (2003c) Two-dimensional gap analysis: a tool for efficient conservation planning and biodiversity policy implementation. *Ambio* 33(8), 527–534.

Angelstam, P., Mikusiński, G. and Fridman, J. (2004a) Natural forest remnants and transport infrastructure – does history matter for biodiversity conservation planning? *Ecological Bulletins* 51, 149–162.

Angelstam, P., Edman, T., Dönz-Breuss, M. and Wallis deVries, M. (2004b) Land management data and terrestrial vertebrates as indicators of forest biodiversity at the landscape scale. *Ecological Bulletins* 51, 333–349.

Angelstam, P., Roberge, J.-M. and Lõhmus, A. *et al.* (2004c) Habitat modelling as a tool for landscape-scale conservation – a review of parameters for focal forest birds. *Ecological Bulletins* 51, 427–453.

Angelstam, P., Persson, R. and Schlaepfer, R. (2004d) The sustainable forest management vision and biodiversity – barriers and bridges for implementation in actual landscapes. *Ecological Bulletins* 51, 29–49.

Angelstam, P., Boutin, S., Schmiegelow, F. *et al.* (2004e) Targets for boreal forest biodiversity conservation – a rationale for macroecological research and adaptive management. *Ecological Bulletins* 51, 487–509.

Angelstam, P., Dönz-Breuss, M. and Roberge, J.-M. (2004f) Targets and tools for the maintenance of forest biodiversity – an introduction. *Ecological Bulletins* 51, 11–24.

Angelstam, P., Ek, T., Laestadius, L. and Roberge, J.-M. (2005) Data and tools for conservation, management and restoration of forest ecosystems at multiple scales. In: Stanturf, J.A. and Madsen, P. (eds) *Restoration of Boreal and Temperate Forests*. Lewis Publishers, Boca Raton, FL, pp. 269–283.

Anon. (2003) *Ecoregion Action Programmes. A Guide for Practitioners*. WWF, Washington, DC.

Anon. (2004) *New perspectives for EU rural development*. The European Commission, Brussels.

Antrop, M. (1997) The concept of traditional landscapes as a base for landscape evaluation and planning. The example of the Flanders region. *Landscape and Urban Planning* 38, 105–117.

Antrop, M. (2004) Why landscapes of the past are important for the future. *Landscape and Urban Planning* 70, 21–34.

Arler, F. (2000) Aspects of landscape or nature quality. *Landscape Ecology* 15, 291–302.

Bender, O., Boehmer, H.J, Jens, D. and Schumacher, K.P. (2005) Using GIS to analyse long-term cultural landscape change in southern Germany. *Landscape and Urban Planning* 70, 111–125.

Bengtsson, J., Nilsson, S.G., Franc, A. and Menozzi, P. (2003) Biodiversity disturbances, ecosystem function and management of European forests. *Forest Ecology and Management* 132, 39–50.

Besseau, P., Dansou, K. and Johnson, F. (2002) The international model forest network (IMFN): elements of success. *The Forestry Chronicle* 78, 648–654.

Björklund, J. (1984) From the Gulf of Bothnia to the White Sea – Swedish direct investments in the sawmill industry of tsarist Russia. *Scandinavian Economic History Review* 32, 18–41.

Boutin, S., Dzus, E., Carlson, M., Boyce, M., Creasey, R., Cumming, S., Farr, D., Foote, F., Kurz, W., Schmiegelow, F., Schneider, R., Stelfox, B., Sullivan, M. and Wasel, S. (2002) The active adaptive management experimental team: a collaborative approach to sustainable forest management. In: Veeman, T.S. *et al.* (eds) *Advances in Forest Management: from Knowledge to Practice*. Proceedings from the 2002 sustainable forest management network conference, University of Alberta, Edmonton, pp. 11–16.

Breitenmoser, U. (1998) Large predators in the Alps: the fall and rise of Man's competitors. *Biological Conservation* 83, 279–289.

Buller, H., Wilson, G.A. and Höll, A. (eds) (2000) *Agri-Environmental Policy in the European Union*. Ashgate, Aldershot, UK.

Busch, D.E. and Trexler, J.C. (2003) The importance of monitoring in regional ecosystem initiatives. In: Busch, D.E. and Trexler, J.C. (eds) *Monitoring Ecosystems*. Island Press, Covelo, pp. 1–23.

Bütler, R., Angelstam, P., Ekelund, P. and Schlaepfer, R. (2004) Dead wood threshold values for the three-toed woodpecker in boreal and sub-alpine forest. *Biological Conservation* 119, 305–318.

Chapin, III, F.S., Peterson, G., Berkes, F. *et al.* (2004) Resilience and vulnerability of northern regions to social and environmental change. *Ambio* 33(6), 244–349.

Clark, T.W. (2002) *The Policy Process. A Practical Guide for Natural Resource Professionals*. Yale University Press, New Haven, Connecticut.

Darby, H.C. (1956) The clearing of woodlands in Europe. In: Thomas, W.L. (ed.) *Man's Role in Changing the Face of the Earth*. University of Chicago Press, Chicago, Illinois.

Dramstad, W.E, Olson, J.D. and Forman, R.T.T. (1996) *Landscape Ecology Principles in Landscape Architecture and Land-use Planning*. Harvard University, American Society of Landscape Architects, and Island Press, Washington.

Egan, D. and Howell, E.A. (2001) *The Historical Ecology Handbook*. Island Press, Covelo, California.

Enoksson, B., Angelstam, P. and Larsson, K. (1995) Deciduous trees and resident birds: the problem of fragmentation within a coniferous forest landscape. *Landscape Ecology* 10(5), 267–275.

Fahrig, L. (2002) Effect of habitat fragmentation on the extinction threshold: a synthesis. *Ecological Applications* 12, 346–353.

Fraser Hart, J. (1998) *The Rural Landscape*. The Johns Hopkins University Press, Baltimore, Maryland.

Fries, C., Johansson, O., Pettersson, B. and Simonsson, P. (1997) Silvicultural models to maintain and restore natural stand structures in Swedish boreal forests. *Forest Ecology and Management* 94, 89–103.

Fries, C., Carlsson, M., Dahlin, B., Lämås, T. and Sallnäs, O. (1998) A review of conceptual landscape planning models for multiobjective forestry in Sweden. *Canadian Journal of Forest Research* 28, 159–167.

Glowaciński, Z. (2000) *Monografie Bieszczadzkie tom IX*. Oficyna Wydawnica 'Impuls', Krakow, Poland.

Good, D.F. (1994) The economic lag of Central and Eastern Europe: income estimates for the Habsburg successor states, 1870–1910. *The Journal of Economic History* 54, 869–891.

Grabherr, G., Koch, G., Kirchmeir, H. and Reiter, K. (1998) *Hemerobie österreichischer Waldökosysteme*. Universitätsverlag Wagner, Innsbruck, Austria.

Gunderson, L.H. and Pritchard, L. Jr. (2002) *Resilience and the Behavior of Large-Scale Systems*. Island Press, Covelo.

Gunst, P. (1989) Agrarian Systems of Central and Eastern Europe. In: Chirot, D. (ed.) *The Origins of Backwardness in Eastern Europe: Economics and Politics from the Middle Ages until the Early Twentieth Century*. University of California Press, London, pp. 53–91.

Gutzwiller, K.J. (2002) Applying landscape ecology in biological conservation: principles, constraints and prospects. In: Gutzwiller, K.J. (ed.) *Applying Landscape Ecology in Biological Conservation*. Springer, pp. 481–495.

Heckscher, E.F. (1941) *Svenskt arbete och liv från medeltiden till nutiden*. Albert Bonniers Förlag, Stockholm.

Höll, A. and Nilsson, K. (1999) Cultural landscape as a subject to national research programmes in Denmark. *Landscape and Urban Planning* 46, 15–27.

Holling, C.S. (1995) What barriers? What bridges? In: Gunderson, L.H., Holling, C.S. and Light, S.S. (eds) *Barriers and Bridges to the Renewal of Ecosystems and Institutions*. Columbia University Press, New York, pp. 3–34.

Humphrey, J., Sippola, A.-L., Lempérière, G., Dodelin, B., Alexander, K.N.A. and Butler, J.E. (2004) Deadwood as an indicator of biodiversity in European forests: from theory to operational guidance. In: Marchetti, M. (ed.) *Monitoring and Indicators of Forest Biodiversity in Europe – From Ideas to Operationality*. EFI symposium No. 51, pp. 181–206.

Hunter, M.L. (ed.) (1999) *Maintaining biodiversity in forest ecosystems*. Cambridge University Press, Cambridge, UK.

Jansson, G. and Angelstam, P. (1999) Threshold levels of habitat composition for the presence of the long-tailed tit (*Aegithalos caudatus*) in a boreal landscape. *Landscape Ecology* 14, 282–290.

Jongman, R.H.G. (2002) Homogenisation and fragmentation of the European landscape: ecological consequences and solutions. *Landscape and Urban Planning* 58, 211–221.

Jonsson, B.G. and Kruys, N. (eds) (2001) Ecology of woody debris in boreal forests. *Ecological Bulletins* 49.

Jordan, A. (2002) *Environmental Policy in the European Union – Actors, Institutions and Processes*. Earthscan, London.

Kohler, R.E. (2002) *Landscapes and Labscapes. Exploring the Lab-field Border in Biology*. The University of Chicago Press, Chicago, Illinois.

Larsson, A. (2004) Landskapsplanering inom jordbrukspolitik. *Agraria* 442. Swedish University of Agricultural Sciences.

Lee, K.N. (1993) *Compass and Gyroscope*. Island Press, Covelo, California.

Martikainen, P., Siitonen, J., Punttila, P., Kaila, L. and Rauh, J. (2000) Species richness of Coleoptera in mature managed and old-growth boreal forests in southern Finland. *Biological Conservation* 94, 199–209.

Mason, W.L. (2003) Continuous cover forestry: developing close to nature forest management in conifer plantations in upland Britain. *Scottish Forestry* 57(3), 141–149.

Matthews, J.D. (1989) *Silvicultural Systems*. Oxford University Press, Oxford, UK.

Mayer, H. (1984) *Die Wälder Europas*. Gustav Fischer Verlag, Stuttgart, Germany.

Mikusiński, G. and Angelstam, P. (1998) Economic geography, forest distribution and woodpecker diversity in central Europe. *Conservation Biology* 12, 200–208.

Mikusiński, G. and Angelstam, P. (2001) Striking the balance between use and conservation in European landscapes. In: Richardson, J., Bjorheden, R., Hakkila, P., Lowe, A.T. and Smith C.T. (Comp.) *Bioenergy from Sustainable Forestry: Principles and Practice*. Proceedings of IEA Bioenergy Task 18 Workshop, 16–20 October 2000, Coffs Harbour, New South Wales, Australia. New Zealand Forest Research Institute, Forest Research Bulletin No. 223, pp. 59–70.

Mikusiński, G. and Angelstam, P. (2004) Occurrence of mammals and birds with different ecological characteristics in relation to forest cover in Europe – do macroecological data make sense? *Ecological Bulletins* 51, 265–275.

Mikusiński, G., Gromadzki, M. and Chylarecki, P. (2001) Woodpeckers as indicators of forest bird diversity. *Conservation Biology* 15, 208–217.

Mikusiński, G., Angelstam, P. and Sporrong, U. (2003) Distribution of deciduous stands in villages located in coniferous forest landscapes in Sweden. *Ambio* 33, 519–525.

Miya, K. (2000) *Maramuresh*. Humanitas, Bucarest.

Møller Madsen, L. (2002) The Danish afforestation programme and spatial planning: new challenges. *Landscape and Urban Planning* 58, 241–254.

Mönkkönen, M. and Reunanen, P. (1999) On critical thresholds in landscape connectivity – management perspective. *Oikos* 84, 302–305.

Nilon, C.H., Berkowitz, A.R. and Hollweg, K.S. (2003) Introduction: ecosystem understanding is a key to understanding cities. In: Berkowitz, A.R., Nilon, C.H. and Hollweg, K.S. (eds) *Understanding Urban Ecosystems. A New Frontier for Science*. Springer, New York, pp. 1–13.

Nilsson, J. and Grennfelt, P. (1988) Critical loads of sulphur and nitrogen. *Nordic Council of Ministers* 15.

Noss, R.F. (1990) Indicators for monitoring biodiversity: a hierarchical approach. *Conservation Biology* 4, 355–364.

Oñate, J.J., Anderson, E., Peco, B. and Primdahl, J. (2000) Agri-environmental schemes and the European agricultural landscapes: the role of indicators as valuing tools for evaluation. *Landscape Ecology* 15, 271–280.

Opelz, M. (2004) *Towards a Carpathian Network of Protected Areas. Final Report*. Bundesministerium für Umwelt, Naturschutz und Reaktorsicherheit.

Osbeck, P. (1996) *Djur och natur i södra Halland under 1700-talet*. Bokförlaget Spektra, Halmstad.

Oszlányi, J., Grzodynska, K., Badea, O. and Sharyk, Y. (2004). Nature conservation in Central and Eastern Europe with a special emphasis on the Carpathian Mountians. *Environmental Pollution* 130, 127–134.

Perzanowski, K., Olech, W. and Kozak, I. (2004) Constraints for re-establishing a meta-population of the European bison in Ukraine. *Biological Conservation* 120, 345–353

Peterken, G. (1996) *Natural Woodland: Ecology and Conservation in Northern Temperate Regions*. Cambridge University Press, Cambridge, UK.

Pimentel, D., Westra, L. and Noss, R.F. (2000) *Ecological Integrity. Integrating Environment, Conservation and Health*. Island Press, Washington.

Powelson, J.P. (1994) *Centuries of Economic Endeavor*, 4th edition. University of Michigan Press, Ann Arbor, Michigan.

Primdahl, J. and Brandt, J. (1997) CAP, nature conservation and physical planning. In: Laurent, C. and Bowler, I. (eds) *CAP and the Regions: Building a Multidisciplinary Framework for the Analysis of the EU Agricultural Space*. Institute National de la Recherche Agronomique, Paris, pp. 177–186.

Primdahl, J., Peco, B., Schramek, J., Andersen, E. and Oñate, J.J. (2003) Environmental effects of agri-environment schemes in Western Europe. *Journal of Environmental Management* 67,129–128.

Rackham, O. and Moody, J. (1996) *The Making of the Cretan Landscape*. Manchester University Press, Manchester, UK, p. 237.

Rametsteiner, E. and Mayer, P. (2004) Sustainable forest management and Pan-European forest policy. *Ecological Bulletins* 51, 51–57.

Redko, G.I. and Babich, N.A. (1993) *Korabelnij les vo slavu flota rossijskovo*. (Ship forest for the glory of the Russian fleet). Arkhangelsk Severo-zapadnoe Knidznoe Izdatelstvo, p. 153.

Roberge, J.-M. and Angelstam, P. (2004) Usefulness of the umbrella species concept as a conservation tool. *Conservation Biology* 18(1), 76–85.

Sandström, U.G., Angelstam, P. and Khakee, A. (2006) Urban planner's knowledge of biodiversity maintenance – an evaluation of six Swedish cities. *Landscape and Urban Planning* 75, 43–57.

Sauberer, N., Zulka, K.P., Abensperg-Traun, M. *et al.* (2004) Surrogate taxa for biodiversity in agricultural landscapes of eastern Austria. *Biological Conservation* 117(2), 181–190.

Sayer, J.A. and Campbell, B.M. (2003) Research to integrate productivity enhancement, environmental protection and human development. In: Campbell, B.M. and Sayer, J.A. (eds) *Integrated Natural Resource Management: Linking Productivity, Environment and Development*. CABI, Wallingford, UK, and Centre for International Forestry Research (CIFOR), pp. 1–14.

Sayer, J.A. and Campbell, B.M. (2004) *The Science of Sustainable Development: Local Livelihoods and the Global Environment*. Cambridge University Press, Cambridge, UK.

Schnitzler, A. and Borlea, F. (1998) Lessons from natural forests as keys for sustainable management and improvement of naturalness in managed broadleaved forests. *Forest Ecology and Management* 109, 293–303.

Scott, J.M., Tear, T.H. and Davis, F.W. (eds) (1996) *Gap Analysis: a Landscape Approach to Biodiversity Planning*. American Society for Photogrammetry and Remote Sensing, Bethesda, Maryland, p. 320.

Scott, J.M., Heglund, P.J., Morrison, M., Haufler, J.B., Raphael, M.G., Wall, W.A. and Samson, F.B. (eds) (2002) *Predicting Species Occurrences: Issues of Scale and Accuracy*. Island Press, Covelo, California, p. 868.

Siitonen, J. (2001) Forest management, coarse woody debris and saproxylic organisms: Fennoscandian boreal forests as an example. *Ecological Bulletins* 49, 11–41.

Smeding, F.W. and Joenje, W. (1999) Farm-Nature Plan: landscape ecology based planning. *Landscape and Urban Planning* 46, 109–115.

Sporrong, U. (1998) Dalecarlia in central Sweden before 1800: a society of social stability and ecological resilience. In: Berkes, F. and Folke, C. (eds) *Linking Social and Ecological Systems*. Cambridge University Press, Cambridge, UK, pp. 67–94.

Tucker, G.M. and Heath, M.F. (1994) *Birds in Europe: Their Conservation Status*. BirdLife International, BirdLife Conservation Series no. 3, Cambridge, UK.

Turnock, D. (2001) Railways and economic development in Romania before 1918. *Journal of Transport Geography* 9, 137–150.

Turnock, D. (2002) Ecoregion-based conservation in the Carpathians and the land-use implications. *Land Use Policy* 19, 47–63.

UNESCO (2002) Biosphere reserves. Special places for people and nature. UNESCO workshops, Paris.

Valve, H. (2002) Implementation of EU rural policy: is there any room for local actors? The case of East Anglia, UK. *Landscape and Urban Planning* 61, 125–136.

Vera, F.W.M. (2000) *Grazing Ecology and Forest History*. CABI, Wallingford, UK.

von Haaren, C. (2002) Landscape planning facing the challenge of the development of cultural landscapes. *Landscape and Urban Planning* 60, 73–80.

von Thünen, J.H. (1875) *Der isolierte Staat in Beziehung auf Landwirtschaft und Nationalökonomie*. Wissenschaftliche Buchgesellschaft, Darmstadt.

Vos, W. and Meekes, H. (1999) Trends in European cultural landscape development: perspectives for a sustainable future. *Landscape and Urban Planning* 46, 3–14.

Vourela, N., Petteri, A. and Kalliola, R. (2002) Systematic assessment of maps as source of information in landscape-change research. *Landscape Research* 17, 141–166.

Whyte, I.D. (1998) Rural Europe since 1500: Areas of retardation and tradition. In: Butlin, R.A. and Dodgshon, R.A. (eds) *An Historical Geography of Europe*. Oxford University Press, Oxford, UK, pp. 243–258.

been evidence that species and ecosystem protection requires a large-scale landscape approach, bringing nature conservation and people together.

Integrating perspectives from natural and cultural resource conservation, coupled with the dynamic qualities of landscapes, poses challenges for management. Adrian Phillips, IUCN's World Heritage Advisor, has noted that the long tradition of

> the separation of nature and culture – of people from the environment which surrounds them – which has been a feature of western attitudes and education over the centuries, has blinded us to many of the interactive associations which exist between the world of nature and the world of culture.
>
> (Phillips, 1998)

Our experience with the Mount Tom forest reaffirms and expands upon the underlying thesis that many types of cultural landscapes must be evaluated and managed as dynamic entities that are continually shaped over time by both natural and human forces.

The Mount Tom Forest as a Cultural Landscape: Defining an Approach

Laying the foundation

The direction for managing the forest as a cultural landscape was underscored in a letter by the then Director of the National Park Service, James Ridenour, appended to the park's 1992 legislation (italics added):

> Important historical and cultural aspects of the Marsh/Billings National Historical Park are the forest management practices instituted by Frederick Billings. Many of the trees on the property were planted under the direction of Billings and represented a major advance in reforestation practices at that time....Active forest management is an important part of not only preserving the resource but of *interpreting the cultural importance of the landscape....*[5]

The park's 1998 General Management Plan (GMP) established the basic management philosophy to guide future decision-making for the park and forest (US Department of the Interior, 2001). The GMP calls for the National Park Service to treat the forest as a cultural landscape and continue to actively manage the forest in order to convey a sense of the site's evolution through the occupancy of the Marsh, Billings and Rockefeller families, and to continue the tradition of professional forest management as an educational demonstration of conservation stewardship.

> This management approach necessarily includes active management, including appropriate harvesting, to preserve the character-defining features of the forest while perpetuating its historic use as a model forest.
>
> Public participation throughout the process will ensure that the public understands the practices of good forest stewardship....A variety of educational programs will be developed that demonstrate the basic principles of forestry. Programming will stress the importance of balance in forest management respecting historic character, natural values, aesthetics, and recreational use.
>
> (US Department of the Interior, 1999, p. 22)

The GMP directed the park to identify and retain features and characteristics that contribute to the forest's historical or ecological significance, and to apply best management practices in the care of the forest (US Department of the Interior, 1999, p. 27). The GMP further recommended 'rehabilitation' as the overall cultural landscape treatment strategy for the park. Rehabilitation is an approach to preservation defined by the National Park Service that allows for repairs, alterations and additions necessary to make the property operational while preserving those portions or features that convey historical and cultural values.

With the general management philosophy established by the GMP, park staff began to grapple with identifying those characteristics of cultural and ecological importance and developing a strategy to retain the character of the forest and work with the dynamic nature of forest change.

Evaluating Cultural Landscapes: Defining Historic Character

The Mount Tom forest is an unusual forest. It represents a remarkable milestone in American conservation – the beginning of a long road from what appeared in the middle of the 19th century to be an endless spiral of forest destruction, stretching from the mountains of New England to the shores of the Pacific. The evolution of that recovery is the result of three generations' commitment, over nearly 130 years, to continuous forest management and ecological succession.

History from the cultural perspective

To understand the cultural significance of the Mount Tom forest, the park worked with a broad set of collaborators to develop several resource inventories and historic context studies.[6] This research provided the framework for understanding the complexities of the park's historical significance and interrelationships between the landscape of Mount Tom and the people who worked with and learned from it.

George Perkins Marsh grew up on the property that is now the national park during a time of enormous social and environmental upheaval. By the mid-19th century, Mount Tom, like thousands of other once-forested landscapes in New England, had succumbed first to clearing for hill farms, potash and firewood and then to further clearing to meet an almost insatiable demand for open land for sheep pasture. The creation of commercial markets for this relatively brief boom in Merino sheep precipitated one of Vermont's earliest environmental catastrophes. Almost 8000 wooden rails were needed to enclose a 40-acre pasture. In a historical blink of an eye, places like Woodstock's Mount Tom were stripped of most of their vegetation, then were quickly eroded and left deeply gullied and infertile. Upland top soil was washed into streams and rivers, threatening drinking water and creating massive fish kills. Meanwhile, struggling lowland villages were afflicted by frequent mud slides and flooding. Both George Perkins Marsh and Frederick Billings witnessed this rapid degradation of the Vermont landscape. Years later, while serving as US Ambassador to Italy, Marsh wrote passionately about the consequences of deforestation and argued for a new ethic of stewardship in his 1864 landmark book, *Man and Nature* (Marsh, 1864).

Frederick Billings, a Vermont native, lawyer, railroad executive and pioneer conservationist purchased the Marsh property including the Mount Tom forest in 1869. Billings was a believer in material progress and sustainable use, an outlook characteristic of American conservation up to the middle of the 20th century. In the west, he had directed efforts to encourage settlement and commerce along the route of the Northern Pacific Railroad, by planting trees, building windbreaks and other measures to stimulate rural development (Auwaerter *et al.*, 2004). In Woodstock, Billings set about creating a farm and forest on the former Marsh property that would serve as a model of land stewardship for the depressed agricultural economy of his home state. He harboured a vision of social improvement and rural recovery based in part on reforestation, agricultural improvement and conservation. As Billings was planting trees by the thousands, he simultaneously developed 12 miles of carriage roads to showcase his pioneer forestry work to the public. He worked to enhance his new estate's productivity and beauty. Landscape architect Robert Morris Copeland, who worked with Billings on the layout of his estate, wrote of a synthesis of the 'useful and the beautiful' (Auwaerter *et al.*, 2004).

When Billings started this bold experiment, forestry was not yet an established profession in America. He used scientific practices borrowed from 19th-century European forestry, drawing heavily on his personal library of German and French forestry texts. The earliest plantations established by Billings on Mount Tom were Norway spruce (*Picea abies*), European larch (*Larix decidua*), and Scots pine (*Pinus sylvestris*) – fast-growing European species thought to be best suited to the New England climate (Fig. 9.1). The use of these species is one demonstration of the dominant influence of European scientific forestry on the nascent profession in America (Auwaerter *et al.*, 2004). In some places on the property Billings simply added desirable native hardwood species into the mix of other trees that were naturally regenerating. In other places, particularly on more subtly graded open land, Billings established plantations (using row plantings of single species) of fast-growing conifer trees.

Billings' scientific forestry programme on worn-out agricultural lands influenced other efforts of forestry conservation throughout Vermont and the New England

Fig. 9.1. This stand of Norway spruce was planted in 1887 as part of Frederick Billings' efforts to create a model forest that demonstrated progressive forestry techniques of the late 19th century. Today, the Mount Tom forest is one of the oldest planned and managed forests in the USA. (Photo featured in *American Forests*, February 1910.)

region. Billings promoted the first state commission to study forestry in Vermont, and was a principal author of its final report which emphasized the role of forestry in the revitalization of rural Vermont (Auwaerter *et al.*, 2004). In the context of American conservation history, his forestry work was far-sighted and pioneering for its time. As the century turned, Frederick Billings' daughters succeeded him in the reforestation work on Mount Tom. The plantations set out by Billings' daughters included both native white pines (*Pinus strobus*) and red pines (*Pinus resinosa*) that would dominate the 20th-century reforestation techniques in the USA. These later plantings were concurrent with the rapid growth of the forestry profession in America and the rise of forest conservation in the public sector, particularly with the establishment of municipal, state and national reforestation programmes.

Reforestation on Mount Tom continued through the mid-20th century. By that time most of the open land had been planted and forestry work shifted to maintaining existing plantations and managing the property for additional recreational uses, such as hiking and skiing. Billings' gently graded carriage roads provide inviting portals into the forest, while overhead, the sky is almost completely blocked by a dense canopy of mature trees. Along the verge on either side of the road, the understorey has been thinned and opened up so one can peer deeply into the forest, a legacy of Rockefeller's emphasis on aesthetics and recreation – hallmarks of his stewardship during the later half of the 20th century.

History from the ecological perspective

While the cultural dimension of the reforestation practices of Billings and his heirs is critical to understanding the Mount Tom forest, this story is incomplete without the ecological history. As the park was developing the baseline information on the cultural significance of the site, simultaneously research was also conducted on the natural processes that have influenced the creation and evolution of the forest (Auwaerter et al., 2004). This research provided the framework for understanding the complexities of the park's ecological significance and interrelationships between ecological change, human activity and the resulting landscape character. These studies illuminated the important role of natural succession in the development of the Mount Tom forest.

While part of Mount Tom was being reforested with thousands of seedlings, other abandoned fields slowly began the transition from field to a mixed northern hardwood forest through natural succession. As this process unfolded, early 'pioneer' tree species such as white pine (*Pinus strobus*), white and grey birch (*Betula papyrifera* and *B. populifolia*), and aspen (*Populus tremuloides*) were the first to colonize the unmanaged, open fields. These trees could sprout and grow in the thick pasture grasses, tolerate nutrient-poor soils, and thrive in the dry, sunny open land. As these 'pioneers' became established and grew, they began to influence the site by adding nutrients to the soil, and forming dense canopies that shaded the grasses and dense herbaceous plants. Under these conditions, the shade-tolerant tree species such as sugar maple (*Acer saccharum*), American beech (*Fagus grandifolia*) and eastern hemlock (*Tsuga canadensis*) became established and to this day dominate the composition of the forest. From Billings' time forward, succession in this part of the forest has been influenced by management. As pioneer species matured, they were harvested or thinned to favour the more shade-tolerant, longer-lived hardwood species. Poor quality and diseased trees were also removed. The resulting managed hardwood forest has readable signs that tell this story, including the high quality of the remaining trees.

Meanwhile, in the decades following the establishment of Billings' first plantations in the 1870s, succession began to influence the plantation development as well. Many of the conifer species used in the reforestation efforts, such as the white pine (*Pinus strobus*), Norway spruce (*Picea abies*) and red pine (*Pinus resinosa*), were selected because they were fast-growing, could tolerate nutrient-poor soils and could compete with the grasses of the agricultural fields. These species quickly established a continuous cover that inhibited the growth of native seedlings.

For a while, these planted trees enjoyed a competitive advantage over other plants that might have naturally colonized the site. However, in order to maintain the health of plantation trees, periodic thinning is required to avoid the potential stagnation of the plantation from overcrowding. The thinning increases the amount of sunlight reaching the forest floor and results in a burst of regeneration from native seedlings. Just like the pioneer species, the plantations enhanced the soil nutrients and created a mature forest canopy, providing the necessary conditions for the establishment of shade-tolerant native seedlings such as sugar maple, beech, yellow birch, white ash and hemlock. After a thinning, plantation trees quickly responded with a growth spurt that once again closed the canopy and repressed the growth of the native hardwood seedlings. However, as the plantations age and the openings created through thinning become larger, the hardwoods become a significant component of the plantation composition (Fig. 9.2).

Man and nature on Mount Tom: an integrated approach to landscape character

It is the interplay between human intention and natural processes that ultimately shapes the character of the forested landscape of Mount Tom. The landscape character can be described as:

- the patchwork configuration of fields, hardwoods, and plantations;
- the diversity of forest architecture including hardwood stands and even-aged conifer plantations of varying ages and levels of maturity; and
- scattered individual 'legacy trees'.

Fig. 9.2. This red pine stand planted in 1911 has a well-established under-storey of naturally regenerated native hardwoods, including sugar maple and beech, that could eventually become a significant component of the over-storey as the plantation continues to mature.

Fig. 9.3. The Mount Tom forest's patchwork represents an evolution in land use. This hay field is one of the remaining agricultural fields that date back to the early 1800s. The trees in the foreground are an allee of sugar maples which line a former farm road that is part of the 20-mile carriage road system developed by Billings. In the background a 1952 red pine stand represents one of the youngest and most dense plantations on Mount Tom. (Photo provided by Nora Mitchell.)

Landscape patchwork

The character of the forest is defined by a mosaic of spaces formed by the interrelationship of natural and planted forest stands, meadows, the Pogue (a 14-acre pond) and vista clearings. Woodland, pasture and water features are woven together by the network of carriage roads. This is a forest of contrast – light and dark, open and enclosed, intimate and expansive. This patch-like character reflects the agricultural origin of the landscape, the influence of late 19th-century landscape design, and over 130 years of continuous forest management (Fig. 9.3).

This land-use history, coupled with natural succession, has created 16 different vegetation communities in the park (excluding the plantations). The current diversity of the forest over a relatively small spatial scale is valuable habitat for many wildlife species.

Forest architecture

The Mount Tom forest has a wide variety of stands. In some areas, such as the mature conifer plantations and hardwood stands, the open under-storey and high canopy offer sweeping views into the forest. In other stands, such as in younger hardwood stands and naturalized plantations, the thick under-storey creates enclosed, intimate forest experiences.

The complex structure of the park's forest provides a diversity of wildlife habitats and other ecological functions. Many forest stands are developing greater vertical diversity as intentional forest thinning and natural ageing of the stands opens up the canopy, increasing light for shade-tolerant trees in the under-storey. Forest inventory and monitoring suggest that low intensity forestry in the park appears to mimic natural disturbances in both plantations and semi-natural stands, supporting increased structural and species diversity within these stands. Some of the park's oldest plantations are starting to develop 'old growth', late-succession structural characteristics.

Legacy trees

This is a forest of big trees. There are sturdy white pines, graceful Norway spruces swaying in the wind high above the forest canopy, giant old-growth hemlocks and allees of stately sugar maples – surviving sentinels still at their posts along the old hill farm roads. These old hemlocks and sugar maples, also referred to as *legacy trees*, are witnesses to the march of history. They have been spared the axe, often growing wildly and defiant with their gnarled trunks and branches in the midst of the otherwise orderly geometry of pine plantations. One can speculate that perhaps these survivors commanded special respect from Billings and his foresters (Fig. 9.4).

These 'legacy trees' also biologically enrich the park's forested ecosystems. They provide an abundance of habitat attributes including cavities utilized by a host of species. Old-growth trees increase the representation of large-diameter trees in mature forests and enhance vertical structure. Diameter distributions of the remnant old-growth trees in the park extend well beyond the sizes reported for natural hardwood stands in the north-east.

The Challenges of Preserving Character in a Dynamic Landscape

In preparing a management approach for Mount Tom, the National Park Service identified the following challenges.

Preserving historic character within the nature of forest change

Forest management needs to work with the long-term nature of forest growth and the natural life-span of trees, in this case, between 100 and 200 years. The character and composition of the forest are the result of both human intervention and natural succession. As such, the cycles of forest change are part of the historic character of the forest. Management needs to work with

Fig. 9.4. The wide-spreading character of the branches of this sugar maple is a testament to the open, agricultural origins of this area that is now a dense red pine plantation. (Photo provided by Christina Marts.)

the dynamics of forest growth and change in a way that continues the legacy of stewardship and retains forest characteristics that illustrate the rich history of forest management on Mount Tom.

Integrating forestry and agricultural practices to retain traditional land uses within the forest

The diversity of forests and fields on Mount Tom is the result of over 130 years of continuous forestry and agricultural practices including thinning, pruning and harvesting. These activities need to continue to be used as both a tool to retain the historic character of the forest and as a demonstration of the processes that have shaped the landscape.

Embracing Billings' legacy of sustainable forestry and public education

Frederick Billings reforested Mount Tom as a model of sustainable, innovative forestry and public education. The park recognizes that forestry and best forest management practices have evolved since the time Billings began his reforestation campaign. Management activities need to model contemporary sustainable management while still retaining characteristics that represent the historical evolution of conservation thought.

A Strategy for Transition and Integration

Retaining forest character

The remaining even-age, single-species plantations on Mount Tom provide an illustrative example of Billings' pioneering reforestation techniques and a continuum of forest management that followed in the 20th century. However, all of these stands face strong competition from native hardwood trees. The distinctive, cathedral character of mature plantation reflects a snap-shot of history, a moment in time,

along an almost irreversible trajectory of forest succession. Eventually all of the historic plantations will reach maturity and will no longer convey the same sense of original planting patterns or species composition that illustrates the historic reforestation efforts. The extent that these softwood plantations might be perpetuated or re-created into the future is problematic and presents a major challenge in managing this cultural landscape.

To re-establish plantations in their existing locations and with the existing even-aged structure would require management techniques including over-storey removal by clear-cutting and the removal of the existing regeneration through aggressive hand cutting or the use of herbicides. These management options would run counter to the momentum of forest succession, would be extremely costly and labour intensive, and could have adverse ecological impacts. This approach to preservation would also be at variance with a long legacy of forward-thinking forest stewardship.

Accepting the reality that the forest on Mount Tom is a dynamic resource with underlying ecological processes at work is a necessary step in conceptualizing strategies for the preservation of the cultural landscape. Thoughtful, unconventional approaches to retaining and enhancing character are being implemented that respect the unique history of the land and its long and intimate connection to a long line of well-intentioned stewards. The US National Park Service has had to work with a philosophical framework that addresses preservation at *both* the broad characteristic and feature level. Preservation in a forested cultural landscape transcends the traditional architectural fixation on 'fabric integrity' and the perpetuation of individual objects or features. As the forest plantations eventually age and decline over time, management emphasis will shift more to renewing broad distinctive patterns and characteristics of the forest as a whole, and the *tradition* of forward-thinking forest stewardship.

For example, new plantings, wherever

they occur, may draw on a 'palette' of historic species and the forest will continue to be managed for its characteristic large trees. New generations of 'legacy' trees will also be cultivated. It may be possible to re-establish a few conifer plantations on a much smaller scale by hand cutting competing hardwoods. (A few new plantations could be established in the few existing open fields and pastures; however, these agricultural lands are also now a valued character-defining feature of the cultural landscape.) Opportunities will also be pursued to either retain and renew the edges of plantations or seek out new locations where scaled-back plantings of new softwoods might be accomplished. The greatest effort will be made along the principal carriage roads where there is the highest concentration of park visitors.

This work is already being started with the realization that action must be initiated long before the historic plantations begin to decline and significantly change. It may take decades if not a generation of work to successfully effect this transition.

Continuing a legacy of education and sustainability

The forest is a place not only to interpret conservation's early history, but also to demonstrate principles of contemporary forest management and sustainability. Forest management is conducted in a way that makes the intent and process of management practices transparent to the public. Programmes and interpretive displays are created in association with management activities to provide further explanation of what is being done and why. Whenever possible, management operations are conducted as public activities, providing hands-on learning opportunities. Educational activities address the complex social, economical and ecological issues associated with cultural landscape in a local and global context.

The park has initiated several projects to continue the tradition of innovative, sustainable forest management.

Certification pilot project

The park is undergoing voluntary assessment under the Forest Stewardship Council certification system to demonstrate and interpret certification as a new chapter in the park's legacy of conservation innovation. Third-party certification is one of the fastest growing new developments in sustainable forestry. The purpose of certification programmes is to provide market recognition of good forest management through credible, independent verification of best forest practices.

Value-added economics

Value is added to forest products through their association with a special place; responsible, sustainable management; and craftsmanship. To interpret these important connections, the park is working with Eastern National, the park's cooperating association, to commission products made by artists from wood harvested in the park, including bowls and pens which are for sale at the park's visitor centre bookstore. Wood from Mount Tom has also been used for furniture for the visitor centre, rehabilitating historic buildings and other park maintenance projects. The park demonstrates and interprets practical, low-impact techniques for harvesting, and on-site milling and drying of lumber. Wherever practical, wood harvested from the forest will be used for value-added purposes, supplying local crafts people and manufacturers (Fig. 9.5).

Crop tree release demonstration

Crop tree management enhances the growth of selected forest trees and improves wildlife habitat, recreational opportunities and forest aesthetics. The demonstration site was created through a public workshop held in cooperation with the USDA Forest Service and the Vermont Department of Forests, Parks and Recreation. The growth of the crop trees will be measured and monitored over time to evaluate the effectiveness of the treatment.

Fig. 9.5. A local furniture maker demonstrates the art of his craft while discussing how local sustainably managed forests inspire and support the tradition of regional wood product craftsmanship and his work. (Photo provided by Rolf Diamant.)

A forest for every classroom: learning to make choices for the future

Inspired by a common vision of students learning from and caring for public lands, the park has joined with Shelburne Farms, the Conservation Study Institute, Green Mountain National Forest and the Northeast Office of the National Wildlife Federation to create *A Forest for Every Classroom*, a professional development programme for educators on place-based education. The public and private partner organizations share a common vision that if today's students are to become responsible environmental decision makers, they must understand the landscapes in which they live, and they must have educational opportunities based on real-life issues that encourage them to practice environmental citizenship in their own communities. The curriculum developed by the teachers integrates hands-on learning about concepts in cultural landscapes, forest ecology, sense of place, stewardship and civic responsibility. Students immersed in the interdisciplinary study of 'place' are more eager to learn and be involved in the stewardship of their communities and public lands.

Adaptive management

A programme of adaptive management is used to understand site-specific change and to evaluate and refine forest management activities. Working with the University of Vermont's Rubenstein School of the Environment and Natural Resources, the park has established 64 permanent plots to assess successional dynamics and structural changes associated with forest stands. This is an important tool for monitoring cultural landscape change over time and assessing overall ecosystem health (forest growth and structural changes, regeneration, biological diversity, forest pests and diseases, invasive plant populations, and water quality). These efforts ensure that ongoing forest management is continuously modified to reflect insights gained from on-site monitoring and knowledge emerging from new research.

Watershed and community connections

Recognizing that the forest's ecological and historical connections extend beyond the park boundary, the park is working with local land owners and community organizations on collaborative projects such as the development of an integrated community trails system. The park also involves the local community, educators, interested professionals and the broader public as active participants in the management of the forest. The park will continue to build a network of partners to enhance research, management and educational efforts related to forest stewardship.

Conclusion

In reclaiming and reoccupying lands laid waste by human improvidence or malice, [man must] become a co-worker with nature in the reconstruction of the damaged fabric.

George Perkins Marsh

In developing the forest management programme for Marsh–Billings–Rockefeller National Historical Park, we were confronted with a tradition of historic preservation that emphasized retaining fabric and features of historic places and an ecological perspective that was ahistorical. However, management of this historic forest challenged these traditions. The forest character we see today is only a snap-shot in time; change is constant, but the nature of that change is infinite. Trees are moving along their own unique trajectories influenced by the dynamics of competition and disease, and the availability of light, soil, water and nutrients. The pull of forest succession never rests. Softwood plantations were planted in the late 19th and early 20th centuries in open, abandoned agricultural fields without the powerful competition from native hardwoods. Today, all but a few of the fields are covered with trees and competition from hardwoods is intense. To maintain new generations of softwoods, exact replicas of today's stands, may well be beyond our reach or simply impossible without clearing land on the scale of early settlement. The response to this challenge is not so much in the preservation of detail, such as *in situ* replication of forest stands and species composition, but rather in the preservation of landscape character and an enduring legacy of responsible, sustainable forest management.

This approach to forest management reflects a new direction in conservation philosophy and practice. There are lessons learned from this storied forest ecosystem where the combination of human intervention and natural processes has created a place of utility and beauty, of nature and culture. These lessons can be applied in other places where natural and cultural systems are not simply parallel, nor distinct resource sets that co-exist. In these cultural landscapes, we need to transcend a traditional historic preservation approach that views landscape as historic artefact and a solely natural ecosystem approach that regards human intervention as undesirable or unnecessary. Instead, we need a historic preservation perspective that incorporates the role and influence of natural succession and ecological processes, and a natural resource conservation perspective that is informed and shaped by a sense of history and stewardship. The dynamic character of these landscapes, by definition, is a product of a long history of very complex interactions. Planning therefore needs to take into account a longer time horizon, thinking about landscape change in terms of hundreds of years. This challenges the traditional perception of the past as separate and distinct from the present. Inherent in the recognition of this broad set of natural and cultural values is the opportunity to manage cultural landscapes in a manner that respects ecology, history, continuing cultural traditions, principles of sustainability and education.

Notes

1. Rolf Diamant is Superintendent of Marsh-Billings-Rockefeller National Historical Park (MBRNHP), US National Park Service, Christina Marts is Chief of Resources Management (MBRNHP) and Dr. Nora Mitchell is Director of the National Park Service Conservation Study Institute.
2. *Cultural Landscapes: The Challenges of Conservation*, World Heritage Papers 7. Paris: UNESCO World Heritage Centre, 2003; Nora Mitchell and Susan Buggey, "Protected Landsdcapes and Cultural Landscapes: Taking Advantage of Diverse Approaches," *The George Wright Forum* 17 (1): 35-46, 2000; *A Handbook for Mangers of Cultural Landscapes with Natural Resource Values*, a web-based publication, 2003, see http://www.nps.gov/csi/csihandbook/home.htm.
3. In the United States, a number of publications have provided multiple tools for identifying, understanding, and managing cultural landscapes. *Preservation Brief 36: Protecting Cultural Landscapes: Planning, Treat-*

ment and Management of Historic Landscapes, by Charles A. Birnbaum. Washington, D.C.: US Department of the Interior, National Park Service, 1994; and *A Guide to Cultural Landscape Reports: Contents, Process, and Techniques,* by Robert R, Page, Cathy A. Gilbert, and Susan A. Dolan, 1998, both offer guidance in analysing, documenting and protecting cultural landscapes. *The Secretary of the Interior's Standards with Guidelines for the Treatment of Cultural Landscapes* edited by Charles A. Birnbaum and Christine Capella Peters, Washington, D.C.: U.S. Department of the Interior, National Park Service, 1996, provides direction for decision-making about cultural landscapes, which is particularly useful for designed historic landscapes.

4. *National Register Bulletin 30: Guidelines for Evaluating and Documenting Rural Historic Landscapes,* Linda Flint McClelland, J. Timothy Keller, Genevieve P. Keller, and Robert Z. Melnick, Washington, D.C.: U.S. Department of the Interior, National Park Service, 1990.

5. Public Law 102-350. August 26, 1992, 102nd Congress, S. 2079.

6. *General Management Plan Marsh-Billings-Rockefeller National Historical Park* (GMP), Washington, D.C.: U.S. Department of the Interior, National Park Service, 1999, 22.

References

Auwaerter, J. *et al.* (2004) *Cultural Landscape Report for the Forest at Marsh-Billings-Rockefeller National Historical Park, Woodstock, Vermont Volume 2: Analysis.* National Park Service, Brookline, Massachusetts.

Auwaerter, J. and Curry, G. (2002) *DRAFT Cultural Landscape Report for the Mansion Grounds: Marsh-Billings-Rockefeller National Historical Park, Volume 1: Site History.* National Park Service, Boston, Massachusetts.

Beck, J. (1994) *Report for the National Park Service based on Interviews Conducted about the Marsh-Billings National Historical Park.* Vermont Folklife Center, Middlebury, Vermont.

Birnbaum, C. A. (1994) *Preservation Brief 36: Protecting Cultural Landscapes: Planning, Treatment and Management of Historic Landscapes.* US Department of the Interior, National Park Service, Washington, DC.

Birnbaum, C.A. and Peters, C.C. (Eds) (1996) *The Secretary of the Interior's Standards with Guidelines for the Treatment of Cultural Landscapes.* US Department of the Interior, National Park Service, Washington, DC.

Foulds, E., Lacy, K. and Meier, L.G. (1994) *Land Use History for Marsh-Billings National Historical Park.* National Park Service, Boston, Massachusetts.

Madison, M. (1999) *Landscapes of Stewardship: the History of the Marsh-Billings Site.* Unpublished. Marsh, G.P. (1864) *Man and Nature, or, Physical Geography as Modified by Human Action.* New York and London.

McClelland, L.F., Keller, J.T., Keller, G.P. and Melnick, R.Z. (1990) *National Register Bulletin 30: Guidelines for Evaluating and Documenting Rural Historic Landscapes.* US Department of the Interior, National Park Service, Washington, DC.

Mitchell, N. and Buggey, S. (2000) Protected landscapes and cultural landscapes: taking advantage of diverse approaches. *The George Wright Forum* 17 (1): 35-46.

Nadenicek, D. (2003) *Frederick Billings: the Intellectual and Practical Influences on Forest Planning, 1823-1890.*

Page, R. R., Gilbert, C. A. and Dolan, S. A. (1998) *A Guide to Cultural Landscape Reports: Contents, Process, and Techniques.*

Phillips, A. (1998) The nature of cultural landscapes – a nature conservation perspective. *Landscape Research* 23 (1) 36.

Plachter, H. and Rossler, M. (1995) Cultural landscapes: reconnecting culture and nature. In: von Droste, B., Plachter, H. and Rossler, M. (eds) *Cultural Landscapes of Universal Value, Components of a Global Strategy.* Jena, Stuttgart and Gustav Fischer Verlag, New York, with UNESCO.

Public Law 102-350. August 26, 1992, 102nd Congress, S. 2079.

UNESCO (2003) *Cultural Landscapes: the Challenges of Conservation,* UNESCO World Heritage Papers 7.

A Handbook for Managers of Cultural Landscapes with Natural Resource Values (2003) see http://www.nps.gov/csi/csihandbook/home.htm.

US Department of Interior (2001) *Report on the Management of the Historic Mount Tom Forest, Woodstock, VT.* US Department of Interior, National Park, pp 4-5.

US Department of Interior (1999) *General Management Plan Marsh-Billings-Rockefeller National Historical Park* (GMP). US Department of the Interior, National Park Service, Washington, DC.

US Department of the Interior (2004) *DRAFT Cultural Landscapes Inventory: Marsh-Billings National Historical Park.* US Department of the Interior, National Park Service, Brookline, Massachusetts.

Wilcke, S., Morrissey, L., Morrissey, J.T. and Morrissey, J. (2000) *Cultural Landscape Report for the Forest at Marsh-Billings-Rockefeller National Historical Park.* Conservation Study Institute and University of Vermont, Burlington, Vermont.

10 Working Forest Landscapes: Two Case Studies from North Carolina

G.B. Blank

North Carolina State University, Department of Forestry,
Raleigh, North Carolina, USA

Introduction

In the USA many opportunistic actions have secured 'development rights easements' or made outright purchases of ecologically valued land parcels threatened by potential development. Actions by land conservation groups at local and regional spatial scales have been effectively expanding the amount of natural landscape protected and accessible by the public for decades (Mann and Plummer, 1996). Such actions are usually directed toward lands showing minimal degrees of prior disturbance, thus preserving important biological values such as biodiversity. However, relatively few places in the landscape of the USA remain absolutely pristine (Dickinson, 2000; Mann, 2002). Thus, as development pressure grows in certain regions, conservation actions increasingly need to focus on remaining large contiguous tracts of natural landscape, no matter what their past use history (Mann and Plummer, 1996). Such large tracts often contain managed forests and tracts harvested but not systematically managed.

At federal and state levels in the USA, a number of programmes exist to provide funding for land conservation. North Carolina, for example, initiated the Clean Water Management Trust Fund to make monies available for acquiring lands to further water quality protection goals. At the federal level, a programme titled 'Forest Legacy' enables participating states to use funds channelled through the USDA Forest Service specifically to acquire forested lands for conservation. States must qualify to participate in the Forest Legacy Program and must develop criteria for determining which lands will become eligible for acquisition. In North Carolina, through a very public process, proponents for joining the Forest Legacy Program declared an intention to secure working forests for long-term protection from development (Blank, 1999).

According to the definition, working forest landscapes can provide wood for extraction, clean water and air, necessary habitat for game and non-game species, recreational space, visual beauty, tranquility and an array of specialty items such as mushrooms, medicinal plants, mistletoe, pine straw, etc. (Blank, 1999). Working forest landscapes produce economic, aesthetic and spiritual values perceived to varying degrees by people from widely varied perspectives. Working forests provide the aforesaid renewable resources and values under direction of a management plan designed to meet defined objectives in a sustainable manner.

This definition emerged from discussions held by the steering committee convened to direct the Forest Legacy application process and ensure broad representation across the range of parties interested in the future of North Carolina's forested landscape. The definition was circulated to directors of land conservation trusts across the state and to interested citizens who participated in the scoping process. Consensus emerged that for the aims of this federal programme as applied in North Carolina, the inclusive language of this definition best addressed the threats facing the state's areas of large contiguous forest. More restrictive definitions were considered unlikely to provide incentive to private landowners to cooperate in the initiative. Two examples of how this definition applies are given here to illustrate implications of the concept of working forests and to consider how historic contexts impact our understanding of landscape management choices.

Landscape Contexts

The 4455 ha of Progress Energy's Harris Lands encompass the 514-ha North Carolina State Harris Research Tract (HRT, http://legacy.ncsu.edu/classes/for784001/harristract.htm) in Wake County, North Carolina (Blank *et al.*, 2002). These lands lie in the southern portion of a region that has experienced sustained urbanization since economic growth began accelerating rapidly during the late 1970s. Growth was spawned by the innovative creation of Research Triangle Park, situated within a geographic triangle formed by the cities of Durham, Chapel Hill and Raleigh. Each of these three cities has a major university located within its boundary: respectively, Duke University, University of North Carolina, and North Carolina State University. In large part because of its proximity to these three universities, Research Triangle Park succeeded beyond most people's expectations in its ability to attract clean industry and research enterprises that, in turn, generated more jobs and economic vitality. Housing and associated service

centres have since sprawled in all directions from the three urban cores. Some small communities peripheral to these urban centres have grown into substantial municipalities in their own right. Cary, North Carolina, for instance, grew from a village of about 4000 people in 1970 to around 96,000 people at the 2000 census. Amidst such growth, the Harris Lands were acquired to provide a security buffer around a nuclear power plant built and activated in the 1980s and to accommodate Harris Lake, needed to provide water for the reactor's cooling tower. Perhaps the most important characteristic of the Harris Lands, the 1619-ha Harris Lake, draws boaters and fishermen from throughout the region. Besides HRT and the lake, the Harris Lands include a regional park and extensive game lands, the latter managed by the North Carolina Wildlife Resources Commission (NCWRC).

Dupont State Forest (DSF, http://www.dupontforest.com) occupies 4171 ha of land in Transylvania and Henderson Counties, in the mountainous western region of North Carolina. Unlike the Harris Lands, DSF lies farther from urban influence, although the North Carolina mountain region includes some counties that lead the state in percentage rates of population growth. Here, pressure comes from retirees and growing service industries catering to outdoor recreation and tourism. Mountain resort and retirement community developments fill the valleys and pepper the hills. Relatively short distances to major cities such as Atlanta, Charlotte and Knoxville make it easy for urbanites to reach second homes and weekend vacation spots, while the Blue Ridge Parkway and Smokey Mountains National Parks attract enormous numbers of visitors to the region.

Land that became DSF was intentionally assembled as a private retreat surrounding an industrial plant built by DuPont de Nemours to manufacture photographic film. Before that, in the 1920s approximately 2025 ha of the property were still owned by 25 individuals, with much of the forest being harvested and large areas cleared for agriculture (Scott and Blank,

2003). The majority owner of the property, Frank Coxe, sold land in 1956 to DuPont Corporation, whose owners believed that the clean air and water of the mountains would improve the film production process. Over time, DuPont acquired additional land parcels around the original 2025 ha. The land was used by employees for recreation and relaxation and used to entertain clients and executives of the corporation, so these activities prompted development of a recreational infrastructure. Most farmed lands were replanted to white pine forest, but most of the land was allowed to regenerate and develop as natural forest. The 4171-ha tract now known as DuPont State Forest was acquired by the State in three major phases, spanning from 1995 to 2000. Under General Statute 113-34, state forests may be used to demonstrate the practical utility of timber culture and water conservation, as refuges for game and for experimental, demonstration, education, park and protection purposes.

Geographic positions and past land uses have, of course, shaped biological conditions and created impacts on the landscapes these forests occupy. Though past activities in some degree define the potential work these forests can do, the resiliency of temperate North American forests allows flexibility and dynamic change under management direction. Currently, management direction of the Harris Lands and DSF will move in different directions because of their change in ownership status and the differing attributes of the forests themselves. However, both forests will still merit being called working forests, though the primary focus of their work may vary in emphasis.

Forest Characteristics

The Harris Research Tract contains remnants of the transitional longleaf (*Pinus palustris*) pine community within mixed pine and hardwood forests that dominate most of the Harris Lands owned by Progress Energy (Blank, 2004). Within these forests, 247 species of plants have been identified

(Parker, 1998) representing the site's transitional position between coastal plain and piedmont regions. The plant community also reveals a number of successional stages and introductions of exotic species because of past disturbance activity. Overall, the piedmont forests in this area once dominated by longleaf pine were altered when farming and logging began with European colonization and settlement in the 1700s and gradually reduced the viability of longleaf trees in the forest canopy (Bode, 1997; Parker, 1998). Thus, these forests show the effects of fire suppression, conversion to non-forest uses and the natural ability of other early succession species to surpass longleaf pine (Parker, 1998). They show evidence of past grazing and timber harvesting in areas continuously forested. Progress Energy intensively manages most of its land in loblolly pine plantations for high-yield fibre and lumber harvests. Restoring the longleaf component in the forest remains in the experimental stage on HRT, with only 69 ha actually being managed to favour longleaf (Blank, 2004), so the amount of longleaf-dominated area that eventually exists will depend on various factors.

The Dupont State Forest consists of mixed pine/hardwood, white pine/hemlock and hardwood forests laced with 145 km of roads and trails and studded with four major waterfalls along a stretch of the Little River. DSF lies in the Blue Ridge Province of the Southern Appalachian Mountains. In fact, DSF abuts and lies immediately west of the Blue Ridge as identified on the Standingstone Mountain quadrangle (USGS 7.5 Minute series, 1965, Photorevised 1990). The forest lies in the French Broad Basin, and the Little River drains a majority of the tract to eventually empty into the French Broad River. The geological character of the land that makes up DSF created a variety of natural terrestrial communities in this location. They subsequently were variously impacted by human uses of the land, probably by Amerindians, certainly by Euro-American settlement late in the 18th century, and then through successive owners to the present. In short, the DSF does not offer a pristine environment.

Rather, it offers examples of past human actions ranging from extractive and utilitarian to custodial and preservative. The result is a mosaic of plant communities in sometimes dynamically changing and sometimes modestly altered environments, depending upon where in the landscape they occur. Fifteen Natural Heritage Program primary areas have been identified in the forest, and forest types include rich cove, acid cove, oak-montaine and Canada hemlock.

Case Similarities

The two cases at hand represent somewhat parallel situations up to a point in their history. Their origins in natural forest subject to random disturbance events can be assured, as can their disturbance by human influences during prehistoric periods, but relatively little detail from that period is available about either site. Frost's (1998) estimates of pre-settlement fire frequencies can only provide general guidance, but they do make clear that fires were considerably more frequent on the Harris Lands than on DSF. Farming and past logging altered forest composition in both of these forested settings, the extent and duration of intensive farming and forest utilization being wider and longer at HRT than at DSF. In both cases corporate ownership implemented conservative management and restoration practices, but such intentional management at DSF preceded HRT by several decades.

Development alternatives at DSF posed the threat galvanizing local citizens and neighbouring residents into a potent force which enabled acquisition of DSF for public use. In Wake County, urban sprawl poses a realistic threat to continued forest integrity and the ability to manage this large block of forest as productive timberland. Clearly the attributes of managed forests can be compatible with suburban neighbours, but the Harris Lands prompt the question of whether the attributes of suburban neighbours can accommodate silviculture practices required to keep managed forests economically viable. Harris Lake

Park and HRT, embedded as they are within the large contiguous forest block, benefit from the buffering effect resulting from Progress Energy's management. Yet designated (multiple) uses will likely change through time across the 445 ha encompassed by the Harris Lands, and the application of fire to manage these forests is always a doubtful option.

The most significant issues at DSF, from the standpoint of remaining a working forest, concern the amount of recreation activity that can be sustained and the relationship of forest management practices to the recreation activity. Certainly as a result of the Primary Areas designation, some limitations on conventional forest harvesting and the extent of recreation allowed in those areas must be maintained. More broadly, the interaction of timber harvesting and recreation users must be addressed carefully. The opportunity to explain natural and human-induced forest succession and the benefits of careful multiple-use management exists because of the numbers of people visiting the forest, but those numbers also pose the potential for varied and conflicting perceptions about appropriate intensity of timber management.

In both cases, rapid growth rates statewide concern members of the conservation community, who see the decline of open space and natural areas generally eroding the capacity for many species of wildlife to sustain their populations, for air to be cleaned and for people to find recreational space. The aesthetic and life quality issues brought about by intensive development in landscapes perceived to be natural spark reactions across the population spectrum. However, local community dynamics can lead to markedly different levels of engagement in what happens to forest parcels.

Case Differences

The most obvious difference between these two cases is that private land at DSF became public but the private lands that became the Harris Lands remain in corporate hands, with the public having leases and rights to use substantial portions of those lands. DSF is managed by a state agency with a clear mandate to balance forest uses but a clear public perception that recreation and aesthetically pleasing outdoor experiences are the primary reason for the land to be accessible. The Harris Lands are managed by several different entities, with the majority being managed for timber production and effectively off-limits to most public access. The areas of public access (Harris Lake, Harris Lake Park, the NCWRC designated game lands) focus uses on various forms of recreation. HRT, managed jointly by North Carolina State University and Progress Energy, remains focused on research and teaching with access by recreation users tolerated, but as yet not actively encouraged.

Terrain features create the most striking difference between the two cases. The high gradient landscape at DSF provides recreational challenges and scenic vistas attractive to a wide range of users. Thus, DSF timber stand management has and will be a secondary focus of the working forest, while management of people in the landscape will remain the primary focus of staff and state-allocated resources. In contrast, the mildly rolling topography of the Harris Lands does not provide scenic vistas, and the creeks feeding Harris Lake are small, relatively unattractive points of recreation interest. On the Harris Lands not primarily devoted to public access, the largest portion of the forested area, silviculture will dominate unless suburban development proves to be an irresistible alternative, but these timber stands are unremarkable aesthetically, so they attract few casual visitors.

The recreational draw of DSF is huge. Significant natural features of waterfalls and granite domes combined with abundant recreational trails at DFS attract thousands of visitors to the forest annually. For, as the DSF Trails Master plan points out,

> one of the main reasons the forest is renowned for its recreation opportunities is the extensive network of trails. With over 145 km of roads and trails meandering through the forest, the opportunity for

different recreation experiences is seemingly limitless. However, in order to maintain and improve the recreation experience on the forest the trail system must be properly managed.
> (Scott and Blank, 2003)

Because the trails and roads link segments of the river and the various waterfalls and climb up and down over a number of steep granite domes and mountain peaks, the variety of vistas and scenic experiences is diverse and appealing to a wide range of visitors with varying physical capabilities. Recreation on the Harris Lands will primarily involve boating, picnics and hunting, though trail development is increasing the attraction for people seeking places to ride horses and bikes. Linkage between Harris Lake Park and HRT may increase this recreation component in the area. HRT is, of course, a research site where restoration of a now somewhat unique plant community calls to question whether restoration of an isolated example amidst the surrounding later succession forest makes sense.

Conclusions

In considering working forests, we must understand the historically important role landscapes played in many dimensions of local residents' everyday lives, including sustained human needs for wood-based resources and the emerging desire for recreational escape. Better knowledge of pragmatic and spiritual attachments that past inhabitants maintained with landscapes of contemporary interest can help us determine where to direct scarce monetary resources in the future. Including working forests among our areas of conservation attention recognizes those historic and continuing interests and needs. Such inclusion can effectively stretch funds available for landscape conservation.

Dedicated volunteers made a difference in acquiring the Dupont State Forest for public recreation and ecological preservation. While potential damage by recreation users is a concern, careful attention to people management in that environment is a clearly articulated goal (Draft EA Document). The Harris Lands, which sit amidst a rapidly urbanizing landscape, have not yet experienced the full impact of development on their borders and are just now beginning to see increased interest in both conservation and recreational opportunities. How they are managed as working forests depends on the management prerogatives of a corporate entity and its relationship with the public.

The two case studies examined here demonstrate different aspects of the working forest definition. Public and private lands of similar size are involved. The intensities of management and recreation use differ in the two cases, but in each instance the heritage of management, the landscape history, is an important factor in shaping the future management direction of these working forests.

References

Blank, G.B. (1999) *Conserving North Carolina's Forests: Assessment of Need for the Forest Legacy Program.* NC Division of Forest Resources, Raleigh, North Carolina.

Blank, G.B. (2004) A case integrating historical ecology to restore a transitional *Pinus palustris* community. In: Honnay, O., Verheyen, K., Bossuyt, B. and Hermy, M. (eds) *Forest Biodiversity: Lessons from History for Conservation.* CABI, Wallingford, UK, Chapter 16.

Blank, G.B., Parker, D.S. and Bode, S.M. (2002) Multiple benefits of large undeveloped tracts in urbanized landscapes: a North Carolina example. *Journal of Forestry* 100(3), 27–32.

Bode, S.M. (1997) Land use and environmental history of the Shearon Harris Tract. Masters thesis. North Carolina State University, Raleigh, North Carolina.

Dickinson, W.R. (2000) Changing Times: the Holocene legacy. *Environmental History* 5(4), 483–502.

Frost, C.C. (1998) Presettlement fire frequency regimes of the United States: a first approximation. In: Pruden, T.L and Brennan, L.A. (eds) *Fire in Ecosystem Management: Shifting the Paradigm from Suppression to Prescription.* Tall Timbers Fire Ecology Conference Proceedings, No. 20, Tall Timbers Research Station, Tallahassee, Florida, pp. 70–81.

Mann, C.C. (2002) 1492. *Atlantic Monthly* 289(3, March), p. 41.

Mann, C.C. and Plummer, M.L. (1996) *Noah's Choice: the Future of Endangered Species.* Alfred A. Knopf, New York.

Parker, D.S. (1998) Using botanical analysis to shape a longleaf restoration project. Masters thesis, North Carolina State University, Raleigh, North Carolina, p. 119.

Scott, J. and Blank, G.B. (2003) *Trails Master Plan for DuPont State Forest: Prepared for NC Division of Forest Resources.* College of Natural Resources, North Carolina State University, North Carolina.

11 Restoration in the American National Forests: Ecological Processes and Cultural Landscapes

N. Langston

Department of Forest Ecology and Management, University of Wisconsin-Madison, Madison, WI, USA

Introduction

While American old-growth forests are commonly thought of as pristine, they are profoundly cultural landscapes, shaped by Indian burning, forest management, industrialization and fire exclusion. Yet, although human efforts have altered American forests in complex ways, the changes that people have brought about have rarely been the changes they had hoped for. Unintended consequences have resulted from each effort to regulate and reshape American forests. While American forests are cultural landscapes, they are also wild in important ways, for they resist the bounds of human control. Professional foresters and the timber industry persist in seeing the forests as under their control, even as environmental groups persist in seeing the same forests as pristine, wild entities best left untouched. Neither perspective is particularly accurate or helpful. If forest conservation is to be successful, both foresters and environmentalists need to recognize the ways that culture has shaped American forests, as well as the ways that wild processes have reshaped cultural landscapes.[1]

American national forests are at a crisis point. Changing societal values and new understandings of ecosystem processes have called into question decades of Forest Service management aimed at regulating the forests for increased timber production. As a result, timber harvests on the 191 million acres of national forests have dropped by 85%, from 12.7 billion board feet harvested in financial year 1987 to 1.8 billion board feet harvested in financial year 2003 (Congressional Research Service, 2000; USDA Forest Service, 2004, www.//fs.fed.us/forestmanagement/reports/sold-harvest/documents/1905-2005_Natl_Sold_Harvest_Summary.pdf). Decades of fire exclusion have made western forests far more susceptible to the threat of stand-replacing fires, and a public outcry against intense wildfires has led to even more confusion about correct forest policy. Changing climate regimes have increased the susceptibility of many public forests to insect epidemics, and millions of acres of forests from Alaska to Wisconsin face a forest health crisis. The national forests, in other words, are a mess.

While most people agree that something has gone badly wrong with management of America's national forests, agreeing on new policies is much more difficult. Restoration of an earlier 'natural' ecosystem is the favoured strategy for many federal agencies who feel that traditional management went badly wrong. Foresters are expected to restore the forests back to the 'historic range of variability', or the landscape before whites arrived (Langston, 1995a). But these restoration goals make problematic assumptions about history, pristine nature, and the role of humans in nature. In this chapter, I will begin by reviewing some of the goals of American forest restoration, and then turn to a case study from the old-growth forests of eastern Oregon to argue that forest restoration should not be based on a pristine myth, but on an understanding of forests as cultural landscapes.

On Restoration and History

American restorationists work within a set of assumptions about pristine nature (Langston, 1999). Their work is ironic at heart, for it uses human labour to erase the physical evidence of human labour, attempting to return an altered landscape to something that appears pristine and free of human presence. According to one recent American textbook in the field, the goal of ecological restoration is 'to take a degraded landscape and return it to its original condition' (Bush, 1997, p. 400). The Society for Ecological Restoration (SER, the international professional society of restoration ecologists) has struggled with the definition over the last several years. In 1990, SER defined ecological restoration as 'the process of intentionally altering a site to establish a defined, indigenous, historic ecosystem. The goal of this process is to emulate the structure, function, diversity and dynamics of the specified ecosystem' (SER, www.ser.org). In 1993, the official SER definition changed to: 'Ecological Restoration is the process of re-establishing to the extent possible the structure, function, and integrity of indigenous ecosystems and the sustaining habitats that they provide'. The National Research Council (1992) focused on the idea of humans as disturbers of ecosystems, defining restoration as 'the return of an ecosystem to a close approximation of its condition prior to disturbance'.

Many European ecologists have disagreed with such interpretations of restoration which stress the return to an original, pre-disturbance, indigenous ecosystem. They argue that such an attempt makes little sense in a world of extensive human manipulations, where no single point in the past can be called original (see Bowler, 1992; Baldwin et al., 1994). Yet most American restorationists agree with the ecologist William Jordan III (1995, Madison, Wisconsin, personal communication) that only returning to a pre-European community can be called restoration; all the rest is mere rehabilitation.

Restoration attempts to use human labour to return damaged landscapes to some earlier point in their history, with the assumption that earlier ecosystems were more sustainable than current ones. Scientifically, this is problematic. As the ecologist John Cairns (1995) argues, stochastic variation due to historical events is critical in the development of ecological communities. This means that it is impossible to predict the endpoint of a community from any set of beginning points, and that therefore it is not possible to recreate any ecosystem from the past, nor to recreate any currently existing reference site. Since every ecosystem constantly changes, it is impossible to determine a baseline for restoration, a normative state deserving to be maintained or restored. Ecosystems are dynamic, rather than static, and disturbance processes operate even in the absence of human intervention. Assuming that all disturbances are harmful and that all human interventions damage an ecological system makes little sense given current ecological understanding of ecosystem processes (Dunwiddie, 1992).

In arguing that restoration should return a site to its 'original' condition, the implicit assumption is that before Europeans altered these landscapes nature was undisturbed by humans. Yet, as environmental historians, palaeoecologists and geographers have demonstrated, nearly all ecosystems on earth have been affected by humans over many thousands of years. Human processes have had profound effects on landscapes that most people now think of as natural. To ignore the roles of people in shaping successional processes is to miss a critical ecological point: namely, that repeated disturbance processes, many of them anthropogenic, shaped the landscapes we wish to restore. Excluding human disturbances as 'unnatural' will ensure that restoration of those communities cannot work.

This chapter proposes a different approach to restoration, arguing that restoration will be most successful when its practitioners recognize that the forests they are trying to fix are cultural landscapes, not purely natural landscapes in need of having human presence erased. I will focus on the three national forests in the Blue Mountains of Oregon and Washington, USA, where millions of hectares have been badly damaged by over-logging, fire exclusion, insect epidemics, climate change and poor management choices. Rather than describing in detail the ecological changes in the Blue Mountains (see Langston, 1995a, for an analysis of these changes), this chapter will focus on the dilemmas of cultural and ecological restoration in the region.

The Blue Mountains

When Euro-Americans first came to the Blue Mountains of eastern Oregon and Washington in the early 19th century, they found a land of lovely open forests full of ponderosa pines five feet across. These were stately giants the settlers could trot their ponies between, forests so promising that people thought they had stumbled into paradise. But they were nothing like the humid forests to which easterners were accustomed. Most of the forest communities across the inland West were semi-arid and fire-adapted, and whites had little idea what to make of those fires.

After a century of trying to manage the forests, what had seemed like paradise was irrevocably lost. The great ponderosa pines were gone, and in their place were thickets of fir and lodgepole. The ponderosa pines had resisted most insect attacks, but the

trees that replaced them were the favoured hosts for defoliating insects such as spruce budworm and Douglas-fir tussock moth. As firs invaded the old ponderosa forests, insect epidemics swept the dry Western forests. By 1991, in the 5.5 million acres of Forest Service lands in the Blue Mountains, insects had attacked half the stands, and in some stands nearly 70% of the trees were infested (Langston, 1995a).

Even worse, in the view of foresters and many locals, was the threat of catastrophic fires. Although light fires had burnt through the open pines every 10 years or so, few exploded into infernos that killed entire stands of trees. But as firs grew underneath the pines and succumbed to insect damage, far more fuel became available to sustain major fires. Each year, the fires seemed to get worse and worse. By the beginning of the 1990s, one major fire after another swept the inland West, until it seemed as if the forests might entirely go up in smoke.

Forest change comes about not just because people cut down trees, but because they cut down trees in a world where nature and culture, ideas and markets, tangle together in complex ways. On one level, the landscape changes resulted from a series of ecological changes. Heavy grazing removed the grasses that earlier had suppressed tree germination, allowing dense thickets of young trees to spring up beneath the older trees. When the federal foresters suppressed fires, the young firs grew faster than pines in the resultant shade, soon coming to dominate the forest understorey. High grading – removal of the valuable ponderosa pine from a mixed-conifer forest – helped change species composition as well. But the story is much more complex than this. Changes in the land are never just ecological changes: people made the decisions that led to ecological changes, and they made those decisions for a complex set of motives.

The story of these drastic landscape changes is, in the simplest version, a story of the land's transformation into a set of commodities that could be removed out of one landscape and moved to another. Indians had certainly altered the landscapes, but when whites showed up they set into motion changes that far outpaced the previous changes. The critical difference was that the Blues finally became a source of resources – timber, gold, meat and wool – to feed the engines of market capitalism.

Before whites came, the Blues were certainly connected to markets outside the region. Local tribes had an extensive set of ties to trading networks that spread west to the Pacific Ocean and east to the Great Plains (Meinig, 1968). Indians did extract elements from the local ecosystem, and in the process, they changed the local ecology to meet their needs, largely through burning. However, their needs did not include removing large quantities of wood fibre for fuel, fertilizer or construction. Indian land use was not necessarily sustainable, nor was it in any kind of inherent balance with the land's limits. Yet it was still fundamentally different from the land use that whites instituted, for it did not include the wholesale extraction of resources and their export elsewhere. Indians who made the Blues their home did not see the land as a set of distinct, extractable resources, as most whites would come to see it even when they had strong emotional connections to the place.

Euro-American settlement in the Blues, as in the West at large, had been driven by a vision of limitless abundance. The forests seemed endless; the land in need of improvement; the world available for the taking; but as the timber industry reached the Pacific, people began to fear that there might be an end in sight. Many worried that if the nation continued to deplete its forests without thought of the future, it might one day find itself without the timber upon which civilization depended. Federal scientists in particular were certain that, because of wasteful industrial logging practices, a timber famine was about to devastate America. By the last decade of the 19th century, the Blues seemed to be in serious trouble. The bunchgrass was largely gone, depleted by intense grazing. Wars between small cattle ranchers, itinerant sheepherders and large cattle operations from

California had left thousands of sheep and several sheep-herders dead. Timber locators and speculators were taking up the best timber land; small mills and miners were illegally cutting throughout the watersheds; irrigators feared that their investments in water projects would be lost (Langston, 1995a). It was in this context that federal foresters came west in 1902 – to save the Blues from unrestricted abuse fostered by the desire for short-term profits.

To restore and protect ponderosa pine forests, early foresters felt they needed to keep out fire, encourage the growth of young trees and replace old trees with young ones. Old growth seemed to threaten the future by taking up the space that young trees needed to grow, and fire seemed even worse, for it actually killed young trees. Since foresters were certain that young trees were the future of the forest, fire and old growth seemed clearly the enemy. To understand these decisions to suppress fire and remove old growth, we need to understand their scientific, cultural and economic contexts. In 1906, the basic premise of the new Forest Service was simple: if the USA was running out of timber, the best way to meet future demands was to grow more timber. More than 70% of the Western forests were old-growth stands – what foresters called 'decadent and over-mature', which meant forests that were losing as much wood to death and decay as they were gaining from growth. Because young forests put on more volume per acre faster than old forests, foresters believed that old-growth forests needed to be cut down so that regulated forests could be grown instead. Regulated forests were young, still growing quickly, so that they added more volume in a year than they lost to death and decay. The annual net growth could be harvested each year, without ever depleting the growing stock.

Scientific forestry seemed impossible until the old growth had been replaced with a regulated forest. For example, in 1911, C. S. Judd, the assistant forester for the Northwest region, told the incoming class of forestry students at the University of Washington that a timber famine was on its way unless the Forest Service did something quickly. Since the forest was running out of trees, the way to fix the problem was to get National Forest land to grow trees faster. As Judd put it, 'the good of the forest ... demands that the ripe timber on the National Forests and above all, the dead, defective, and diseased timber, be removed.' The way to accomplish this was to 'enter the timber sale business' and heavily promote sales. This would get rid of the old growth, freeing up land to 'start new crops of timber for a future supply' (Judd, 1911, unpaged document). Foresters saw old growth not as a great resource, but as a parasite, taking up land that should be growing trees.

The unregulated forest was something to be altered as quickly as possible for moral reasons, to alleviate what one forester, Thorton Munger, termed 'the idleness of the great areas of stagnant virgin forest land that are getting no selective cutting treatment whatsoever' (Munger, 1936, unpaged document). The problem was not just with old growth or dying timber; the problem was with a forest that did not produce precisely what people wanted – a recalcitrant, complex nature marked by disorder and what the forester George Bright called 'the general riot of the natural forest' (Bright, 1913, unpaged document).

This logic shaped a Forest Service that, in order to protect the forest, believed it necessary to first cut it down. Beginning in 1902, across the 5.5 million acres of public forests of the Blue Mountains, federal foresters focused on liquidating old-growth pine to make a better nature. By replacing slow-growing 'decadent' forests with rapidly growing young trees, the Forest Service hoped that the human community and the forest itself would become stable and predictable. Foresters believed that disease, dead wood, old growth and fire all detracted from efficient timber production. In other words, they were assuming that the role of the forest was to grow trees as fast as it could, and any element that was not directly contributing to that goal was bad. Whatever was not producing timber competed with trees that could be producing

timber, foresters believed. Any space that a dead tree took up, any light that a fir tree used, any nutrients that an insect chewed up – those were stolen from productive trees. If timber trees did not use all the available water, that water was wasted. If young, vigorous pine did not get all the sun, that sun was lost forever. These assumptions made it difficult for foresters to imagine that insects, waste, disease and decadence might be essential for forest communities; indeed, that the productive part of the forest might *depend* on the unproductive part of the forest.

Liquidating Old Growth

Cultural ideals alone are not enough to transform forests: technology, markets and political conditions all play important roles as well. Until World War I, for all the foresters' desire to cut old growth, the Forest Service sold little timber in the Blue Mountains (Langston, 1995a; see also Skovlin, 1991). Forest Service timber was inaccessible, prices were set so high that few contractors were willing to invest, and the industry still had enough private stock to make sales of federal timber unattractive. After the war, however, markets for public ponderosa pine opened up, since there were few remaining accessible stocks on private land, and the Forest Service began to heavily push sales of ponderosa pine in the Blues. This in turn enabled them to seriously begin the campaign to regulate the forests by liquidating old growth.

The Forest Service believed that to ensure local prosperity, old-growth forests needed to be converted to regulated forests that could produce harvests forever; but to regulate the forests, planners needed markets for that timber, and they needed railroads to get the timber out to the markets. Railroads were extraordinarily expensive, particularly after World War I. Financing them required capital, which often meant attracting investment from midwestern lumber companies. These companies were only going to be interested in spending money on railroads if they were promised sales large enough and rapid enough to cover their investments. The results in the Blues, as across the West, often damaged both the land and the local communities that depended on that land.

Throughout the Blue Mountains in the 1920s, Forest Service planners encouraged the construction of mills which had annual milling capacities well above what the Forest Service could supply on a sustained-yield basis. On the Malheur National Forest alone, for example, two large sales during the 1920s offered over 2 billion board feet of pine, out of only 7 billion in the entire forest. Two mills followed – one capable of processing 60 million board feet a year, and another that could process 70 to 75 million board feet each year. With mill capacities reaching 135 million board feet a year, it would take only 15 years – not the 60 years of the cutting cycle – to process the two billion board feet in these sales, and only 52 years to process all the ponderosa in the entire forest.

Even though the Forest Service sales programme started out conservatively, it quickly gained a momentum that seemed to overwhelm the good sense of foresters. Throughout the 1920s, foresters set up plans knowing that harvests would drop by at least 40%, leading to probable mill closures in the 1980s (Langston, 1995a). This, unfortunately, is exactly what happened. Harvests collapsed at the beginning of the 1990s – not because of environmentalists or spruce budworm, but because planners set it up that way in the 1920s, figuring it was a reasonable price to pay for getting forests regulated as fast as possible.

The training of early foresters was heavily influenced by European silviculture, which had as its ideal a waste-free, productive stand: nature perfected by human efficiency. Early Blue Mountains foresters believed that to make the forests sustainable they needed first to transform decadent old growth into vigorous, regulated stands. Yet until World War I they never tried to implement these ideals, largely because there were few markets for the trees. It was neither economically nor technologically feasible to cut the forests

heavily enough to bring about intensive sustained-yield forestry. After World War I, however, the Forest Service established extremely high rates of ponderosa pine harvests, creating the ecological and economic conditions that directly led to the forest health crisis of the 1990s. Why did the Forest Service promote such high harvests? Desire for profit, power struggles, bureaucratic empire building – all of these played an institutional role, but none of them can explain the motivations of individual foresters. To make sense of their decisions, we need to examine the links between ideals and material reality in American forestry. Federal foresters shaped the western landscapes according to a complex set of ideals about what the perfect forest ought to be. In turn, these visions were shaped by available logging technology, developing markets for forest products, the costs of silvicultural practices, and what the historian Rich Harmon (1995, E-6) has called 'the unrelenting pressures...aimed at government officials to make public resources available for private profit.'

After World War II, managers became ever more enamoured of intensive forestry. No-one had yet proven any of the claims of intensive forestry; no-one had managed to regulate a western old-growth forest, but the Forest Service was optimistic all the same – surely, someday soon, with the help of loggers, silviculturists would be able to transform all the western forests into vigorous young stands growing at top speed (Hirt, 1994). When that day finally came, the Forest Service estimated that loggers could harvest 20 billion board feet a year forever (Wilkinson, 1992). There hardly seemed to be an end in sight to what managers thought forests could eventually produce.

The forest health crisis changed all this. Just before the Forest Service published the 1991 Forest Health report, loggers had harvested over 860 million board feet a year of timber from the Blues – nearly 600 million of this from federal lands. By 1993, however, harvests had slowed to a trickle. A lot of money, a lot of timber and a lot of jobs were at stake. In an unusual admission of guilt and confusion, the Forest Service stated that this crisis was caused by its own forest management practices – yet no-one could agree exactly which practices caused the problems, much less how to restore the forests.

Restoration and Cultural Landscapes in the Blue Mountains

Most people now agree that a forest health crisis threatens the Blues, but few people agree on the solution. Many environmentalists argue that the best way to restore the forest is to leave the land alone, stop logging and let nature heal itself. Natural processes, they say, will heal the forests better than human intervention ever could. Yet this perspective overlooks the fact that these are no longer natural forests. Logging, road building, fire suppression and grazing have degraded the soil- and water-holding capacities of these forests and increased fuel loads dramatically – and the result is a forest much less resilient to disturbance (Perry, 1994). If we simply removed ourselves from these forests at this point, letting the forests burn might prevent the re-establishment of ponderosa pine forests for centuries (Agee, 1994). Leaving these forests alone may seem like the most natural thing to do, but, ironically, it would lead to highly unnatural effects, since we have so radically altered the forest communities.

For many foresters, restoration means intensive management, not an end to management. Their ideal past is one of wide open stands, with few trees per acre – a past they hope to return to with the help of heavy salvage logging. Because many pre-settlement mixed-conifer communities used to be open and park-like, proponents of salvage logging have argued that we should log out the dense under-storey now present in these forests. After the catastrophic wildfires of 2002 and 2003, Congress passed the Bush Administration's 'Healthy Forests Initiative' (H.R. 1904), which hoped to save the forests from fire by using intensive logging to restore pre-settlement forest structure.

Definitions of forest health are at the root of these justifications for salvage logging, and these definitions reflect long-held cultural ideals of what a virtuous forest should look like. According to the Idaho Policy Planning Team, the best measure of forest health is when mortality is 18.3% of gross annual growth – the definition offered by the Society of American Foresters (O'Laughlin et al., 1993). By this definition, intensively managed industrial forests in Idaho are in a much healthier condition than non-industrial forests, and old growth is in the worst condition of all, since mortality and growth are nearly equal. Therefore, the Idaho report concludes, intensive, industrial management is what keeps forests healthy. Early foresters justified liquidating old-growth pine forests for exactly this reason – so young, healthy, rapidly growing forests could take their place.

Salvage logging tries to restore the forests by focusing on just one element, the ecological changes in tree structure, ignoring the policies and the cultural ideals that led to the changes. It ignores the ideological basis of forest health problems, and so it ends up with a proposal that repeats the same errors that created the changes. Salvage logging ignores the political forces that led to forest devastation: namely, an economic and political system which made forests into storehouses of commodities to feed distant markets and fill distant pockets. It also gets the ecology wrong, since it does not realize that ideology and politics shape the ways one sees ecology. For example, at the heart of the desire to save the forests with intensive management is the belief that by making current forest over-storeys look like they used to look, we will make fires behave as they used to behave. One hundred years ago, when light fires burnt frequently in some mixed-conifer forests, those forests were open, with minimal fuel loads, little organic matter on the ground, and few firs in the under-storey; but after years of fire suppression and intensive management, the forest is a different place, a landscape that is as much cultural as natural. Even light fires may now have surprising effects. After decades without fire, increased litter has led to cooler microclimates near the forest floor and increased soil moisture. Root structures have changed in response, with more roots clustering close to the surface. In those conditions, even a very light fire may singe tree roots, killing old ponderosas if the soil moisture is low (Harrington and Sackett, 1992). The important point here is that history matters: the world has changed, so that simply rearranging the trees will not return a forest to its earlier condition.

What we need to restore forest health is a new vision of restoration and its relation to history. The goal of restoration should be not to bring humans back to the pristine, wild past, but instead to do the opposite: to restore elements of the wild back into cultural, managed landscapes. This may sound quixotic, but several private foresters in the region are trying to do just this. Bob Jackson and Leo Goebel work a forest site that lies on a moist north slope near the town of Joseph in the Wallowa Mountains of eastern Oregon. Over the past 40 years, after working for the Forest Service and Boise Cascade and growing disgusted with them both, Jackson and Goebel have developed an alternative vision of good forestry built out of their experience working in the woods and out of their passion for a particular place (Langston, 1995b).

On their land, the most valuable species were high-graded off about 70 years ago and soil organic matter was badly depleted by clear-cutting. Jackson and Goebel's primary goals have been to restore the soil fertility by nurturing dead wood, and to restore a variety of species native to the site – ponderosa pine, larch, grand fir, and Douglas fir. Growing soil means growing diversity, they argue, not just in trees, but in insects, birds and spiders, and microbes and dead wood. When they are in the woods, one of their primary concerns is counting spiders, since they think many of the spider species only return when the soil is in better condition. They hate clear-cutting, feeling that while it might bring in more money all at once, short-term profit comes at the cost of soil, young trees and organic matter. Instead, they selectively

harvest, waiting until each tree is at least 18 inches in diameter. To increase growth rates, they thin young trees by hand, opening up space and light for the trees they leave behind. To get the long, knot-free lengths that bring in the best money, they do what is called 'limbing', which is a labour-intensive effort that involves cutting off low branches while the tree is still growing. To control insect damage, Jackson and Goebel grow as many different tree species as possible and keep the dead wood thick on the ground. By doing their own work, they can keep skid trails, yarding sites and roads down to about 5% of each harvest area, reducing soil compaction. In the Forest Service that figure is 20%. All these practices require a great deal of careful hand labour, and extensive knowledge about the forest itself. Few contractors could afford to pay people to take this much care for the land; Jackson and Goebel do it because they have a great deal of attachment to both the place and to their craft. Although they work the land intensively, the forest looks much like old growth – multi-layered, multi-aged, with numerous trees over 18 inches in diameter, a rich soil, abundant snags and a forest floor thick with dead wood. Trees do not grow in rows and there is nothing neat or tidy about the place, but it is a productive working forest all the same.

Jackson and Goebel's sustainable forestry work has managed to bring together political factions in the area who normally refuse to speak to each other. In 1994, a leader of an environmental group was burned in effigy by representatives of the local county movement, yet both these groups now agree that what Jackson and Goebel are trying to do is the best hope for the region's troubled forests. Groups in the area with very different political goals – from the Indian tribes to ranching and timber industry groups, and environmentalists – have managed to collaborate on a watershed plan proposing that Jackson and Goebel's sustainable forestry practices be applied to small private forests throughout the county (Wallowa County Commissioners, 1995).

Jackson and Goebel's decision to restore forest productivity by suppressing fire, increasing soil organic matter and managing for a mixed-age, mixed-species forest makes sense for their particular place, given their specific goals of making a living here without destroying the forest's ability to persist. Many details of the Jackson and Goebel model would be different in other, much drier inland forests, where fire suppression is not a viable option. Yet the basic framework of the Jackson and Goebel model does apply to other forests. Theirs is one example of a general principle that can be adapted to other forest communities on many different, particular sites. They have turned the industrial forestry model on its head: instead of transforming decadent old forests into young intensively growing forests, they have turned cutover forests into something much more like old growth – and made a living out of it as well.

What matters for forest persistence in the inland West may be exactly what large-scale forestry has tried to remove, and what Jackson and Goebel have encouraged – death and decay, the dark stinky unnerving heart of the wild forest. They have shown that you do not need to trade off this wild core for a living. The choice is not necessarily between untouched forests and industrial monocultures; nor is the choice between keeping people out and the kind of boom and bust economy that industrial logging has fostered in the Blue Mountains ever since the first mill went up. The Forest Service thought science would let its foresters leap past the constraints of a local place – in this case, a cold, high land with fragile soils, fires and floods, insects and droughts, a place of extremes. Jackson and Goebel have done well not by trying to eliminate those constraints, but by restoring them, blending human culture and care with wildness.

But what can wildness mean in this intensively humanized context? What makes their forest different from industrial tree farms? The critical difference is the presence of functioning communities, where ecological processes function with some autonomy. In contrast, many industrial forests are designed so that ecological interrelationships are fragmented to the point that they do not function without extensive inputs of petrochemicals. Trees exist in isolation, each one cut off from potentially competing plants by herbicides. Managers line these trees up in rows and begin to think that nature is just a collection of parts. From these machine-like forests, one learns a kind of contempt for nature; one starts believing that people can actually control both the trees and the forest.

Functioning communities do something else: they teach us the limits to human control and omniscience. A restored forest, while not entirely wild, can tell two major interconnected stories, one about change, and another about the links between people and the land. Restorations at their best do not erase human history, but instead they point out the different ways people have altered the landscape, while also showing the ways the land has affected people by setting ecological constraints. What you learn when you walk in the woods with Jackson and Goebel is that all the cultures who have depended on the Blues forests have changed them in different ways, reshaping them to fit their own needs and desires; but for all the stories they wrote upon the land, none of them ever controlled the forest. People can study ecological communities, change them, pull them apart and try to restore them, but they never have full control over ecological processes. These are lessons that both restorations and environmental histories can teach – lessons about the limits to human control that we badly need to learn.

Managers have always hoped that they can engineer the forest to produce what people desire, but the forest is far too complex for this. No matter how many facts we accumulate and how many theories we test, we will never have the knowledge to manipulate natural systems without causing unanticipated changes. When we manage ecosystems, all we are really doing is tinkering with processes we are just beginning to understand. There is no doubt that we can push succession in different directions – but rarely are those directions the ones we intended. The more managers alter a forest, the less they can predict the paths that succession will take. Each road we build, each stand we cut and replant with another species, each application of herbicide and pesticide adds another confounding layer of possibility. This is startling, since the changes managers have made in the forest have been aimed at making succession *more* predictable, not less – making more of what we want, and less of what we do not want.

Conclusions

Much as we try, we cannot actually substitute our version of nature for the nature out there – instead, we can only play around with it a bit, tugging on this process, pushing a little at that other process, adding our own agents of mortality (loggers) on to the agents of mortality that are always going to be out there – decay, insects, fire and wind. Given the limits of our present understanding of forest complexity, health problems cannot become the justification for wholesale applications of thinning, burning and salvage. We know little about how these forests function now, much less how they functioned in the past, so we need to recognize the limits to our knowledge and control.

Across the West, the places where we should be considering restoration are not the wilderness areas or roadless areas – places where many managers now call for intensive logging in the name of forest health. Instead, we should focus on the places that have already been intensively transformed to fit human ideas of what a civilized forest should be. Those are the areas most in need of restoration. Rather than trying to return landscape to an imagined original condition, restoration does best when it offers a way of working with the continuum of humanized cultural landscapes that occupy much of the planet – from reserves that have been minimally influenced by industrial society, to urban landscapes where trees grow inside metal cages in the sidewalk. Restoration can return elements of wildness to all these managed landscapes, without attempting to hide the fact that they are cultural landscapes that may benefit from continued human intervention.

Note

1. This chapter is based in part on Langston, 1995a and Langston, 1999.

References

Agee, J.K. (1994) *Fire Ecology of Pacific Northwest Forests.* Island Press, Washington, DC.

Baldwin, A.D. Jr, De Luce, J. and Pletsch, C. (eds) (1994) *Beyond Preservation: Restoring and Inventing Landscapes.* University of Minnesota Press, Minneapolis.

Bright, G. (1913) Relative merits of western larch and Douglas-fir in the Blue Mountains, Oregon. Forest Service Research Compilation Files, National Archives, Region VI, Entry 115, Box 135.

Bowler, P. (1992) Shrublands: in defense of disturbed land. *Restoration and Management Notes* 10, 144–149.

Bush, M. (1997) *Ecology of a Changing Planet.* Prentice Hall, Upper Saddle River, New Jersey.

Cairns, J. (1995) *Restoration Ecology: Protecting our National and Global Life Support Systems.* CRC Press, Ann Arbor, Michigan.

Congressional Research Service (2000) Congressional Research Service Reports, memo – timber harvesting and forest fires, 22 August 2000. Ross W. Gorte, p. 3. Available at: http://ncseonline.org/NLE/CRS/abstract.cfm?NLEid=670

Dunwiddie, P. (1992) On setting goals. *Restoration and Management Notes* 10, 116–119.

Harmon, R. (1995) Unnatural disaster in the Blue Mountains. *Portland Oregonian*, 24 December, E-6.

Harrington, M.G. and Sackett, S.S. (1992) Past and present fire influences on southwestern ponderosa pine old growth. In: *Old Growth Forests in the Southwest and Rocky Mountain Regions. Proceedings of a workshop.* USDA Forest Service General Technical Report GTR-RM-213.

Hirt, P. (1994) *A Conspiracy of Optimism*. University of Nebraska Press, Lincoln, Nebraska.

Judd, C.S. (1911) Lectures on timber sales at the University of Washington, February 1911. Forest Service Research Compilation Files, National Archives, Region VI, Entry 115, Box 136.

Langston, N. (1995a) *Forest Dreams, Forest Nightmares: The Paradox of Old Growth in the Inland West*. University of Washington Press, Seattle.

Langston, N. (1995b) A wild, managed forest. *The Land Report (The Land Institute)*, Summer 1995.

Langston, N. (1999) Environmental history and restoration. *Journal of the West* 38, 45–54.

Meinig, D. (1968) *The Great Columbia Plain; a historical geography, 1805–1910*. University of Washington Press, Seattle.

Munger, T.T. (1936) Basic considerations in the management of ponderosa pine forests by the maturity selection system. Umatilla National Forest Historical Files, Supervisor's Office, Pendleton, Oregon.

National Research Council (1992) *Restoration of Aquatic Ecosystems: Science, Technology, and Public Policy*. National Academy Press, Washington, DC.

O'Laughlin, J., MacCracken, J.G., Adams, D.L., Bunting, S.C., Blatner, K.A. and Keegan, C.E. III (1993) *Forest Health Conditions in Idaho: Executive Summary*. Idaho Forest, Wildlife and Range Policy Analysis Group Report 11, Moscow, Idaho.

Perry, D. (1994) *Forest Ecosystems*. Johns Hopkins University Press, Baltimore, Maryland.

Skovlin, J. (1991) Fifty years of research progress: a historical document on the Starkey experimental forest and range. USDA Forest Service General Technical Report PNW-GTR-266.

Wallowa County Commissioners (1995) *Wallowa County Watershed Plan*. Enterprise, Oregon.

Wilkinson, C. (1992) *Crossing the Next Meridian: Land, Water, and the Future of the West*. Island Press, Washington, DC.

12 Land-use and Landscape Histories: the Role of History in Current Environmental Decisions

S. Anderson

President, Forest History Society, Durham, NC, USA

Introduction

People hold varied ideas about the value of history. They range from George Santayana's 'Those who cannot remember the past are condemned to repeat it' to Henry T. Ford's 'History is more or less bunk.' (The Quotation Page, www.quotationspage.com/subjects/history, 2004). The challenges to history's value mainly revolve around concerns about who writes the history and how accurate it is, and hence, how valuable the information can be to present-day decisions. In natural resource management, and particularly the field of forestry, professionals are trained to 'read the landscape' with an eye towards what happened on the land previously that may have an influence on the success of current management decisions. For example, knowing that a forested area was previously in agriculture could have important implications regarding the capacity of the soil to support certain species, both nutritionally and physically.

So, what lessons can land-use and landscape histories offer? The following is a brief and almost sure to be incomplete list. Land-use and landscape histories can:

- help us to understand the diversity of land uses in the past;

- provide information regarding vegetation types and patterns over time;
- help us to decipher what land uses and vegetative patterns were adapted to specific climates;
- help us estimate the human impact on native flora and fauna;
- help us understand the effect of non-human disturbances on the forest (Foster *et al.*, 1998) and the interaction of those disturbances with those caused by human shaping;
- help us understand the effect of changing ownership regimes and private property rights;
- provide models of sustainable and non-sustainable land use (that is, where traditional landscape use has existed for an extended period of time, it could represent potential models for sustainable approaches) (Phillips, 1998);
- help us understand our present decisions and motivations (Cronon, 2000), a critical item needed to ensure conservation efforts across the globe (Mitchell and Buggey, 2000);
- help us to overcome the contradiction between the static character of protection or preservation measures, and the dynamic processes of landscape evolution (Willis and Garrod, 1992; Cook, 1996).

© CAB International 2006. *The Conservation of Cultural Landscapes*
(ed. M. Agnoletti)

174

Assuming that understanding the history of a landscape, whether it is an individual tract of land or a greater region, has the potential to provide such insights and understanding, then it would follow, as a minimum expectation, that resource professionals should see it as a desirable precursor to forest management decisions. This chapter allows the assumption that landscape history and land-use histories can productively inform present management decisions, and then looks at it through the lens of forest certification, a rapidly developing effort to encourage sustainable forestry worldwide.

A Note on Definitions

Land-use histories are self explanatory. They are simply the record of how humans used the land, usually associated with a particular ownership. When we use the term landscape history, though, varied concepts about the area of land in question come into play. Such uncertainty in definitions, while a limitation for certain purposes, does not necessarily devalue a term for all purposes. For instance, Donald Floyd suggests that 'sustainability', while possibly never attainable, is a worthwhile goal and something to be strived for (Floyd, 2002). It is limited because the notion of sustainability is much in the eye of the beholder. Nevertheless, the term confers a general understanding that something is ongoing over time. At the very least, it becomes the basis to bring people together for discussion. Just as our notion of sustainability can be a useful term, discussions about landscapes can be valuable, even though it can also be a slippery term. The landscape we see or shape is constructed from what we value, what is important to us.

One of the most encompassing, brief definitions of 'landscape' is provided by the Forest Stewardship Council. It defines a landscape as 'a geographical mosaic composed of interacting ecosystems resulting from the influence of geological, topographical, soil, climatic and human influences in a given area'. Most definitions of landscape include some reference to human and/or non-human activities in a particular area as 'influencing' or 'shaping' the land. This evolutionary aspect of 'landscape' is indicated by the origin of the word. It is both noun and verb. *Webster's International Dictionary* defines landscape as: (i) 'a portion of land or territory that the eye can comprehend in a single view including all the objects so seen'; or (ii) 'the landforms of a region in the aggregate especially as produced or modified by geologic forces' (but not limited to them). As a verb, landscape means 'to arrange and modify the effects of natural scenery over a tract of land as to produce the best aesthetic effect with regard to the use to which the tract is to be put.' The *Merriam Webster's Collegiate Dictionary*, 10th edition, simplifies all this and says landscape is 'a particular area of activity.'

Basically, we see that landscape includes the notion of 'shaping' something. The term 'landscape' combines 'land' with a word of ancient Germanic origin, the verb 'scapjan', which means to work, to be busy, or to do something creative – mostly with a plan or design in mind (Haber, 1979). 'Scapjan' became 'schaffen' in German, which means to shape.

Certification Protocols

One of the most visible developments during the last decade of programmes intended to foster sustainable forestry has been the development of certification protocols. These are programmes that attempt to describe what should be the elements of a well-managed forest and then provide for verification. When a forest receives its certification, the wood or other products that result from that forest can carry a label stating that it came from a responsibly managed, well-managed or sustainably managed forest. Certification protocols, however, have generally eliminated the use of the word 'sustainable' because of its uncertainties. The word is used in this chapter because it generally aligns with the values of resource managers.

Interest in certification was initially driven by concerns over the exploitation of tropical forests and reported losses of some species from these forests (Vogt *et al.*, 2000). Following reports in the 1983 FAO Yearbook that tropical moist forests were being lost at a rate of over 11 million ha annually, environmental organizations began boycotting the international tropical timber trade. When these efforts failed to reduce deforestation rates, certification was introduced as a market-based effort to foster sustainable management of forests including aspects such as human rights of indigenous populations, alleviating poverty levels, and respect for conservation legislation.

Three of the world's predominant forest certification protocols are the Forest Stewardship Council's (FSC) standards and policies, the Sustainable Forestry Initiative (SFI) Standard supported by the American Forest and Paper Association, and the Pan-European criteria supported by the Pan European Forest Certification (PEFC) Council. The organizations supporting each protocol have their own geographical scope and approach to sustainability. As of 29 August 2002 the organizations reported on their web sites the data shown in Table 12.1 (American Forest and Paper Association, 2002; Forest Stewardship Council, www.fscus.org/html/about_fsc/index.html; Pan European Forest Certification, www.pefc.org/Ramme2.htm).

The following represents a cursory survey of the specifics of three forest certification protocols as they relate to history and landscape.

Forest Stewardship Council (FSC)

The Forest Stewardship Council promotes responsible forest management globally by providing overarching standards for certification organizations. It essentially accredits or certifies the certifiers. The organization was founded in 1993 by environmental groups, the timber industry, foresters, indigenous peoples and community groups from 25 countries. The FSC in the USA is backed by 14 major environmental organizations including the Nature Conservancy, World Wide Fund for Nature, Sierra Club and others. It is also supported by businesses such as Home Depot and Lowes (home improvement centres), although arguably some of this support seems to be instigated by fringe environmental groups practising a sort of environmental blackmail.

At the time of writing this chapter, in the USA there were two organizations that had been identified as accredited certifying organizations by the FSC. These include Scientific Certification Systems (Forest Conservation Program) and the Rainforest Alliance (Smartwood Program). By using their own assessment methods, the groups determine whether a given operation meets the FSC standards. The operation can then advertise their products as coming from a well-managed forest.

The Forest Stewardship Council defines forest stewardship in a set of global principles and criteria. These principles and criteria, which apply to all forests worldwide, are used to ensure that certified forests according to the standards are managed in an ecologically sound, socially responsible and economically viable manner. While these terms essentially describe the concept of sustainability, the words 'sustainable forestry' are notably absent from their dialogue.

The FSC's principles and criteria apply to all tropical, temperate and boreal forests. Many of these principles and criteria apply

Table 12.1. Component statistics for three forest certification protocols as of 29 August 2002.

	Forest area (million ha)	Number of certificates issued	Number of countries
FSC	29.1	437	57
SFI	45.7	205	2
PEFC	43.2	146	9

also to plantations and partially replanted forests. More detailed standards for these and other vegetation types may be prepared at national and local levels. The FSC identifies 10 principles:

- Principle 1. Compliance with laws and FSC principles.
- Principle 2. Tenure and use rights and responsibilities.
- Principle 3. Indigenous people's rights.
- Principle 4. Community relations and worker's rights.
- Principle 5. Benefits from the forest.
- Principle 6. Environmental impact.
- Principle 7. Management plan.
- Principle 8. Monitoring and assessment.
- Principle 9. Maintenance of high conservation values.
- Principle 10. Plantations.

Principles 3, 4, 6, and 7 of the FSC standards contain criteria that relate loosely to land-use and landscape history. They are presented below for the reader to become more familiar with the language and level of detail used in describing various criteria.

Principle 3. The legal and customary rights of indigenous peoples to own, use and manage their lands, territories and resources shall be recognized and respected.

3.1 They shall control forest management on their lands.
3.2 Management shall not threaten or diminish resources or tenure rights.
3.3 Sites of special cultural, ecological, economic, or religious significance to indigenous peoples shall be clearly identified in cooperation with such peoples, and recognized and protected by forest managers.
3.4 They shall be compensated for the application of their traditional knowledge regarding the use of forest species or management systems.

Principle 4. Forest management operations shall maintain or enhance the long-term social and economic well-being of forest workers and local communities.

4.1 The communities within, or adjacent to, the forest management should be given opportunities for employment, training, and other services.
4.2 Forest management should meet or exceed all applicable laws and/or regulations covering health and safety of employees and their families.
4.3 The rights of workers to organize and voluntarily negotiate with their employers shall be guaranteed as outlined in Conventions 87 and 98 of the International Labour Organization (ILO).
4.4 Management planning and operations shall incorporate the results of evaluations of social impact. Consultations shall be maintained with people and groups directly affected by management operations.
4.5 Appropriate mechanisms shall be employed for resolving grievances and for providing fair compensation in the case of loss or damage affecting the legal or customary rights, property, resources, or livelihoods of local peoples. Measure shall be taken to avoid such loss or damage.

Principle 6. Forest management shall conserve biological diversity and its associated values, water resources, soils, and unique and fragile ecosystems and landscapes, and, by so doing, maintain the ecological functions and the integrity of the forest.

6.1 Assessment of environmental impacts shall be completed – appropriate to the scale, intensity of forest management and the uniqueness of the affected resources – and adequately integrated into management systems. Assessments shall include landscape level considerations as well as the impacts of on-site processing facilities. Environmental impacts shall be assessed prior to commencement of site-disturbing operations. (6.2–6.10 refer to specific management practices).

Principle 7. A management plan – appropriate to the scale and intensity of the operations – shall be written, implemented and kept up to date. The long-term objectives of management, and the means of achieving them, shall be clearly stated.

7.1 The management plan and supporting documents shall provide:

- management objectives;
- description of the forest resources to be managed, environmental limitations, land use and ownership status, socio-economic conditions and a profile of adjacent lands;
- description of silvicultural and/or other management system ...;
- rationale for rate of annual harvest and species selection;
- provisions for monitoring of forest growth and dynamics;
- plans for identifying and protection of rare, threatened, and endangered species;
- maps describing the forest resource base including protected areas ...;
- description and justification of harvesting techniques and equipment to be used.

(7.2–7.3 deal with revising the plan, training workers, and confidentiality of information).

It is a stretch to interpret that there is any concern of consequence indicated in the above that would indicate to a reviewer, landowner, or consumer that knowing the land-use history or landscape history is an important component of the FSC standards. Criteria 3.3.3 cites that significant cultural areas shall be protected. Common sense indicates that this would include significant historic sites but the emphasis is on protection of individual areas, most likely of limited extent. Criteria 4.4.4 indicates that management plans should incorporate analysis of social impact, yet this is not specific to changes in historical land use. As well, 4.4.5 vaguely mentions 'customary rights', but does not clarify that loss of historical land uses may or may not be important. Criteria 6.6.1 articulates that landscape-level considerations should be made in order to protect biological diversity. While a landscape view is recommended with regards to environmental impact assessments, no mention of landscape history is made. Under 'Principle 7. Management Plan' where there is great opportunity to cite the relevance of land-use and landscape histories, the criteria are silent.

Sustainable Forestry Initiative (SFI)

The Sustainable Forestry Initiative was adopted by the American Forest and Paper Association (AF&PA) in October 1994 and officially launched in 1995. It is a set of principles, objectives and performance measures that integrates the perpetual growing and harvesting of trees with the protection of wildlife, plants, soil and water quality and other conservation goals. As of 2002 there were 45 organizations listed on the SFI web site that support the Sustainable Forestry Initiative including such environmental organizations as the Conservation Fund, the Wildlife Society, and Ducks Unlimited (Haber, 1979).

The association represents individuals and companies that are actively involved in management of forest lands. During the first year after SFI was established 17 of about 200 member companies were suspended because they failed to meet the requirements of the SFI. A few members also resigned. Initially, verification was limited to reporting systems to the AF&PA. In 1998, the SFI programme added voluntary verification options that allowed first-, second- and third-party approaches for programme participants to declare their conformance with the SFI standards. As of March 2004, the SFI program listed 16 third-party certifiers in 15 different organizations.

In July of 2000, the Sustainable Forestry Board was chartered as an independent body to oversee development and continuous improvement of the SFI programme standard, associated certification processes, and procedures and programme quality control mechanisms. In September 2002, the Sustainable Forestry Board was recognized by the Internal Revenue Service as a 501(c)3 not-for-profit organization under the US tax code. The 15-member Board consists of one-third SFI programme participants, one-third conservation and environmental community interests, and one-third broader forestry community representation.

The SFI standard identifies six principles for sustainable forestry and then articulates 11 objectives under those principles with performance measures and core indicators within each objective. The SFI principles include:

1. Practice sustainable forestry.
2. Use responsible practices.
3. Forest health and productivity.
4. Protecting special sites.
5. Legal compliance with laws.
6. Continual improvement.

SFI Principle 4 concerning protecting special sites states: 'To manage forests and lands of special significance (e.g. biologically, geologically, culturally or historically significant) in a manner that takes into account their unique qualities.' Similar to the FSC standards, this principle seems to be directed at particular historical sites rather than any broader focus on land-use history as an important precursor to sustainable management.

The principles are followed by 11 objectives that are considered the substance of the programme. They are intended to translate the principles into action by providing a roadmap to practice sustainable forestry and visibly improve performance. The main objectives include:

1. Broaden the practice of sustainable forestry.
2. Ensure prompt reforestation.
3. Protect water quality.
4. Ensure wildlife habitat.
5. Minimize the visual impact of harvesting.
6. Protecting special sites.
7. Promote efficient use of forest resources,
8. Cooperate with procurement chain to promote SF.
9. Publicly report progress.
10. Support public outreach and public input.
11. Continual improvement in practicing SF.

Only objectives 4 and 6 include performance measures that recognize historical or landscape components.

Objective 4. Manage the quality and distribution of wildlife habitats and contribute to the conservation of biological diversity by developing and implementing stand- and landscape-level measures that promote habitat diversity and the conservation of forest plants and animals including aquatic fauna.

    4.1 Performance measures.
        4.1.1 Programme participants shall have policies to promote habitat diversity at stand- and landscape-levels.

(Core SFI indicators – seven core indicators and six other indicators are listed, none of which specifically address landscape-level requirements).

        4.1.3 Programme participants shall apply knowledge gained through research, science, technology and field experience to manage wildlife habitat and contribute to the conservation of biological diversity.

(Core SFI indicators – three core indicators make no mention of landscape or historical land-use although 'other SFI indicators' under this performance measure allows a programme participant to show involvement:

1. Participation in cooperative ecological landscape planning efforts where available.
2. Professional expertise available to assist in developing wildlife, aquatic, or biodiversity programmes, or staff allocated to wildlife, aquatic, or biodiversity research.
3. Participation in appropriate agreements with external parties on wildlife habitat management, federally threatened and endangered species conservation, landscape planning, or conservation of biological diversity.)

Objective 6. Manage programme participant lands of ecological, geological, cultural, or historical significance in a manner that recognizes their special qualities.

    6.1 Performance measures.
        6.1.1 Programme participants shall identify special sites and manage them

in a manner appropriate for their unique features.

(Core SFI indicators:

1. Written policy to identify, map and manage special sites.
2. Obtain existing natural heritage data and cooperate with experts in identifying or selecting sites for protection of significant ecological, geological, cultural or historic qualities.
3. Map and catalogue existing sites.)

Still, this represents cursory attention to history and focuses solely on selected special sites. This is confirmed by the definition of 'culturally' presented by SFI, which states 'Culturally: special sites of importance because of their significance as examples of Native American peoples (e.g., Indian burial mounds)'. The overall approach to forest management in SFI has no mention of the importance of assessing land-use history or landscape history.

Pan European Forest Certification

The Pan European Forest Certification (PEFC) Council was officially launched in Paris on 30 June 1999. It was developed in response to the lack of flexibility of the Forest Stewardship Council protocol to account for different cultural and ecological differences between European countries. The PEFC scheme, a voluntary private sector initiative, will provide assurance to the customers of woodland owners that the products they buy come from independently certified forests managed according to the Pan-European criteria as defined by the resolutions of the Helsinki and Lisbon Ministerial Conferences of 1993 and 1998 on the Protection of Forests in Europe.

The Third Pan-European Ministerial Conference on the Protection of Forests in Europe, held in Lisbon, on 2–4 June 1998 declared their commitment to adopt the six Pan-European criteria for sustainable forest management, published as part of Annex 1 of the Resolution L2. These criteria, which were adopted together with the Pan-Euro-

pean Indicators (Annex 3 of the PEFC Technical Document), were previously adopted by the expert level of follow-up meetings of the Helsinki Conference in Geneva, on 24 June 1994 and in Antalya, on 23 January 1995. Therefore, in some publications these six criteria are also referred to as the six Helsinki criteria. The six Pan-European criteria read as follows:

- Criterion 1: maintenance and appropriate enhancement of forest resources and their contribution to global carbon cycles.
- Criterion 2: maintenance of forest ecosystem health and vitality.
- Criterion 3: maintenance and encouragement of productive functions of forests (wood and non-wood).
- Criterion 4: maintenance, conservation and appropriate enhancement of biological diversity in forest ecosystems.
- Criterion 5: maintenance and appropriate enhancement of protective functions in forest management (notably soil and water).
- Criterion 6: maintenance of other socio-economic functions and conditions.

Under each criterion there are one or more 'Concept Areas' which have one or more quantitative and/or descriptive indicators associated with them that constitute the performance measures under PEFC. Criterion 6 related to 'other socio-economic function and conditions' mentions protecting culturally valuable sites and could be construed as recognizing the value of historical land use. Descriptive indicator number 2, below, does, however, seem to indicate that conserving culturally valuable *landscapes* can be an important criterion.

Concept area: cultural values

Descriptive indicators (examples):

1. Existence of a *legal/regulatory framework*, and the extent to which it: provides for programmes and management guidelines which recognise cultural heritage in relation to forestry.
2. Existence and capacity of an *institutional framework* to: develop and maintain pro-

grammes to conserve culturally valuable sites and landscapes.

3. Existence of *economic policy framework* and *financial instruments*, and the extent to which it: provides for sufficient financial incentives for acknowledgement of cultural values in forest management planning.

4. Existence of *informational means* to implement the policy framework, and the capacity to: conduct studies on proportion of culturally valuable sites and sites with special visual value.

The Third Pan-European Ministerial Conference on the Protection of Forests in Europe, held in Lisbon, on 2–4 June 1998, declared their commitment to endorsing the voluntary Pan-European Operational Level Guidelines for sustainable forest management. These Pan-European Operational Level Guidelines were previously adopted by the fifth Expert Level Preparatory Meeting of the Lisbon Conference on the Protection of Forests in Europe, in Geneva, on 27–29 April. The Operational Level Guidelines form a common framework of recommendations that can be used on a voluntary basis and as a complement to national and/or regional instruments to further promote sustainable forest management at the field level, on forest areas in Europe. The Pan-European Operational Level Guidelines for Sustainable Forest Management state:

Criterion 6. Maintenance of other socio-economic functions and conditions

6.1 Guidelines for Forest Management Planning

c. Sites with recognised specific *historical*, cultural or spiritual significance should be protected or managed in a way that takes due regard of the significance of the site.

6.2 Guidelines for Forest Management Practices

a. Forest management practices should make the best use of local forest related experience and knowledge, such as of local communities, forest owners, NGOs and local people.

Criterion 4. Maintenance, conservation and appropriate enhancement of biological diversity in forest ecosystems

4.1 Guidelines for Forest Management Planning

a. Forest management planning should aim to maintain, conserve and enhance biodiversity on ecosystem, species and genetic level and, where appropriate, *diversity at landscape level*.

4.2 Guidelines for Forest Management Practices

d. *Traditional management systems* that have created valuable ecosystems, such as coppice, on appropriate sites should be supported, when economically feasible.

The PEFC, similar to the SFI and FSC guidelines, also recognizes the importance of specific sites of historical significance. It mentions making use of local knowledge, but in the end also fails to specifically cite the value of land-use or landscape histories as necessary information that should be collected in preparation of forest management plans and decisions. We see, however, the first evidence that 'traditional management systems' may have value and should be part of the management analysis.

These efforts may eventually be strengthened by the support provided by the European Landscape Convention. The Council of Europe's Committee of Ministers adopted the European Landscape Convention (Council of Europe, 2000) and opened it for signature on 20 October 2000 during the ministerial conference on landscape protection in Florence, Italy. The convention aims to encourage public authorities to adopt policies and measures at local, regional, national and international levels for protecting, managing and planning landscapes throughout Europe. It provides for a flexible approach to landscapes and calls for various types of action, ranging from strict conservation through protection, management and improvement to actual creation.

The idea to draft a new legal text for better management and protection of the continent's landscapes was first proposed by the Council of Europe's Congress of Local and Regional Authorities (CLRAE) in 1994. It received strong political support from both the Parliamentary Assembly and the Committee of Ministers as part of the Council's work on natural and cultural heritage, spatial planning, environment and local self-government. The convention proposes legal and financial measures at the national and international levels, aimed at shaping 'landscape policies' and promoting interaction between local and central authorities to protect landscapes. It sets out a range of different solutions which states can apply, according to their specific needs. The convention will come into force three months after ten Council of Europe member states have ratified it.

Conclusion

It is apparent through this survey of three major certification protocols that as of autumn 2002, none of them have made specific efforts to articulate the importance of looking at land-use histories as a guide for land management decisions, nor even as a major component of a management plan. The protocols also do not fully identify landscape histories or culturally important landscapes as a core component of future management decisions. The failure to address coherently landscape or land-use history may very well be a growing weakness that will have to be reconciled if the public is to have any confidence in the protocols to recognize well-managed forests and, ultimately, sustainable management. Of the three protocols examined, the Pan European Forest Certification standards comes the closest to addressing these issues but does so incompletely. The PEFC also may have the best opportunity of the three to strengthen its landscape and land-use history language as the European Landscape Convention, if approved, will provide legal impetus to move discussion of landscape history to the forefront.

References

American Forest and Paper Association (2002) 2002–2004 Edition: Sustainable Forestry Initiative (SFI) Program. Available at: http://www.afandpa.org/forestry/sfi/Standard_02_04.pdf Accessed 1 July 2002.
Cook, R.E. (1996) Is landscape preservation an oxymoron? *The George Wright Forum* 13(1), 42–53.
Council of Europe (2000) *European Landscape Convention and Explanatory Report*. Council of Europe, Strasbourg, 19 July.
Cronon, W. (2000) Why the past matters. *Wisconsin Magazine of History* Autumn 2000, 3–13.
Floyd, D.W. (2002) *Forest Sustainability: The History, the Challenge, the Promise*. Forest History Society, Durham, North Carolina.
Forest Stewardship Council web site (2002) http://www.fscus.org/html/about_fsc/index.html (accessed 29 August 2002).
Foster, D., Knight, D.H. and Franklin, J.F. (1998) Landscape patterns and legacies resulting from large, infrequent forest disturbances. *Ecosystem* 1, 497–510.
Haber, W. (1979) Concept, origin and meaning of "Landscape". In: Von Droste, B., Harols Plachter, H. and Rossler, M. (eds) *Cultural Landscapes of Universal Value: Components of a Global Strategy*. Gustav Fischer Verlag, Jena, Stuttgart.
Mitchell, N. and Buggey, S. (2000) Protected landscapes and cultural landscapes: taking advantage of diverse approaches. *The George Wright Forum* 17(1), 35–46.
Pan European Forest Certification (2002) http://www.pefc.org/Ramme2.htm (accessed 29 August 2002).
Phillips, A. (1998) The nature of cultural landscapes – a nature conservation perspective. *Landscape Research* 23(10), 21–38.
Vogt, K.A., Larson, B.C., Gordon, J.C., Vogt, D. and Franzeres, A. (2000) *Forest Certification: Roots, Issues, Challenges, and Benefits*. CRC Press, Boca Raton, Florida.
Willis, K.G. and Garrod, G.D. (1992) Assessing the value of future landscapes. *Landscape and Urban Planning* 23, 17–32.

Through findings deriving from analysis of reports, publications, cartographic studies and planning proposals, a statistical analysis based on EUROSTAT data, a visual analysis of case studies and through interviews and discussions with experts all over Europe (see Meeus *et al.*, 1990) combined with the above listed selection criteria, Meeus (1995) defines 30 pan-European landscapes, which Stanners and Bourdeau (1995) group into eight distinctive landscape types in the 'Dobrís Assessment on Europe's Environment'. These landscape types are shown in Table 13.3.[2]

Meeus' landscape typology is not only the most cited, but probably also the most criticized approach within the circle of experts (see, for example, Schenk, 1997a, b;

J. Vervloet and T. Spek, 1998, unpublished material; W. Vos, 1999, unpublished material). They express disapproval of the fact that the characterization is dominated by natural aspects, whereas influences by man are, if at all, of secondary importance. The only 'artificial' landscapes are polders, some deltas and the Spanish Huertas, while towns and cities, for example, are regarded as landscape-deteriorating instead of seeing them as a part of landscape. Trade and traffic landscapes, mining and industrial landscapes as well as urban cultural landscapes are not mentioned at all.

This imbalance between natural and anthropic criteria does not only become obvious in the classification, but also in the choice of terminology: in northern Europe,

Table 13.3. Meeus' pan-European landscape types and distinctive landscape types in the 'Dobrís Assessment'.

Distinctive landscape types (Dobrís Assessment, Stanners and Bourdeau, 1995)		Landscape types (Meeus, 1995)	
I	Tundra	1	Arctic tundra
		2	Forest tundra
II	Taiga	3	Boreal swamp
		4	Northern taiga
		5	Central taiga
		6	Southern taiga
		7	Subtaiga
III	Uplands	8	Nordic highlands
		9	Mountains
IV	Bocage	10	Atlantic bocage
		11	Atlantic semi-bocage
		12	Mediterranean semi-bocage
V	Open fields	13	Atlantic open fields
		14	Continental open fields
		15	Aquitaine open fields
		16	Former open fields
		17	Collective open fields
		18	Mediterranean open land
VII	Regional landscape	19	Coltura promiscua
		20	Montados/dehesa
VIII	Artificial landscape	21	Delta
		22	Huerta
		23	Polder
VII	Regional landscape	24	Kampen
		25	Poland's strip fields
VI	Steppic and arid landscapes	26	Puszta
		27	Steppe
		28	Semi-desert
		29	Sandy desert
–	Do not appear	30	Terraces

mainly natural criteria are used and types are named 'tundra' and 'taiga', while in central and southern Europe the land-use type, formal shaping or the like are regarded, which results in types named 'open fields' or 'bocage'. This leads to another point of critique: Meeus' classification is not consistent. Heather landscapes are classified in landscape type number 8 'Nordic highlands' by Meeus, whereas one can find them also in lowland areas of central Europe (but with a different degree of human influence and history), for example, Lüneburger Heide in northern Germany. Landscapes that fit in neither one of the other categories are called 'regional landscapes'.

'Terraces' – Meeus' (1995) landscape type number 30 – do not appear as a proper type in the Dobříš Assessment, although they can be considered as an important element of the European cultural heritage, not to mention that there are also different (sub-)types of terraces. This lack is due to the scale of the examination. Social and historical aspects influence cultural landscapes on an often very small dimension, which is not even roughly represented in the chosen presentation scales of about 1:35 million. Moreover it could be counterproductive to the original intention, if the responsible planners and policy-makers conceive that Meeus' 30 types and eight types in the Dobříš Assessment (which are politically relevant) represent the real diversity of European cultural landscapes. The same is true for the illustrations and descriptions, which represent stereotypes rather than reality. Landscapes are romanticized and idealized. Landscape values and functions are seen as described too positively, although negative performance is one part of history which resulted in the actual landscape as well.

The conclusion of this discussion ought to be to concern 'landscapes' not as entities but as spatial constructions for a specific purpose mostly at a regional scale; and in the German-speaking countries it is better to speak of 'Kulturlandschaften' (cultural landscapes), if you would like to signal that you are mainly interested in the cultural aspects of regions, especially in any kind of spatial cultural heritage (Schenk, 2002).

European Landscape Convention

One milestone in European cultural landscape management has been set by the European Landscape Convention. It is based on the 'Carta del paisaje mediterráneo', which was signed in 1993 in Siena (Italy) by the regions Andalusia (Spain), Languedoc-Roussillon (France) and Veneto (Italy). About the same time, the Dobříš Assessment (Stanners and Bourdeau, 1995) and the World Conservation Union (IUCN, 1993) in its publication *Parks for Life: Actions for Protected Areas in Europe* recommend drawing up a 'European Convention on Landscapes', which would involve the Council of Europe. After having set up an *ad hoc* working group composed of members of the Congress of Local and Regional Authorities of Europe (CLRAE) and other international, national and regional bodies, the draft European Landscape Convention is introduced by the CLRAE. Its main objective is that 'public authority concern for landscapes will become a political priority issue, since landscape quality is a key factor in the well-being of European citizens and the strengthening of a European sense of identity' (Council of Europe and Congress of Local and Regional Authorities of Europe, 1998).

On 20 October 2000 the European Landscape Convention was signed in Florence by 18 countries, and entered into force on 1 March 2004. Seventeen countries have ratified the Convention so far (as of April 2005: Armenia, Belgium, Bulgaria, Croatia, Czech Republic, Denmark, Ireland, Lithuania, Moldova, Norway, Poland, Portugal, Romania, San Marino, Slovenia, the former Yugoslav Republic of Macedonia and Turkey).

The convention's objective is to enhance landscape protection, management and planning and to organize a European cooperation on landscape issues. Measures to be realized on national levels are:

- awareness-raising;
- training and education;
- identification and assessment of landscapes;
- identification of landscape quality objectives; and
- implementation.

The Landscape Award (see below) has been included in the convention. The European Landscape Convention shares a broad definition of landscapes by, in its Article 2, referring to 'natural, rural, urban and peri-urban areas ... land, inland water and marine areas'. Besides, it 'concerns landscapes that might be considered outstanding as well as everyday or degraded landscapes' (Council of Europe, 2000).

The Convention is supposed to be put into practice by the citizens of Europe, and not just something 'to dream of as a theoretical possibility in front of a computer screen' (Council of Europe and Congress of Local and Regional Authorities of Europe, 1998). Nevertheless, it is held in quite general terms in order to take into account individual states' traditions, organizations and practice in the field.

Council of Europe Activities: Recommendation (95) 9, Landscape Award and European Diploma

Recommendation No. R (95) 9 of the Committee of the Ministers to member states on the integrated conservation of cultural landscape areas as part of landscape policies sets out principles for cultural landscape area conservation and managed evolution within the context of general landscape policy. The recommendations refer in particular to areas susceptible to damage, destruction and transformations harmful to the balance of the environment and 'especially concern the conservation of cultural landscape areas' (Council of Europe, 1995). It proposes an integrated approach, 'reflecting all the cultural, historical, archaeological, ethnological, ecological, aesthetic, economic and social interests of the territory concerned' and concerted action by all the concerned parties. Article 4 describes the

process of identifying and appraising cultural landscape areas. One condition is that a multidisciplinary approach should be adopted. The member states themselves determine the level at which the identification process should be carried out, and operations should be conducted by competent authorities with assistance of appropriate, independent experts and with the participation of the local communities. The same multidisciplinary approach is proposed for strategies for action. National governments are responsible for making the necessary institutional provision and for providing an adequate legal or regulatory framework. All policies should draw on the principles of sustainable development. Strategies should be devised at the administrative level consistent with the landscape identification and appraisal procedures.

In 1965, the Committee of Ministers or the Council of Europe institutionalized the *European Diploma* for different types of landscapes, reserves and natural monuments. Those need to be of exceptional European interest from the biological, geological or landscape diversity point of view. Although the focus lies on natural aspects, cultural qualities are included as well. Proposals are handed in as priority lists by the member states of the Council of Europe. They are reviewed by an *ad hoc* group of specialists, and an independent expert carries out an on-the-spot appraisal. The final decision about the award is taken by the Committee of Ministers of the Council of Europe. The diploma is awarded for 5 years and can then be renewed. When a site is undergoing degradation, the diploma may be withdrawn, which happened in one case, the French Pyrenees National Park, in which the situation degraded dramatically with the expansion of tourism. Currently 65 diplomas in 25 Council of Europe member states and Belarus are awarded, to among others the Swiss National Park and the German–Luxemburg Nature Park. For 2005 the applications of Gran Paradiso National Park in Italy and Piatra Craiului National Park in Romania were being examined for admission to the network (see Council of Europe, 2005).

Originally being focused on strict and often static conservation policies, the approach today is dynamic. In the vast majority of the sites in Western Europe that have been awarded the diploma, a (sometimes significant) human influence exists (Ribaut, 1998). Hacourt (1993) describes primarily positive effects deriving from the diploma, like the creation of jobs, measures concerning aesthetics and the environment and the extension of scientific research and information politics. Another positive aspect he mentions is that both the award and extension of the diploma contain recommendations or conditions to be followed by the responsible management agency. On the other hand, tourist numbers increase, which is considered to be a threat to the protected zone.

The Council of Europe Landscape Award is dedicated to local and regional authorities as well as NGOs which have taken initiatives for the conservation, management and/or development of landscape quality. The initiatives have to fall into one of the following categories:

- awareness, education and participation;
- scientific and technical activities; and
- protection, management and planning.

A process is to be stimulated, in which high-quality landscape management is initiated, enhanced and recognized (Council of Europe, 2003).

The Landscape Award was introduced in 1997 under the 'Europe – a common heritage' campaign. With the European Landscape Convention entering into force, its Article 11 is regulating the award process. A committee of experts is identifying the necessary criteria.

Cultural landscapes in European spatial planning

The overall framework for European spatial planning is, although without obligation, set by the European Spatial Development Perspective (ESDP). The ESDP is the result of a process, lasting several years, of intense discussion among the member states of the European Union (EU-15) and with the Commission on Spatial Development (CSD) in the EU. Starting from 1993 the main principles and important analyses were developed until in 1997 the Dutch presidency presented a first official draft, referred to as the 'Noordwijk Paper'. This draft was revised in Glasgow in 1998 and finally adopted in 1999 in Potsdam (see BMBau, 1995; European Commission, 1997, 1998; Council of Europe, 1999).

In the beginning of the process three operational objectives for spatial development are identified for an ESDP: development, balance and protection. The protection objective aims at – among others – preserving '...cultural identity, the heritage of European rural and urban settlements, and the diversity of landscape' (BMBau, 1995). As one sphere of activity the 'wise management and sustainable development of Europe's natural and cultural heritage' is suggested. This section of the paper proposes to set up an inventory of the European heritages on the basis of consistent criteria taking into account the diversity of national and regional context. It is stressed that 'at the same time, special attention has to be paid to the preservation of "cultural landscapes", which form an important part of regional cultural identity in Europe.' Describing policy aims and options for the EU territory, one subchapter of the ESDP is dedicated to the 'creative management of cultural landscapes' (Council of Europe, 1999). The way in which agriculture is practised is seen as a major threat to the cultural landscapes, and the increasing uniformity in landscapes and the loss of biodiversity is considered. Therefore, the ESDP suggests putting a small number of places under protection as 'unique examples of historical cultural landscapes'. The following policy options are suggested for cultural landscapes. Naturally, those policy options are kept in general terms.

- 'Preservation and creative development of cultural landscapes with special historical, aesthetical and ecological importance.

- Enhancement of the value of cultural landscapes within the framework of integrated spatial development strategies.
- Improved co-ordination of development measures which have an impact on landscapes.
- Creative restoration of landscapes which have suffered through human intervention, including recultivation measures.'

Also within the ESDP process, in 1997 a concept for a Study Programme on European Spatial Planning (SPESP) was developed in order to set up a European Spatial Planning Observatory Network (ESPON) (Ministerium für Raumordnung Luxemburg, 1997). In one of three programme sections indicators that reflect European spatial differentiation were to be developed. One of the seven working groups within this section worked on cultural assets, which included cultural landscapes.[3] Parallel to the development of indicators the question of the assessment of landscapes was treated.

Concerning indicators, in a first step a distinction between 'significance degree indicators' and 'endangering degree indicators' is made. Significance indicators stand for all the intrinsic properties of single cultural elements and of their context, as well as objects, activities and facilities that make them considerable and culturally significant. Endangering degree indicators represent all those conditions and activities, as well as objects and facilities whose existence, absence or inadequacy determine a condition of imbalance leading to situations of degradation and refer mainly to probable future development of the cultural landscape. Then five categories of indicators are identified:

Category I: Physical Geographical Features
Category II: Human Geographical respectively Economic Functional Features
Category III: Special Agricultural Features
Category IV: Special Legislation Instruments
Category V: Cultural Significance Values.

For the description of cultural landscapes, two alternative ways are followed: a direct one by making use of remote sensing data, in which above all physiognomy of the landscapes is regarded, and a rather indirect one making use of statistical data, which are collected within administrative borders. One problem lies in the availability of the data which are relevant for cultural landscapes. Those data collected by EUROSTAT often are not detailed enough. Data collected on local and regional levels are not comparable. This is why the selection of criteria has to follow a pragmatic way: included are indicators such as agricultural production, landscape dissection by traffic or population growth. Another problem of the statistical approach is the fact that cultural landscapes do not necessarily coincide with administrative borders. As discussed above, there remains the problem of illustrating the complex structure on a European scale, where a huge amount of simplification is necessary.

Within the successor ESPON 2006, which is carried out within the framework of the Interreg-III-Programme, cultural landscapes serve as frame-giving topics within the working group which deals with the role and spatial effects of cultural heritage and identity. This project was launched in October 2004.

Pan-European Biological and Landscape Diversity Strategy

The Pan-European Biological and Landscape Diversity Strategy was set up for a period of 20 years (1996–2016) and focuses on contributing to the realization of the Convention on Biological Diversity. In four 5-year action plans the ambitious objectives are formulated in the form of Action Themes. All actions are coordinated by the European Centre for Nature Conservation (ECNC). Action Theme 4 aims at the conservation of landscapes by treating the following issues (ECNC, 1997): development of landscape indicators, links with corresponding topics, priorities for landscape management, strategies to solve conflicts, information policies, and proposals for integrated actions.

Special attention is paid to 'soft' measures such as cross-sector participation, specification of communication mechanisms or learning and exchange of information. The landscape concept is interpreted in multiple ways.

Cultural landscapes within UNESCO World Heritage

Within UNESCO World Heritage, cultural landscapes receive special attention: two decades after the World Heritage Convention entered into force in 1972, in 1992 the operational guidelines were adapted in so far as cultural landscapes of outstanding universal value can be included in the World Heritage List. Three categories of cultural landscapes are distinguished (UNESCO, 2003):

1. Cultural landscapes designed and created intentionally by man.
2. Organically evolved cultural landscapes (these fall into the two subcategories relict/fossil landscape and continuing cultural landscapes).
3. Associative cultural landscapes.

The Sintra (Portugal) was included in the list in 1995 as the first European cultural landscape. One year later, in Vienna an expert meeting took place, which worked explicitly on *European* cultural landscapes. Meanwhile at least 35 cultural landscapes are inscribed in the World Heritage List, of which 22 are located in Europe (as of April 2004; the new design of the World Heritage Centre website does not allow search by heritage types any longer. For detailed discussion of the problems of identification of cultural landscapes on the World Heritage List refer to Weizenegger, 2000).

Cultural Landscape Protection and Management – the German Perspective

The legal and institutional frame of cultural landscape conservation in Germany

The legislature of the Federal Republic of Germany demands explicitly or implicitly

in a number of laws and decrees that planning should deal with the historical heritage of our landscapes, particularly in the sectors of building conservation, nature protection and spatial planning. In practice, a complex network of institutions and activities for 'historical cultural landscapes' has developed, which can hardly be seen through. In this chapter the specific contributions of nature protection and building conservation as well as spatial planning towards the conservation of the assets of cultural landscapes will be presented.

Historical landscapes and nature protection

For a long time since its first publication in 1968 the law of nature protection had been the only one which explicitly demanded the protection of historical landscapes. In the present edition, that of 2002, the text reads as follows:

> Historical landscapes and parts of them with a specific character ought to be conserved. This includes landscapes or parts of landcapes with particular significance for the character or beauty of cultural monuments, historical buildings or archaeological sites which are under protection or considered to be protected.
>
> (Historische Kulturlandschaften und -landschaftsteile von besonderer Eigenart, einschließlich solcher von besonderer Bedeutung für die Eigenart oder Schönheit geschützter oder schützenswerter Kultur-, Bau- und Bodendenkmäler, sind zu erhalten.)

However, the legislator omitted to install a specific category of protected landscape areas comparable to 'nature protection areas'. That is the reason why this passage had been rather unknown, especially in the lower levels of nature protection administrations, for a long time.

Nowadays the situation has changed radically. In fact, nature protection movements are just and once again exploring the 'cultural landscape' for two reasons:

1. The acceptance of 'classic' strategies of single-species protection is declining in the arena of politics as well as in the public. So the involved institutions are looking for new fields of activity, and one area is the 'historical landscape'.

2. There is a general return to the roots of nature protection which are based on the holistic movement of the 'Heimatschutz' concept (the protection of the homeland) of the late 19th and early 20th century. This approach includes nature, landscape and monument protection. Although it was widely misused at the time of National Socialism (and there is no doubt that a lot of leading members of the German nature protection movement in the first half of the 20th century had been very familiar with the ideas of National Socialism), it can now be seen as an appropriate answer to the complex influences on modern landscapes.

As a result, an increasing number of projects initiated by nature protection administrations are dealing with 'cultural landscapes'.

Historical landscapes and monument conservation

There is no doubt that the main task of the monument administration is to take care of the built cultural heritage. Nevertheless, landscapes created by past generations should also be a part of the administration's tasks of conservation. Unfortunately, most colleagues working in this area are educated in the arts – and the subject 'landscape' is rather out of their minds. In spite of this, a common paper of the German Denkmalschützer on the relations of monument preservation and cultural landscape was published in 2000 (Gunzelmann, 2001). However, this paper is hardly accepted in some parts of Germany. Due to the federal structure of Germany, every federal state has its own laws and practice in monument protection. Thus, landscape protection is often a matter of a single person's engagement in monument protection. Especially in Bavaria and in Baden-

Württemberg, two historical geographers succeeded in 'infiltrating' these administrations. They were able to put the ideas of historical landscape preservation into practice by a broad interpretation of formulas in federal laws talking of 'ensembles' and similar terms. In Bavaria you can find some examples of so-called 'Denkmallandschaften' ('landscapes with important historical landscape elements and structures'), like historical vineyards, channels or railways, and the 'Denkmalpflegerische Erhebungsbogen' (a standardized questionnaire concerning historical sites) is nowadays a common instrument for pre-investigations of historical structures at the beginning of village renewal programmes. However, in comparison to the general practice of German monument conservation there are only a small number of projects dealing with cultural landscapes – but it is a growing number none the less.

Historical landscapes and regional planning

In the amendment of the Federal Spatial Planning Act ('Raumordnungsgesetz') from 1 January 1998 it is said that (in principle 13) 'grown landscapes' ('gewachsene Kulturlandschaften') should be protected in their characteristic features including the monuments of nature and culture close to them ('Die gewachsenen Kulturlandschaften in ihren prägenden Merkmalen sowie mit ihren Kultur- und Naturdenkmälern sind zu erhalten'). This article relates to the ideas of the EU of the cultural and natural heritage as an important value for regional development; its first implementation in Germany took place in the so-called 'UVPG' (Gesetz zur Umweltverträglichkeitsprüfung; law on environmental impact assessment) in the early 1990s. However, a large conference in 2000 concerned with the term 'cultural landscapes' showed that the majority of German spatial planners were not familiar with the idea of protecting historical structures in processes of regional planning. Nobody actually knows how to deal with historical landscapes in this context. This is why a working group of the 'Akademie für Raumordnung und Landschaftsplanung' (ARL; Academy for Spatial Research and Planning, Hannover) is looking for ways of implementation.

The concept of historical geography

It has been shown that the dilemma of historical landscape preservation is that it sits on the fence in terms of legal regulation in Germany. For this reason, it is indicated to show paths (including the approaches of the UNESCO and the EU) towards common strategies according to the concept of cultural landscape care (CLC). In this context, landscape conservation is understood as an interdisciplinary concept for the spatial management of the historical cultural heritage. In general, landscapes should be seen as archives of nature and human history and as an important basis for sustainable regional development.

Cultural landscape conservation can be understood as a concept overwhelming the different approaches of planning and handling the cultural heritage in our landscapes. It is based on reflections on what is important in historical landscapes for the present and future societies. Figure 13.2 shows the process of CLC.

First of all, an overview of the present historical structures and elements in our modern landscapes is needed. In Germany we are talking about 'Landschaftskataster', cadastral inventories of historical elements and structures in catalogues combined with texts, photos and maps. In the Rheinlande (the western part of Nordrhein-Westfalen) the historical geographers in Bonn (Burggraaff and Kleefeld, 1998) are involved in building up a huge inventory based on a geographical information system (GIS), the so called KulaDIG.

Second, a broad discussion on the values of these structures and elements is necessary. That demands measures of values. The most important values in the

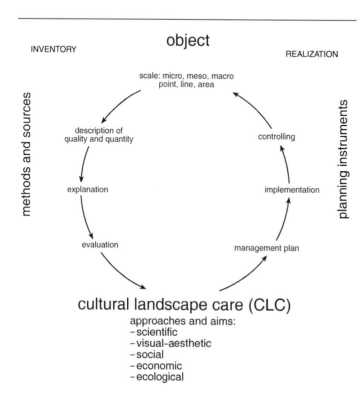

Fig. 13.2. The concept of cultural landscape care (CLC) as a circle of discussions. Source: W. Schenk 2002 after H.-R. Egli, 1996 (see Schenk *et al.*, 1997).

concept of CLC are the age of landscape elements or structures, their specificity and rarity relating to the regional context, their aesthetic quality and their importance for the regional identity. These criteria are a mixture of nature and monument conservation issues as well as regional planning concerns. In the federal state of Saarland this approach has been applied very successfully in a model project.

The third step is to bring together all the related institutions, societies and persons to discuss strategies of landscape management. Once again, the main idea is to use the heritage in our landscapes for regional development – not to put the landscape under a 'cheese cover'. It is very important to stress that cultural landscape care expressly accepts the evolution of landscapes, if historical assets, considered as potential for future development, are not destroyed.

Conclusion

There are many examples of the realization of the concept of cultural landscape care (Schenk *et al.*, 1997; Schenk, 2003). However, some of these examples revealed a general problem of CLC: a lot of people do not know about the value of historical landscapes, so you need a basic education about that. Effective ways to inform the public about the assets of their cultural landscape heritage include writing 'landscape guides' in the form of booklets or information sheets, installing 'landscape museums', producing films about regions with a rich heritage, or to offer field trips guided by local experts, in some cases educated in special seminars. If you are able to explain the landscape as a matter of our own, it is very impressive to feel the warm interest of people in the history and values of their landscapes.

Notes

1. 'Landscapes and Sustainability: A European Workshop on landscape assessment and policy tools', March 1999, organized by the European Centre for Nature Conservation and the Countryside Commission.
2. In the 1998 edition of *Europe's Environment* (EEA, 1998), the chapter on landscapes is no longer included.
3. The results are published in German and English (BBR, 2001a, b).

References

BBR (Bundesamt für Bauwesen und Raumordnung) (ed.) (2001a) *Kriterien für die räumliche Differenzierung des EU-Territoriums: Kulturerbe*. Forschungen, 100.1, Bonn, Germany.

BBR (Bundesamt für Bauwesen und Raumordnung) (ed.) (2001b) *Criteria for the Spatial Differentiation of the EU Territory: Cultural Assets*. Forschungen, 100.2, Bonn, Germany.

BMBau (Bundesministerium für Raumordnung, Bauwesen und Städtebau) (ed.) (1995) *Principles for a European Spatial Development Policy*. BMBau, Bonn, Germany.

Burggraaff, P. and Kleefeld, K.D. (1998) *Historische Kulturlandschaft und Kulturlandschaftselemente*. Angewandte Landschaftsökologie, 20, Bonn, Germany.

Council of Europe (1995) Recommendation No. R (95) 9 of the committee of ministers to member states on the integrated conservation of cultural landscapes areas as part of landscape policies (Adopted by the Committee of Ministers on 11 September 1995 at the 543rd meeting of the Ministers' Deputies).

Council of Europe (1999) ESDP – European Spatial Development Perspective. Towards Balanced and Sustainable Development of the Territory of the European Union. Council of Europe, Luxembourg.

Council of Europe (2000) European Landscape Convention. Available at: www.nature.coe.int/english/main/landscape/conv.htm

Council of Europe (2003) Council of Europe Landscape Award. Available at: www.coe.int/t/e/Cultural_Co-peration/Environment/Landscape/Presentation/6_Content/04landscape_award.asp

Council of Europe (2005) The European Diploma of Protected Areas. Available at: www.coe.int/t/e/Cultural_Co-operation/Environment/nature_and_biological_diversity/ecological_networks/the_european_diploma

Council of Europe and Congress of Local and Regional Authorities of Europe (1998) Fifth Session (Strasbourg, 26–28 May 1998). The Preliminary Draft European Landscape Convention. Explanatory Memorandum. CG (5) 8, Part II.

ECNC (European Centre for Nature Conservation) (1997) Action Theme 4: European Landscapes. Draft Action Plan for European Landscapes. Tilburg (unpublished).

EEA (European Environment Agency) (1998) *Europe's Environment: The Second Assessment*. European Environment Agency, Luxembourg and Oxford.

European Commission (1997) European Spatial Development Perspective. First official draft. Presented at the informal meeting of Ministers responsible for spatial planning of the member states of the European Union. Noordwijk, 9 and 10 June 1997, Luxembourg.

European Commission (1998): European Spatial Development Perspective (ESDP). Complete draft. Meeting of Ministers responsible for spatial planning of the member states of the European Union. Glasgow, 8 June, 1998, Luxembourg.

Gunzelmann, T. (2001) Erfassungen zur Kulturlandschaft innerhalb der Denkmalpflege. In: Kommmunalverband Hannover (ed.) *Kulturlandschaften in Europa. Regionale und internationale Konzepte zu Bestandserfassung und Management*. Beiträge zur regionalen Entwicklung 92, Hannover, pp. 57–69.

Hacourt, H. (1993) Das europäische Diplom. *Naturopa*, 71, 12–13.

IUCN (The World Conservation Union) (1993) *Parks for Life: Report of the IVth World Congress on National Parks and Protected Areas*. IUCN Gland, Switzerland.

Job, H., Metzler, D. and Weizenegger, S. (2000) Strategien zur Sicherung des europäischen Natur- und Kulturerbes. *Informationen zur Raumentwicklung*, 27(3/4), 143–155.

Meeus, J.H.A. (1995) Pan-European landscapes. *Landscape and Urban Planning* 31, 57–79.

Meeus, J.H.A., Wijermans, M.P. and Vroom M.J. (1990) Agricultural Landscapes in Europe and their Transformation. *Landscape and Urban Planning* 18, 289–352.

Ministerium für Raumordnung Luxemburg (1997) Konzept zur Errichtung eines ,Netzwerk Europäisches Raumplanungsobservatorium' (NERO). Konzeptpapier für die Sitzung in Echternach. Echternach.

Ribaut, J.-P. (1998) Outstanding European landscapes. A conservation tool: the European Diploma. *Naturopa*, 86, 23.

Schenk, W. (1997a) Wie man 'wertvolle Landschaften' macht – Geographische Kritik an einer Karte der '30 Landschaften Europas' und am zugehörigen Kapitel in 'Europe's Environment – The Dobříš Assessment'. *Kulturlandschaft* 7, 1, 33–37.

Schenk, W. (1997b) Kulturlandschaftliche Vielfalt als Entwicklungsfaktor im Europa der Regionen. In: Boesler, K.A. and Ehlers, E. (eds) *Deutschland und Europa. Historische, politische und geographische Aspekte*. Festschrift zum 51. Deutschen Geographentag Bonn 1997, 'Europa in einer Welt im Wandel'. *Colloquium Geographicum* 24, 209–229, Bonn.

Schenk, W. (2001) Kulturlandschaft in Zeiten verschärfter Nutzungskonkurrenz. In: Akademie für Raumforschung und Landesplanung, ARL (ed.) *Die Zukunft der Kulturlandschaft zwischen Verlust, Bewahrung und Gestaltung*. Forschungs- und Sitzungsberichte ARL 215, Hannover, pp. 30–44.

Schenk, W. (2002) 'Landschaft' und 'Kulturlandschaft' – 'getönte' Leitbegriffe für aktuelle Konzepte geographischer Forschung und räumlicher Planung. *Petermanns Geographische Mitteilungen* 146(6), 6–13.

Schenk, W. (2003) Historische Kulturlandschaften als Faktor der Regionalentwicklung. *Local Land & Soil News* 7/8, III/IV, 16–18.

Schenk, W., Fehn, K. and Denecke, D. (eds) (1997) Kulturlandschaftspflege. Beiträge der Geographie zur räumlichen Planung. Gebrüder Borntraeger, Stuttgart, Berlin.

Stanners, D. and Bourdeau, P. (eds) (1995) *Europe's Environment – The Dobříš Assessment*. European Environment Agency, Copenhagen.

UNESCO (United Nations Educational, Scientific and Cultural Organisation) (2003) Operational Guidelines. Available at: www.unesco.org/whc

Weizenegger, S. (2000) An appraisal of approaches dealing with European cultural landscapes – with special reference to the UNESCO World Heritage Convention. Diploma Thesis, University of Trier, Germany.

14 The Restoration of Forest Landscapes through Farmland Afforestation Measures in Spain

C. Montiel Molina

Complutense University of Madrid and Department of Regional Geography and Physical Geography, Faculty of Geography and History, Ciudad Universitaria, Madrid, Spain

Introduction

Farmland afforestation is one of the agri-environmental measures that accompanied the Common Agricultural Policy (CAP) of 1992. The objective of these European measures was to promote early retirement systems for farmers and farm workers, to favour the recreational, environmental and forestry use of land withdrawn from agricultural production and to support the protection of the environment, landscape and natural resources.

The Regulation (EC) No 2080/92, which established the support system for forestry measures in agriculture, was applied in Spain following the enactment of the Royal Decree 378/93,[1] of 12 March, which established a support system to promote forestry investment on farms and actions to develop and exploit forests in rural areas. It is this EC regulation that has had the greatest territorial and socio-economic impact in Spain, partly because it has been the best financed. Its application was carried out through the National Farmland Afforestation Programme, complemented by 17 Regional Programmes, whose initial implementation period (1993–1997) was extended to include the years 1998–2000.

The development of these actions has since continued with the application of Regulation (EC) No 1257/99 on support for rural development, which was the culmination of the CAP reform process begun in the 1980s and insisted on the importance of agri-environmental instruments and particularly the afforestation of farmland, establishing the need to maintain the support system for forestry measures. The application of this second regulation was regulated in Spain by the Royal Decree 6/2001, of 12 January 2001, on the promotion of farmland afforestation.

During the period 1993–1999, 435,737 ha of Spanish farmland were afforested with EC financial support. Thus, the Farmland Afforestation Programme has not only contributed to achieving the goals established in the EC regulations (changing land use and assigning duties to farmers additional to production), but has also brought about meaningful changes in the landscape. The latter must be assessed differently according to the environmental and socio-economic conditions in which the measures were taken.

Most of the afforested areas consisted of low-productivity farmland that had been abandoned since the 1960s, where the spontaneous regeneration of vegetation had already consolidated the restoration dynamics of forest ecosystems. In general, the planting done so far has led to a diversification of the landscape in rural areas and has enhanced our forest heritage. None the less, the EC regulations do not respond to the specificity of the Mediterranean regional context and this has had notable consequences for the landscape.

The Failure of EC Regulations to Respond to the Specificity of the Mediterranean Region

The National Programme which regulates the implementation of the EC regulations regarding farmland afforestation in Spain uses the phrase *'superficies agrarias'* ('agrarian areas') which has a Mediterranean meaning that is difficult to translate into other languages. In Spanish, the term *'agrario'* ('agrarian') encompasses agricultural, stock-breeding and also forestry activities. This, together with the specificity of the Mediterranean countryside, where agrosilvopastoral land uses often coexist or take place successively, introduces an atmosphere of conceptual confusion which allows all kinds of rural lands to be eligible for the incentives, including forestlands themselves, which should not *a priori* be the target of this programme (Montiel *et al.*, 2003). In fact, the Royal Decree 152/1996 provides a list of agricultural areas that are eligible for support which includes cork oak forests and 'open woodland and meadows, provided that the crowns of the trees do not cover more than 20% of the surface area and that it is mainly used for pasture.'

It is clear that the original EC philosophy, linked to the CAP Reform (to reduce surplus and diversify agricultural incomes and functions), is difficult to implement in Mediterranean Europe, mainly due to the low forest productivity of the Mediterranean countryside. In the specific case of Spanish regions, most of the afforested plots consisted of land which had already been abandoned and were at different stages of spontaneous evolution towards

the re-establishment of a forest canopy when the plantings were carried out. Thus, many afforestations have been carried out on forest ecosystems (meadows, broom fields, esparto fields) and all this has been possible due to confusing regulations which allow the transformation of 'open woodland' and 'uncultivated grassland', whilst at the same time specifying that 'under no circumstances may it be abandoned land', meaning, therefore, that its use must be generating earnings.

In reality, the use of the Spanish word *'agrarias'* ('agrarian') to designate the land to which the Afforestation Programme may be applied, instead of the word *'agrícolas'* ('agricultural'), is a national response which aims to adapt the European regulation to the problems and specificity of the Mediterranean region. The real socio-economic, environmental and landscape problem in rural Mediterranean areas is not the intensification of agricultural production, but instead the abandonment of less productive dry lands. Consequently, the main challenge involves regaining the management of these areas, something which is only possible by valuing agroforestry systems based on a criterion of multi-functionality.

Although the current legal framework, defined on the basis of Regulation (EC) No 1257/1999, corrects some of the problems that arose from the previous regulation regarding the failure of EC provisions to respond to the specificity of the Mediterranean region, certain problems of formulation and interpretation still exist. This is demonstrated by Article 25 of Regulation (EC) No 1750/1999 which establishes provisions for the application of Regulation (EC) No 1257/1999:

> Agricultural land eligible for support for afforestation according to Article 31 of Council Regulation (EC) No 1257/1999 shall be specified by the Member State and shall include in particular arable land, grassland, permanent pastures and land used for perennial crops, where farming takes place on a regular basis.

This definition, which is specific and unequivocal for regions in north-west

Europe and central Europe, is however ambiguous and creates conceptual problems in the Mediterranean context. Therefore, the Royal Decree 6/2001, for the purpose of implementing these incentives, considers 'those areas which are not officially classified as woodland and have been regularly used for agricultural or stock-breeding activities during the last 10 years, including:

- land occupied by wood crops;
- land occupied by herbaceous crops;
- family vegetable gardens or orchards;
- natural meadows;
- pastures;
- fallow land; and
- uncultivated grassland.

It is true that the wording of the Royal Decree 6/2001 replaces the term '*explotaciones agrarias*' ('agrarian farms'), used in the Royal Decree 378/93, with that of '*tierras agrícolas*' ('agricultural land'). However, the list of property categories that are entitled to receive afforestation incentives still includes 'uncultivated grassland', based on an objective technical definition which conceals the landscape and territorial reality of such lands in Spain. According to the Royal Decree 6/2001, uncultivated grassland is understood to be 'agricultural land which is technically and financially oriented towards stockbreeding'. However, the fact is that, in most cases, and depending on recent socio-economic dynamics, uncultivated grasslands constitute consolidated forest ecosystems (bushes and scrubland), where stock-breeding activities ceased some time ago and whose current situation is the result of a process of abandonment which began in the 1960s. The fact that the majority of the afforestation projects have focused on this kind of land has had territorial and landscape consequences which highlight the specificity of the Mediterranean region and the need to take this singularity into account when defining territorial policies.

The failure of the EC regulations to adapt to the specificity of the Mediterranean region contrasts, on the other hand, with the criteria and indices used in the budgetary distribution process,[2] which have been criticized for favouring Mediterranean regions at the expense of Atlantic regions, which are more productive in forestry terms. However, it is clear that there has been insufficient coordination between the incentive-awarding EC initiative and the territorial actions that it involves, both at a regional and local level. The need to assimilate and adapt an originally foreign approach to the territorial dynamics of the Mediterranean context has generated operational and efficiency-related problems. Furthermore, the failure of the European regulation to take into account the specificity of the Mediterranean region has led to the development of regional regulations that define a permissive framework for the acceptance of applications which has manifested itself, in practice, in the type of land use affected by afforestation and the effect of these incentives on the landscape. The actions have been implemented almost exclusively in areas where agrarian activity no longer existed and which were often in an advanced state of naturalization.

Spanish Rural Landscapes

According to the European Landscape Convention, 'landscape means an area, as perceived by people, whose character is the result of the action and interaction of natural and/or human factors'. In the light of this definition, forestland is a natural and cultural landscape, which means that its dynamic is not only natural, but also social and cultural.

The Mediterranean landscape, recognized for both its singularity and its extraordinary diversity, 'is the most directly perceptible expression and reflection of the transformation of our territories' (Arias and Fourneau, 1998, p. 9). Its specificity is a consequence of the material and immaterial components which define it. The former are related to natural aspects, particularly the Mediterranean climate, and the latter are related to cultural aspects, arising from human involvement throughout history, through the exploitation and management of the land.

The presence of the sea and, above all, the architecture of its basin, surrounded by mountains which give it a closed, complex and fragmented character and turn said regions into mountainous backdrops topped by flood plains, is the first element to determine the territorial characteristics of the Mediterranean context. The climate, marked by many days of sunshine and scarce and irregular rainfall, together with the mountainous topography, is responsible for the predominance of poor soils (except on the alluvial plains) and sclerophyllous vegetation, suited to such environmental conditions. The climate is, in fact, one of the most influential factors in the success or failure of farmland afforestation due to the possibility of drought, the most serious climatic risk suffered by Mediterranean regions.[3]

The natural components of the landscape thereby define an ecosystem which is characterized by biodiversity and fragile ecological balances whilst, at the same time, possessing an extraordinary capacity for natural regeneration which may, none the less, be affected by inappropriate human actions of a repetitive and frequent nature. Indeed, Mediterranean landscapes have a distinctly human component.

Humans have been present and active ever since the origins of these ecosystems, acting as a constituent element and agent of the latter.

The Mediterranean agrarian landscape is a mosaic that has been built and shaped by human activities, exploitation and management throughout history. Society has helped to generate biodiversity by diversifying territorial uses and valuing agricultural and forest land for the purpose of exploiting resources. Both farmland (unirrigated and irrigated) and forestland are *constructed landscapes* which possess strong cultural associations (Fig. 14.1).

The origins of the Mediterranean cultural landscape are to be found in traditional methods of spatial organization. Thus, the practice of transhumance in the stock-breeding of sheep, pigs and goats has given rise to a complex agrosilvopastoral system consisting of meadows. Similarly, the adaptation of agricultural practices to the scarcity of water resources, through arboriculture (olive, almond, carob and fig trees), and the need to create and retain soil has led to hillside terracing. Furthermore, the property structure and various subjective factors related to the mentality of Mediterranean farmers are also constituent elements which have determined the dynamics of the landscape.

Fig. 14.1. Agroforestry landscape in El Solsonés (Catalonia).

Biological and landscape diversity in the Mediterranean countryside is even greater when an active human presence exists in harmony with the dynamics of the natural ecosystem. Any alteration of this balance, as a consequence of the abandonment or disorganized intensification of human intervention, will trigger off a trivialization and degradation of the biodiversity and landscape, respectively. As a result, policies which are aimed at reusing and revaluing abandoned land offer an opportunity for the restoration of rural landscapes.

Furthermore, the socio-economic evolution that the Mediterranean forests experienced throughout the 20th century was characterized by the transition from uses linked to the primary sector in the first half of the century, to the tertiarization of uses and functions after the rural exodus which began in the 1960s. Since the 1980s, their valuation has been fundamentally related to their landscape significance, to the conservation of biodiversity and to the demand for open-air recreational areas. In this respect, actions which are aimed at restoring forest landscapes arouse a great deal of not only natural and cultural, but also socio-political interest.

The National Farmland Afforestation Programme

The Farmland Afforestation Programme, which resulted from Regulation (EC) 2080/92, enables eligible applicants to receive incentives over a period of 20 years. The quantity and duration of the incentives, together with other secondary factors, serve to explain the territorial importance that the programme has had throughout Spain, particularly in the Mediterranean region.

The support system is implemented through programmes which cover several years and specify the technical aspects, such as the quantity and duration of the incentives, conditions that must be fulfilled in order to be eligible, evaluation regulations (environmental and territorial), con-

trol procedures, etc. In February 1993, the national framework programme was created, complemented by 17 regional programmes (July 1993),[4] which may, in turn, be implemented through afforestation programmes that reflect environmental diversity, natural conditions and agricultural structures.

The Spanish case is just one of the different national responses that have been given to the enactment of Regulation (EC) 2080/92, since one may speak of a national (or even regional) specificity in the implementation of the measures (Barrué-Pastor et al., 1995). In fact, various studies have highlighted the existence of marked differences in the implementation and results of the National Farmland Afforestation Programme in different Spanish regions (Bona et al., 1997; Gómez-Jover and Jiménez, 1997; Montiel et al., 2003). This is not only the result of ecological, historical and socio-economic contrasts, but also, and above all, differences in the decisions and resources applied in each case by the regional administrative body implementing the measure.

In general, the results of the National Farmland Afforestation Programme have been the outcome of a disorganized set of uncoordinated, one-off actions of complex viability, due to the lack of a clearly defined overall strategy. The geographical distribution of the afforestation projects in different Spanish Mediterranean regions and provinces has not been effectively supervised or determined by regional administrative bodies; instead it has followed the spontaneous response of agrarian landowners to the successive annual programmes. In general, the management of the programme has been characterized by a lack of planning and an absence of technical and territorial criteria (Montiel et al., 2003). None the less, the size of the total afforested area indicates that we are looking at a sectorial initiative with undeniable (but as yet difficult to assess) repercussions for the territory and landscape (Table 14.1).

The National Farmland Afforestation Programme promotes a reforestation process which has achieved figures of over

Table 14.1. Planned objectives and actual results of the Farmland Afforestation Programme (1993–1999).

Autonomous region	Planned area (ha)	Afforested area (ha)	% afforested/planned
Andalusia	250,000	154,072	61.62
Aragon	30,880	4,761	15.4
Asturias	15,080	7,705	51.1
Balearic Islands	6,080	949	15.6
Canary Islands	5,560	–	0.0
Cantabria	3,560	492	13.8
Castilla–La Mancha	126,113	64,474	51.1
Castilla y León	110,000	89,018	80.9
Catalonia	23,720	1,640	6.9
Extremadura	89,000	41,687	46.8
Galicia	65,000	24,866	38.2
Madrid	10,000	6,019	60.2
Murcia	9,080	6,573	72.4
Navarra	4,500	1,447	32.2
La Rioja	4,000	1,369	34.2
Com. Valenciana	23,520	6,251	26.6
Basque Country	30,500	24,414	80.0
Total	806,593	435,737	54.0

Source: Directorate General for Rural Development (Ministry of Agriculture, Fisheries and Food) and Autonomous Government of Andalusia (Montiel *et al.*, 2003).

70,000 ha per year. To put it another way, and taking one of the greatest landscape transformations of contemporary history as a reference, reforestation is taking place at a pace close to that achieved by the public forestry actions carried out between 1950 and 1970, a period of maximum intensity for reforestation processes in Spain (Gómez and Mata, 1991, p. 41). However, in contrast to the unified nature of the public actions carried out between 1940 and 1980, which had clearly defined objectives (the harvesting of direct products, wood, resin, pine cones, etc., and the protection of river basins), the process that we are now analysing has arisen from a combination of many unconnected private initiatives, with guidelines provided by different administrative bodies (from the European Union to regional governments) with a complete lack of territorial objectives and only vague environmental indications (recommendations regarding the most appropriate species). Furthermore, and due to various reasons which shall be explained later, the operations carried out will have problems in guaranteeing their future viability and do

not ensure a promising future for all the new forests.

Whilst the approach and initial achievements of the programme are debatable from a territorial point of view, the situation is also rather limited from a social perspective. The social objectives that are attributed to the programme, reflected in the Royal Decrees 378/1993 and 152/96 through the idea of injecting money into rural areas, have not been fulfilled in a satisfactory way. For various reasons, most of the farm owners who have benefited from these subsidies in Spain have been, in many cases, either people who are not directly linked to the rural context or public entities (local councils). In general, the programme has not achieved full acceptance amongst the farmers themselves and much less so a replacement of agricultural activities (Gómez-Jover and Jiménez, 1997). In many cases, farmland afforestation has represented an alternative for already-abandoned agricultural land. Most of the areas affected are, therefore, unproductive spaces whose owner is neither directly nor principally linked to the primary sector (Montiel *et al.*, 2003).

Territorial Valuation of Farmland Afforestation: Landscape Transformation

Despite arising from a sectorial agrarian approach, the implementation of the Farmland Afforestation Programme and the other agri-environmental measures of the CAP have had important territorial consequences of a general nature in Spain, since they represent interesting instruments for the management of rural areas. Their socioeconomic and landscape results have been highly conditioned by the regulatory formulation with which the Royal Decree 378/93 has been adapted in each region and by the implementation of the incentives at a regional level.

In any case, no landscape objectives or criteria have been applied in the implementation of the National Farmland Afforestation Programme. The actions carried out in the different Spanish regions have not considered landscape structure nor country planning criteria; instead they have been the combination of a disorganized collection of individual plantations.

The rural landscapes most affected by farmland afforestation projects in Spain have been *mid-mountain* agroforestry landscapes and abandoned dry regions inland. In fact, most of the afforestation projects have been carried out in mid-mountain areas, which contain the best-conserved cultural landscapes and most of the protected natural territories to be found in urban fringe areas (Fig. 14.2). In inland Spain, the Farmland Afforestation Programme has often been used to regenerate highly evolved meadows with open formations and old trees (Fig. 14.3). In other cases, old agricultural enclaves have been forested, thereby causing a loss of landscape diversity and open spaces which previously played an important role in preventing forest fires (Fig. 14.4).

Most of the planting has been done, above all, on the large farm estates located in these mid-mountain areas, with the aim of diversifying functions and improving rural heritage. In the region of Valencia, for example, the afforestation work has been concentrated on the larger farm estates in

Fig. 14.2. Afforestation carried out on the 'El Canchal' and 'Casablanca' farm estates in the Regional Park of Cuenca Alta del Manzanares, Madrid (aerial photograph HNM 533-K-13, domestic flight of August 1984).

inland valleys and high plains, rather than coastal areas, where agriculture has a greater economic and territorial importance. In the drier regions, there also exist examples of afforestation on impoverished hill slopes where the planting work has not only restored the forest landscape but has contributed to controlling the risk of erosion (Fig. 14.5).

A large number of afforestation projects have also been carried out in abandoned dry regions inland. As a result of their management being neglected for decades, these landscapes are usually highly degraded. They include a wide variety of landscapes, depending on the morphology and use of

Fig. 14.3. Regeneration of a meadow through afforestation with *Q. ilex* and *Q. suber* (Soto del Real, Madrid).

Fig 14.4. Gall oak afforestation of an agricultural enclave in a leafy woodland area (Hinojosa, Soria).

Fig. 14.5. Restoration of degraded forest landscapes in the region of Huéscar (Granada).

Fig. 14.6. Afforestation of abandoned farming terraces in Morella (Castellón de la Plana).

the land, including open countryside (herbaceous crops) and terraces (wood crops). The afforestation work has been conditioned in both cases by farm and property structures, characterized by the fragmentation and smallholdings that are typical of the Mediterranean region, and for this reason the actions have generally been small scale (Fig. 14.6).

In contrast to mid-mountain agro-forestry landscapes and dry inland regions, the more natural high mountain landscapes have barely been affected by farmland afforestation, partly because they constitute the most forested areas in the country and partly because of the determining nature of their topography.

With regard to the agrarian land uses that have been replaced by afforestation, most of the planted areas are old pastures which have been unproductive for over 10 years, some of which already display a consolidated forest regeneration process. After pastures, the most frequent property category to figure in the applications for farmland afforestation is dry land. Together, pastures and dry lands represent 70% of the afforested area and if they are combined with 'scrubland' and 'arable or worked land with holm oaks in dry regions', they account for 90.15% of the afforested area.

Therefore, the areas affected by the afforestation work generally consisted of unproductive land. The afforestation of agricultural land is uncommon. This is understandable if we remember that a change from agricultural land use to forest plantations in Mediterranean regions brings with it a considerable reduction in the value of the land and a great loss in financial returns, due to the low productivity of the Mediterranean countryside. In contrast, afforestation is an interesting alternative for low-value, unproductive land since it provides an income through EC incentives and improves the area's heritage. In this respect, the afforestation of agrarian land has provided an opportunity for heritage valuation and landscape improvements on rustic properties, which has been used, above all, by large landowners and local councils.

Suitability of the Chosen Reforesting Method and Species, with Regard to the Geo-ecological Characteristics of the Land

Farmland afforestation can serve to improve the rural heritage of the affected estates and can help to control erosion and increase biodiversity in the medium and long term. However, such actions have a great impact on the landscape as a result of the lack of territorial planning in their implementation (except in the distribution of EC funding between autonomous regions).

In general, the European agri-environmental measures have not been applied in accordance with landscape criteria. The results of the National Farmland Afforestation Framework Programme have been the outcome of a disorganized set of uncoordinated, one-off actions of complex viability, without a clearly defined overall strategy. The geographical distribution of the actions and afforested areas in the different Spanish Mediterranean regions and provinces has not been effectively supervised or determined by the regional administrative bodies; instead it has followed the spontaneous response of agrarian landowners to the successive annual programmes.

When assessing landscape impact, one of the most interesting parameters is usually the species and the afforestation method used (preparation of the land, planting and maintenance work). In this respect, the most frequently chosen species stand out as one of the most striking differences between plantations carried out by farmers and those carried out by the city-dwelling owners of large country estates. Whereas the former usually choose *quercineas* – more demanding, but with higher incentives and a greater environmental value from a subjective point of view – the owners of large properties prefer to use species of the *Pinus* genus, which have less value and fewer incentives, but are less demanding and have greater guarantees in terms of implantation and development. Farmers tend to seek the highest incentives and are not concerned about the commitment involved since they have sufficient time and interest. In the case of large country estates, however, the aim is to limit risks and guarantee benefits with maximum returns and minimum effort.

When it comes to selecting the species, purely financial criteria have taken priority over silvicultural criteria in almost all cases, favouring the species that receive the highest incentives and particularly those present in the region. This is particularly accentuated in areas with large reforested areas due to a certain social rejection of conifers, pines in particular. A detailed study of the current species distribution shows that there is no correspondence between the reforestation methods, species and seasonal conditions of the afforested lands; however, it cannot be said that the general situation is one of inappropriate species selection. Only in regions where environmental conditions greatly limit afforestation (for example, the semi-arid area of Granada) could one say that the choice has been completely inappropriate.

The preparation of the land, planting work and maintenance of the afforested area is all fundamental in guaranteeing the success of an afforestation project. If we start by recognizing that many factors can influence the success of a plantation, we cannot deny that the key to success or failure in Mediterranean afforestation projects is to be found, in the majority of cases, in the guaranteed completion of plantation maintenance work. This is another reason why the Farmland Afforestation Programme fails to adapt to the territorial reality of the Mediterranean context, where plantations need more continued support. In order to ensure the success of the afforestation, it should be subject to maintenance work (and therefore financial support), preferably for a period of 10 years.

Furthermore, maintenance work is a determining factor in the landscape results of the afforestation. Afforestations carried out by farmers tend to show an excellent state of growth. However, the maintenance work that determines this situation, almost always typical of agricultural farming techniques, forms a landscape which could be described as a 'cultivation of forest trees', thereby creating a distinctly regular and homogeneous landscape that distances itself from the final objective, 'to reconstruct a forest with a diversity of species and structures, forming a heterogeneous and structured plant population' (Simón, 1997).

The ecological and landscape results are very similar in plantations carried out on large, productive farming estates which have their own permanent staff. The plantations are usually a complete success in such cases; but they suffer similar limitations as a result of maintenance work that is unsuited to forestry objectives. However, when it comes to large country estates used for recreational and hunting purposes, whose owners live in the capital or some other town or city and have their main interests elsewhere, the most frequent results are a high number of missing trees, from the second or third year onwards, due to neglect of the plantation. In these afforestation projects, maintenance work is limited to the replacement of missing trees. After the failure of the plantation, carried out on land already populated by a variety of bushes, one can normally observe the continuity of pre-existing forest dynamics, whilst hardly a trace of the action remains, except for the protective plant covers.

In general, farmland afforestation causes a great impact on the landscape as a result of the introduction of new species in local or regional settings and through the creation of 'patches of forest' in municipalities and regions with a low proportion of forested land. The visual impact of the plantation is usually highly noticeable (above all due to the use of protective covers) but it is not necessarily negative. In some cases, it has allowed the improvement and recovery of degraded land located close to urban areas (i.e. land adjacent to the rubbish tip in the city of Soria).

Effects upon the Biological and Landscape Diversity of Forestlands

The plantations carried out have not always been successful; however, they have contributed to a diversification of the land's colonizing species, particularly through the introduction of leafy species and, above all, the promotion of forestry activities in these areas.

One of the first questions to be asked when evaluating the results of the Farmland Afforestation Programme, as a complementary measure of the CAP Reform that is currently included in Rural Development Programmes, as in the case of other silvi-

cultural actions, is whether or not it has contributed to the objectives established in the Forestry Strategy for the European Union[5] and incorporated into the Spanish Forestry Strategy and the Spanish Forestry Plan: the multi-functionality of forests, sustainable management and the conservation and increase of biodiversity. In this respect, it is also fundamental to consider the species included in the appendices of the different regional orders, as well as the value of the financial support assigned to each group.

In contrast to the preferential use of conifers in the reforestations carried out in Spain throughout the 20th century, farmland afforestation has mainly used leafy species, above all *quercineas*. It is interesting to note that the lists provided in the different regional regulations give preference to native species and not only the inclusion of trees, but also several types of bush. However, despite the considerable progress made in the knowledge and results of nursery plant production, the actions have been conditioned by the characteristics of the affected land, depending on its prior use, and the species available in the nurseries.

The contribution of afforestation to an increase in biodiversity also depends on the location and size of the plantations. In the Madrid region, for example, the effects would be potentially more positive in the mountains or on foothills, rather than in the southern and south-west sectors which is where the work has in fact been concentrated, with very small-scale actions (Montiel *et al.*, 1999). In general, even in the case of actions covering more than 25 ha on large estates in the Madrid region, we cannot talk of an increase in forest perimeters, the creation of new expanses of forest or landscape diversification. On the contrary, the total area forested by the applicants contrasts with the small size and irregular perimeter of the large number of plots into which the action is often fragmented. In reality, plantations are usually located on the farm's most peripheral land and respond to a strategy of 'gap-filling' which, in many cases, favours homogenization and trivialization rather than landscape diversity. This situation, however, varies in other provinces where it is possible to find a tendency, albeit partial, towards a gradual increase in the species used and an integration of forested plots into the surrounding areas (i.e. the region of Huéscar, Granada).

Of particular interest among the plantations that have successfully improved estate landscapes are those carried out on the peripheral land of large farms devoted to agricultural production (hillsides and slopes), which have contributed not only to landscape diversification, but also to soil binding and protection against erosion. Special mention must be given to the afforestation work carried out in the triangles formed by the irrigation pivots on maize plantations in Las Vegas, an area in the region of Madrid. This is another case of 'gap-filling'; however, it takes a rational approach based on profitability and improvement. Furthermore, given the limited territorial presence of woodland in this area, the contribution of these plantations to biodiversity must be positively valued, although the use of regular distances and their monospecific nature may affect the increase in biodiversity. None the less, afforestation work in the Mediterranean context cannot aim to achieve an immediate increase in biodiversity, since the real difficulties that exist in establishing many forest species mean that the choice of species is itself limited. The contribution of afforestation to biodiversity must be assessed according to the medium-term effect that the new expanse of forest will have on the biodiversity of associated communities: flora, fauna, soil organisms, etc. The importance of the forest areas created by artificial expanses of conifers in many nature reserves in Andalusia is one of the best illustrations of the effects of reforestation on the biology and landscape of an area (Álvarez, 2001).

In general, we can identify a set of factors which have a greater or lesser determining influence on both the success of the plantation and the quality of the environmental and landscape results:

- the integration of forested plots into the estate, in order to prevent them from becoming forest islands located inside farms;
- planning processes which study the ecological suitability and objectives of the chosen species, as well as all other decisions (reforestation methods, preparation procedures, etc.); and
- the peculiarity of the Mediterranean climate, which has not been taken into account in adapting the European regulations to the regional specificity. This has resulted in establishment limitations (drought), slow growth, difficulties in obtaining direct products, etc, and has led to an inappropriate interpretation and implementation of the incentives.

Conclusions

The conceptual confusion which characterizes the Spanish regulations regarding farmland afforestation highlights the difficulty of applying the spirit of the CAP reform (to reduce surpluses and diversify agricultural income and functions) in Mediterranean Europe, due to the low forest productivity of the region. The failure of the European Regulations to take into account the specificity of the Mediterranean region has led to the development of national and regional regulations in Spain which define a permissive framework for the acceptance of applications. As a result, the ecological, socio-economic and landscape results of farmland afforestation in Spain have been extremely mixed.

It is not profitable for landowners in Mediterranean regions to change from agricultural land use to forestry plantations, due to the low productivity of the Mediterranean countryside in timber-yielding terms. As a result, the afforestation work has been focused on 'uncultivated grassland' used for extensive stock-breeding in traditional farming systems or less productive agricultural land in dry regions which was abandoned in the 1960s due to a lack of profitability.

In short, farmland afforestation in

Spain has represented an alternative for already-abandoned land which has favoured the recovery of degraded forest landscapes, some of which were affected by the risk of desertification. In this respect, afforestation work carried out on the peripheral land (hillsides and slopes) of large farms devoted to agricultural production is of particular interest. In other cases, afforestation has allowed the land and cultural heritage of terraced landscapes to be conserved, whilst also recovering the original forestry use of arable land at a time when there exist great demographic and socio-economic pressures. However, farmland afforestation has also contributed to the homogenization and trivialization of the forest landscape when implemented on areas of land which are located in the heart of a forest and previously served a fundamental purpose in ecological and landscape terms, as an element of diversification, complementarity and contrast.

In short, farmland afforestation offers interesting opportunities for the restoration of forest landscapes in areas which have become degraded as a result of the abandonment and demographic exodus that took place in rural areas in the mid-20th century. These territorial policies may provide an alternative for the recovery of forestlands, but their effects depend largely on the degree to which the European regulations are adapted, addressed and applied to the specificity of the Mediterranean context.

Notes

1. Later modified by Royal Decree 152/96.
2. Criteria and indices applied by the autonomous regions:
   - Surface Area of Usable Agricultural Land (SAU)
   - Indication of active production and agricultural production (calculated according to Final Agricultural Production, population working in the sector and population contributing to the Agricultural Social Security System)
   - Erosion rate
   - Inverse of average agricultural productivity per hectare.
3. The Royal Decree 152/96 which modifies Royal Decree 378/93 echoed this problem, indicating the need to 'take into account the years in which prolonged droughts occur, capable of destroying most of the plantations if the necessary measures are not taken', and gave rise to a general modification process similar to that of the regional orders.
4. These were all approved by the European Commission Decision of 27 April 1994.

References

Álvarez, M. (2001) *Paisaje Forestal Andaluz. Ayer y Hoy.* Junta de Andalucía-Ibersilva, Seville, Spain.

Arias, J. and Fourneau, F. (1998) *El Paisaje Mediterráneo.* Universidad de Granada-Junta de Andalucía, Granada, Spain.

Barrué-Pastor, M., Billaud, J.P. and Deverre, C. (1995) *Agriculture, Protection de l'Environnement et Recomposition des Systèmes Ruraux: les enjeux de l'article 19.* INRA, Ivry-sur-Seine, France.

Bona, L., Aramburu, M.P. and Cifuentes, P. (1997) Seguimiento del Programa de Reforestación de Tierras Agrarias en su contexto medioambiental y socioeconómico. In: Puertas Tricas, F. and Rivas, M. (eds) *Actas del I Congreso Forestal Hispano-Luso y II Congreso Forestal Español IRATÍ-97.* Gráficas Pamplona, Pamplona, t. VI, pp. 21–26.

Gómez, J. and Mata, R. (1991) Actuaciones forestales públicas desde 1940. Objetivos, criterios y resultados. *Agricultura y Sociedad* 65, 15–64.

Gómez-Jover, F. and Jiménez, F.J. (1997) *Un Programa de Forestación de Superficies Agrarias (Legislación y aplicación).* MAPA, Madrid, Spain.

Montiel, C., Ferreras, C. and Álvarez, P. (1999) El Plan de Forestación de Superficies Agrarias de la Comunidad de Madrid 1993–1997: Valoración territorial y paisajística. In: *El Territorio y su Imagen. Actas del XVI Congreso de Geógrafos Españoles. Málaga, 9–12 diciembre 1999.* Universidad de Málaga-AGE, Málaga, Spain, pp. 179–190.

Montiel, C., Galiana, L. and Navarro, R. (2003) Participación de las sociedades rurales en la forestación de tierras agrarias. In: García, J.S. and Vázquez, C. (eds) *Las Relaciones entre las Ccomunidades Agrícolas y el Monte.* Universidad de Castilla–La Mancha, Cuenca, Spain, pp. 93–124.

Simón, E. (1997) Proyectos de forestación de tierras agrarias. In: Orozco, E. and Monreal, J.A. (eds) *Forestación en Tierras Agrícolas.* Edic. de la Universidad de Castilla–La Mancha, Cuenca, Spain, pp. 21–42.

PART III

Case Studies

Investigations into landscape end up influencing many different aspects of planning and management. The case presented by Latz in Chapter 16 presents a very interesting situation on the estate of Spannocchia, in Siena province (Italy), where farming activities are still going on and the owners are very interested in managing the land, not only preserving historical values, but also taking advantage of the opportunities offered by the 'added value' that landscape represents for typical products and for a different way of managing farmlands. It may be surprising, but in Tuscany there is no policy of considering landscape as an important resource for rural development, but rather as a minor element in the competitiveness of the whole system. Spannocchia is instead an example of how this resource can be considered a central element in the rural economy, although it suffers from the lack of economic incentives given for conservation and development and from reduced attention to inappropriate policies negatively affecting its quality. Spannocchia is also an important case because a part of the property is included in a natural protected area, showing the problems of creating a network of areas to preserve 'nature' in cultural landscapes – an interesting case also for many other countries in the world.

In this range of different situations, the Temple Valley in Sicily, a UNESCO World Heritage site, is a place where unregulated development has strongly and negatively affected a landscape that should have been submitted to very strict regulations. The landscape of the 'Mediterranean gardens' resulting from the Arab influence in Sicily and restored by the project presented in Chapter 18, had been left to progressive decay for years in many parts of southern Italy. They are an example of the positive results of the multicultural influences affecting many countries of the world; in this case they have created landscapes now presented as a distinctive cultural feature of the area.

Wider issues are presented by Rotherham (Chapter 15) and Johann (Chapter 17). The first shows the problem arising after 200 years of intensive industry with a legacy of dereliction and pollution, where key sites have been identified for either conservation or restoration and habitat creation, with an interesting reintroduction of traditional systems, an activity suggesting actions to recover many areas in the world degraded by intensive industrial activity. The decline of industry and the serious crisis of agriculture create many abandoned areas where restoration projects can be undertaken embedding such projects in the cultural history of the region, through reinstatement of sympathetic and traditional management such as grazing by rare-breed livestock, as also described in Chapter 5. There is in fact the chance to develop restoration, not only by re-creating natural habitats, but also by restoring cultural landscapes especially in areas where the impacts of human utilization are etched deeply into their fabric. In such cases, there is a need to address the conflicts between contemporary sustainable landscapes and their ecology and a need to recognize and conserve the historic archives that these areas represent.

The case of the Viennese forest (Chapter 17) is instead a valuable example of an important issue concerning the use and management of landscapes in the suburbs of urban areas, often affected by stronger conflicts compared to rural territories, due to the contrasting interests of different economic and social groups. There are few capital cities owning a landscape of variety and equivalent extension comparable to that of Vienna. With more than 40,000 ha, 20% of it situated in the urban area, the forest has been dedicated totally to public welfare. Besides showing the use of historical investigation, this case shows how individual communities can succeed in opposing intervention by local authorities and even by powerful market forces – a real issue of modern times.

15 Historic Landscape Restoration: Case Studies of Site Recovery in Post-industrial South Yorkshire, England

I.D. Rotherham

Tourism, Leisure and Environmental Change Research Unit, Sheffield Hallam University, Sheffield, UK

Introduction

The idea of an informed historic context to site restoration has been addressed in a number of papers by Rotherham and colleagues (e.g., Rotherham and Avison, 1998; Rotherham, 2002). In post-industrial South Yorkshire, England, this is especially significant. A region of environmental diversity with a rich cultural and natural heritage, over a period of around 500 years its landscape has changed to a point often beyond recognition (Rotherham, 1996a, 1999). Extensive wetlands in uplands and lowlands have been drained and 'improved' to facilitate agriculture – grazing in the uplands and cereals in the lowlands. Much of the area was progressively urbanized from the 1700s onwards, and large tracts of land taken for massive industrialization – particularly steel and coal (Harrison and Rotherham, 2006; Rotherham, 2002; Rotherham *et al.*, 1997).

Now, with industry in decline and agriculture in serious economic crisis, many sites are abandoned and derelict (Beynon *et al.*, 2000; Handley and Rotherham, 2000; Rotherham and Cartwright, 2000). This creates major opportunities to re-establish sites, to restore lost wildlife habitats and to recover degraded ecosystems (Rotherham and Lunn, 2000). For maximum success and to embed such projects in the cultural history of the region, knowledge of the former landscapes (such as from Scurfield and Medley, 1952 and 1957; and Scurfield, 1986) is used to better inform the renewal of the region's environment. Techniques being applied include both site restoration and recovery – through reinstatement of sympathetic and traditional management such as grazing by rare-breed livestock. Along with this some sites are created within new landscapes, but where possible using seed and materials from donor sites across the region. The sites described include a diversity of landscape types and ecological communities.

Rotherham *et al.* (2000b) discuss the approaches taken to restore a major heathland area at Wharncliffe. This site can be dated back to a substantial Romano-British quern-stone factory perhaps 1700 years ago. Now with removal of encroaching scrub and secondary woodland, with controlled heather burning and cutting, and spraying of bracken, the site is being managed to safeguard heathland wildlife and the rich archaeological resource. To control the re-establishment of woodland and scrub the site is being grazed by rare-breed livestock.

Management here has to balance the ecological requirements of key species with the physical recognition and protection of the historic landscape. In some situations the two are not compatible and compromise is sought.

Along the banks of the River Rother lies the dereliction of 200 years of intensive industry leaving a legacy of decay and pollution. Here, key sites have been identified for either conservation or restoration and habitat creation. The techniques applied include major creation of large-scale wetlands on post-industrial areas, down to small-scale creation of complexes of ponds within a matrix of created and relict wet grasslands and marshes. Once created, sites are then subject to agreed conservation management plans. Where the aim is restoration through management rather than intervention with habitat creation, the approaches taken include re-establishing hay cutting and follow-up with grazing by rare-breed livestock (sheep and cattle) suited to the rough wet conditions. Along with this the existing drainage of the sites is being reversed to allow them to become much wetter. Site recovery is being monitored with key indicator species (plants and animals) recorded to help assess whether the restoration is succeeding (Handley and Rotherham, 2000; Rotherham et al., 2000a).

A target wildlife habitat and landscape has been that of unimproved grasslands. These include traditional grazing pastures and hay meadows. In particular, the sites have been riverine wet meadows and marshes, and relict meadows now within the suburban area (see Rotherham, 1999, for example). Relict but abandoned grasslands in urban green space and parks have been brought back into traditional management, utilizing cuttings where possible, and appropriate grazing with livestock.

Finally, along with the restoration and recovery of heathlands, wetlands and grasslands, are presented examples of woodlands brought into conservation management. This may include traditional coppice work and the creation of conservation glades in amenity woods. It is within the woodland projects that there is considerable scope for tension between different interest groups and lobbies. These woodlands mix both relict ecology and the cultural landscapes of former management systems. The balance of restoration and recovery to conserve and benefit different aspects of the woodland resource may be problematic. The wooded landscapes present a palimpsest of history over more than 3000 years. Deciding which period to conserve or enhance raises fundamental issues about restorative ecology, about conservation and about priorities. There are serious concerns about the lack of understanding of the interactions between wooded sites and other landscapes, and of the importance of antiquity in woodland conservation. The planting of trees to 'create' woodland may destroy other communities of great value and may not generate the type of environment hoped for by its creators. It is important that these projects are effectively constructed in terms of ecology and, for example, likely re-colonization rates of key species (see, for example, Vickers and Rotherham, 2000 and Vickers et al., 2000).

The techniques and case studies are presented along with an assessment of drivers for change, and triggers and barriers for appropriate and sustainable landscapes in the future. The balance between 'restoration' and 'creation' in the wider context of landscape 'renewal' is considered in the light of the case-study sites and projects. Furthermore, it is important for long-term sustainability that these initiatives have resonance with local communities and that they are economically viable.

The Case Studies

Four case-study areas are discussed. These include specific individual sites, and in some cases groups of sites. All are located in the Yorkshire and Derbyshire region of northern central England. They include a riverine meadow landscape (Woodhouse Washlands), a dry heathland and ancient woodland (Wharncliffe Heath and Wood), an acidic grassland and relict woodland (Westwood), and a group of ancient coppice woods (Ecclesall Woods, Gleadless Valley Woods and Owler Carr Wood).

Broad conclusions and common threads are drawn from the different examples. All the areas reflect the inextricable links between landscape history, site utilization and subsequent abandonment, and then recognition by conservationists and a desire to restore or re-create in part the former interests. There are key issues that arise in terms of economic history and in the relationships between the environmental resource and the local people or community that value and utilize this. In former times these landscapes were exploited but conserved – essential for sustainable living. Today they are valued for leisure, for recreation and for conservation.

The individual projects demonstrate very tangibly the huge potential for landscape recovery on these areas, but along with this they highlight some causes for concern and tension.

Westwood Opencast Coaling Site

The case-study site is on Coal Measures Series geology at an altitude of around 170 m, and sloping gently to the southwest. It is about 500 m long by 180 m wide, totalling around 10 ha in a greenbelt area

with a history of previous mining use and subsequent abandonment (Fig. 15.1).

The site has public access and is now managed by appropriate and self-financing, low-key grazing. Historically, the area was a part of the ancient woodland of Westwood and associated with Tankersley Park, a major medieval deer park to the east, and Wortley Park to the west (Jones, 1984). During the late 1980s there was increasing pressure on green space in the area for industrial parks and business parks. The region has substantial areas of derelict and despoiled industrial land, but some of these areas are unexpectedly rich in wildlife and may have a significant conservation value.

There was a proposal to open-cast mine this area primarily for the 'pillars' of coal remaining from previous mining activity. In the case of this particular area, interest and awareness were heightened by both an active and environmentally aware parish council, and an active local conservation group. Local action was to assess the site and to design and oversee the restoration. The result was a scheme with major conservation and community benefits. The compromise to allow mining, but with a comprehensive conservation plan and restoration, achieved a positive result. The scheme secured the effective restoration of the site using local wildflower seed, and the creation of new conservation features (ponds and heath). A major element of the

Fig. 15.1. Open-cast coal mining in the study area.

project was the protection and then restoration of the remnants of the ancient Westwood. These occurred as isolated relics around the site perimeter, and the lane-side hedgerows.

Here, site history has a bearing on the potential 'archaeology' and landscape historical features, and also on the contemporary ecology. Indeed, the diversity and conservation value of the flora and fauna relate directly to the history (recent and long-term) of the area. The western edge of the site is formed by Westwood Lane, the boundary of the parishes of Tankersley and Wortley, probably dating back to the 12th century. It is likely that Westwood Lane and its ditched bank and hedge are of similar age or older. This was given a high priority in the assessment of the site. A map of the area in 1772 shows the site as part of the ancient Westwood, which covered a considerable proportion of Tankersley Parish. The surviving remnants of this predominantly mixed sessile oak woodland lie to the south of this site and have parts that are of high floristic interest with plants typical of ancient woods. The western boundary is probably a remnant of these ancient communities.

Prior to 1906 Westwood had been felled in the northern portion of the site. There is evidence in the form of collapsed workings that this was done by early miners working the shallow seam as drift mines or by bell-pits. The wood may in fact have been used in the local iron-works that sprang up in the 18th and 19th centuries, taking advantage of the Coal Measures ironstone that was extensively mined at this time. It is likely that Westwood was felled up to the boundary woodland present in the 1980s.

Following these small-scale workings, the site was open-cast during the war years around 1942–1943. Probably due to the necessities of wartime, the site was restored only minimally with a thin layer of rather poor soil. Inspection of aerial photographs from 1971 shows that nearly 30 years on there was little regeneration. The woodland that had occupied this lower part of the site was probably clear-felled during the

wartime open-casting. Three small hillocks at the south-western corner of the site were probably from tipping of overburden that was not back-filled. The site passed into agricultural management as low-grade pasture. Small pockets of botanical indicator species did exist where the coaling operations had not disturbed them, particularly in the woodland remnants, but trampling and overgrazing by livestock significantly affected even these.

In conclusion, because of the previous open-cast operations on this site, most features of archaeological or historical interest had long since been removed, and the main thrust of the conservation plan was to safeguard what remained of the ecology and use this as the basis for recovery. A preliminary survey was undertaken and the area divided into main zones and features: woodland relics, hedgerows and drystone walls, grassland and ponds. National Vegetation Classification (NVC) surveys (Rodwell, 1992) were not carried out since the preliminary guidance was only just becoming available. Similarly, this project pre-dated both 'Natural Areas' and 'Biodiversity Action Plans'. If undertaken today, all these influences would be taken into account.

According to Rotherham and Lunn (2000), the results of restoration have been surprisingly good. Despite initial problems with sub-contractors in the replanting of woodland and establishment of hedgerows, the project's success has been spectacular. The woodland areas have recovered substantially, and there is already evidence of old woodland species such as wood sage recolonizing into the new areas. The ponds and ditches are very attractive and very successful in terms of their flora and fauna (with Odonata and both great crested newt and grass snake colonizing from a site close by).

The grasslands established well and quickly, with the heath area lagging perhaps 4–5 years behind. (By 1999, young heather plants and a range of *Cladonia* lichens and wet heath bryophytes were establishing well. Gorse and broom began to establish quite quickly.) The grassland was sampled for NVC classification 3 years after establishment. This produced two main 'communities'. The first came out as NVC MG9, whilst the 'hay meadow' area did not match, probably being a mixture of communities still in flux and missing some expected constants of, for example, MG5.

The site was established as a community open space with agricultural management, and so an attractive landscape with reinstatement of public rights of way was important. The grassland is now cut for hay in late July/early August and then grazed by cattle. The acid grassland/heathland is uncut, but grazed. There is no management of the ponds and ditches. The woodlands and hedges are presently unmanaged, but, importantly, the woodland areas are no longer grazed. In the short to medium term, effective management of the grassland areas was seen as a priority. By establishing economically effective management, but led by conservation guidelines, this is essentially a sustainable regime. There has been no input to the project by any formal conservation body, and the current management does not cost anything to either the local authority or the region's conservation bodies.

The key to success here was the effective involvement of local people. They were very much involved with the project from the outset. The company involved was sufficiently enlightened to recognize the benefits of this involvement, and the need for the consultants as the interface with local people. The scheme has been very successful. One cause for concern, as is so often the case in such situations, is that no funding was set aside for monitoring and review. The valuable lessons to be learnt from such projects are too often lost for the contribution of a minimal amount of finance. Perhaps less than 1% of the project costs would suffice, but without statutory 'encouragement' developers seem to baulk at the idea!

Wharncliffe Heath and Woods

Wharncliffe Heath and Crags is a recently established, major urban nature reserve described by McCarthy *et al.* (1993). It is located on the most easterly of the Peak District Edges, though outside the Peak National Park – this is now a Yorkshire Wildlife Trust Nature Reserve, managed by the Sheffield Wildlife Action Partnership (SWAP) (McCarthy, 1996a,b, 2000). The site incorporates a Geological Site of Special Scientific Interest, Scheduled Ancient Monuments, an early medieval deer park, and ancient woodlands, all surrounded by and overlooking contemporary housing expansion, road and rail networks, and a mixture of heavy industry and major industrial dereliction. The landscape here has a long history of utilization and subsequent abandonment, with impacts of industry and urbanization, and now management as a wildlife site.

The site is hugely important in terms of regional biodiversity, being a vital stronghold for locally rare species such as nightjar, red deer, green tiger beetle, adder and grass snake. The importance of these species, and of local community support in helping to secure vital grant aid, and a balance of appropriate management and low-key access and promotion, have been major issues here.

Wharncliffe Heath and Wood has long been known as one of the finest wildlife areas in the Sheffield district. Being in the core zone of the South Yorkshire Forest, the Forest Team commissioned a comprehensive biological survey in 1993 to be carried out by the Sheffield City Ecology Unit. It was accepted for Countryside Stewardship funding by the then Countryside Commission. This community/environment is not typical of the Sheffield area today, but historically was widespread and common. Unmanaged, much of the interest was under serious and imminent threat.

The core area is a mosaic of deciduous, semi-natural and plantation, woodland, heathland and acid grassland. The area is floristically rather impoverished due to poor management and former high levels of air pollution. In most of the woods there remain patches of much richer woodland flora. These are usually along streams and woodland edges. However, the faunal interest of the complex is considerable. Several bird species are at the edge of their ranges, and Wharncliffe Heath is one of the few remaining intermediate moorland habitat-types in the region, supporting the area's largest breeding population of nightjar (*Caprimulgus europaeus*), with locally and regionally significant breeding populations of reptiles and amphibians. Wharncliffe Woods have been extensively researched for their invertebrate fauna and are notable for a diversity of insect species, in particular the Coleoptera (beetles) and Syrphidae (hoverflies). Wharncliffe Wood and Heath, together with the Greno Wood area, are listed as Grade B (regionally significant) on the English Nature Invertebrate Site Register. The area is also of special archaeological and geological significance. Land tenure of the case-study site is shared between two landowners. Wharncliffe Chase, together with some adjacent farmland and woodland, is owned and managed by Wharncliffe Estates Ltd, whilst Wharncliffe Wood and Wharncliffe Heath are owned by the Forestry Commission and managed by Forest Enterprise.

The past and present land uses are of interest with evidence of early human activity dating back to around 7500 BC. Close to the confluence of the Don and Little Don rivers at Deepcar is a nationally important Mesolithic site, probably used as a summer camp by hunters following seasonal animal migrations to the uplands of the Pennines. The Wharncliffe area was intensively used during the Romano-British period, and settlements from this time have been excavated and the remains of buildings and field boundaries have been identified under what is now the Wharncliffe Chase boundary. The Crags were quarried extensively during this period for quernstones (hand mills for grinding grain) which, it is thought, gave rise to the original name, 'Quern Cliff'. The disc-shaped base stones and beehive-shaped rotating stones can still be found at the base of the crag, mostly as unfinished artefacts. During the early medieval period the first of a series of enclosures of the Chase took place. These were carried out between the early 13th and

late 16th centuries and the area was subsequently developed as a deer park. The remains of two villages, the inhabitants of which were evicted during the enclosures, can still be found within the present Chase boundary. There is much evidence of past industrial use, particularly within Wharncliffe Wood, the most notable being mining. The local coals and their associated ganister and fire-clay have been extensively worked and several old drift mines and adits are in evidence on the talus slope below Wharncliffe Crags. Grenoside sandstone was quarried throughout Greno Wood and the eastern Chase, and was renowned as a high-quality building stone.

Organized exploitation of the woodland resource has probably been ongoing for at least 600 years. Management probably took the form of 'coppice-with-standards' until the 19th century when the planting of conifers such as Scots pine and larch was carried out. By the early part of the 20th century traditional woodland management had all but ceased. Both landscape and vegetation have been considerably modified during the 20th century. Large areas of Wharncliffe Wood were felled during World War I, during the General Strike in 1926, and to a lesser extent during World War II. Further destruction of the woodland resource took place during the period up to 1954, when woodland adjacent to the railway was destroyed in a series of severe fires caused by the sparks from passing steam trains. Since then, the dominant land use has been commercial forestry, with blanket afforestation of the southern section of Wharncliffe Woods. This has reduced what was described in 1903 by W.G. Smith (former President of the British Ecological Society) as 'probably one of the finest oak woods in the country' to a much simplified community (Rotherham, 1995, 1996a). More recently, the area has experienced intensive and largely unplanned recreational use with walking and jogging, but also more intensive and organized pastimes such as mountain biking, orienteering, horse riding, clay pigeon shooting and fox hunting.

In the 1990s, an initiative was developed to address key issues of site restoration and conservation. Establishing the Heath as a nature reserve was perhaps the most significant. Here, the desired management objectives were easy to agree, but more difficult to achieve. A programme of controlled burning, of birch cutting and more recently of grazing with rare-breed livestock was initiated to bring the site back to heathland in a mosaic of woodland. There have been some difficulties and compromises. This is a sensitive area for wildlife, recreation, scientific geological interest and archaeology. Recent surveys on behalf of English Heritage have identified the area as the most important Romano-British quern factory in the UK and possibly in Europe. This has serious implications for some of the proposed nature conservation management.

The proposals for the Chase also highlight the need for a careful and holistic approach to historic sites even when they appear to be in desperate need of management. The Chase was seen as a huge opportunity to re-establish heathland close to the Crags and the Heath itself. Grant aid in the form of Countryside Stewardship was available to both areas, and has underpinned much of the work. On the Chase, grant was taken up to decrease the levels of grazing stock and hopefully tackle the expansion of bracken.

However, there was also the chance to re-establish heather in a more radical approach to the site. Site survey indicated a total absence of heather on the Chase, but expanding bracken and poor acidic grassland. This seemed an ideal opportunity to intervene and re-seed heather or even to encourage natural regeneration. It was believed that heather had been present not that long ago. The Chase has a number of stone and sod revetted mounds, described on the Ordnance Survey maps as 'butts', and it was well-known that the Earls of Wharncliffe had used the site for shooting. The conclusion was that these were indeed shooting butts for grouse shoots, and therefore the site must have been heath or moorland within the last 100–150 years.

This interpretation was completely wrong. Samples of soil taken to examine dormant seed banks produced no heather at all. This seemed perplexing, but perhaps the heather was lost too long ago. Detailed research suggested a totally different conclusion and therefore alternative recommendation for the site. The 'butts' are not grouse butts at all, but are medieval pillow mounds for keeping rabbits, and of great historic interest. The use for shooting was for target practice by soldiers from the estate during wartime training and not after grouse at all! It is probably at least 500 years since heather formed the dominant community on the Chase, and perhaps not even then.

Restoration of this historically important and complex area to heathland would have been both misconceived and difficult to achieve. Management now seeks to decrease grazing, allowing the Chase to recover any diversity 'naturally', and allowing some areas to recover from drainage operations that have substantially dewatered the entire site. Heathland restoration is targeted at the Heath and then perhaps at areas of the wood below the main Crags.

Woodhouse Washlands

The Woodhouse Washlands Nature Reserve is a major Yorkshire Wildlife Trust project of around 70 ha in the shadow of the former Orgreave Colliery, in the heart of urban, industrial South Yorkshire (Fig. 5.2). It is one of the last significant remnants of low-lying riverine wetland that were formerly extensive along the Rother Valley. These were perhaps up to 30–40 km in length and up to possibly 2 km width in a much wider valley bottom. Significant areas remained intact but increasingly degraded until the 1950s (McCarthy, 1994 and 1995). This case study exemplifies the historic context of abandonment and dereliction of the Rother Valley washlands, their eventual recognition in the River Rother Wildlife Strategy (Anon., 1994b), and now piecemeal recovery. The triggers for practical conservation action are noted by Rotherham et al.

Fig. 15.2. Coot feeding young at Woodhouse Washlands.

(2000a). These are examined along with the constraints on the present restoration, as a strategic floodplain site, and as a major amenity resource for local people.

The initial results of long-term monitoring and the recovery of key species are most encouraging with key plants of the ancient meadows recovering and re-establishing across the site. Effective and strategic grant-aid, targeted business support, and local community action in this urban setting, have been vital in the recovery of this site over a 10-year period.

The area remained undeveloped and to some extent protected by designation as strategic floodland by the various agencies with authority over the years. This was Yorkshire Water Authority until the late 1980s, and then the National Rivers Authority, now the Environment Agency. All these bodies, as essentially absentee landlords, managed the site by means of tenant farmers. The area was intensively grazed so that when adopted as a nature reserve in the mid-1990s, the grass sward was very species-poor and only 3–5 cm in length. Simple, superficial drains had been cut across the area and were very successful in de-watering much of the site.

This is an ancient site with a long history of human impact and massive change. There are, or were, tiny areas of degraded but relict wildlife habitat surviving. It is a

greatly degraded site on the urban fringe and has experienced problems of poor management and low landscape value. However, it does have a superb wildlife resource, including relicts and Local Red Data Book species, and along with this, substantial local community interest (Rotherham and Whiteley, 1995). There were and still are serious limits on what could be done: these include the constraints of a statutory flood-control area, plus the fact that the prime target areas are wetlands, but the site no longer floods. It is a regionally unique site, urban fringe and so important to many people, an important link in a strategic green corridor, and linking laterally into the Shire Brook Countryside Management area.

Two further developments were critical in the conservation of this site. First, the South Yorkshire Forest Partnership was established and was able to target funds and strategic recognition towards the area. Second, the Countryside Stewardship Grant Aid Scheme came online in time to support much of the necessary work. These two effectively oiled the wheels of action once the area had been identified in the Sheffield Nature Conservation Strategy (Bownes *et al.*, 1991) and the River Rother Wildlife Strategy (Anon., 1994b).

The overall aims are to conserve the remaining relict wildlife habitat, restoring the degraded wildlife habitats and increasing desired wildlife habitats. The experience of managing the reserve and charting the changes provides a deeper understanding of how and where on the reserve those aims may be achieved. Key components of future work are to develop the monitoring systems to pick up future changes, to encourage responsible local community involvement and to work in partnership with other stakeholders. The management of the reserve is also set into the 'biodiversity' context and the targets set in the local BAPs (Biodiversity Action Plan) used where appropriate. A challenge for a large and varied site such as the Washlands is to accurately assess the potential for achieving a range of different BAP targets. Several questions remain around the initial viability and then sustainability of any wildlife habitat or species management, and how the reserve fits into the bigger regional and national pictures.

In order to fulfil the need for information, and to inform the monitoring and management process, a strategy for vegetation monitoring has been developed. As well as providing objective, scientific information for the site's managers, this also provides a further opportunity to involve local people, students and others in the process.

The site is a complex mix of different, though sometimes closely related plant communities/vegetation types. To further complicate the situation, some of these communities are in a state of rapid flux, related to both the changed management of recent years, and also perhaps to the varying weather patterns of recent years. To address this problem, and to monitor changes, to inform management and to record the status of species considered of nature conservation importance on the site, a mixed strategy was proposed. This approach assumes that the present lack of dedicated resources for monitoring will continue, and that work will be carried on by volunteers, student researchers and professionals donating their time and support. Through the monitoring programme, a comprehensive package of managed restoration (with grazing by livestock, reduction in drainage, and with targeted hay cutting) has been achieved, along with a programme to create up to 50 ponds over the next 20–30 years. Around 15 new ponds have already been constructed and the impacts on wildlife species have been immense.

In conclusion, Woodhouse Washlands is a deceptively complicated site that still has potential for further improvement as a biodiversity conservation resource. It has had, and will continue to have, a range of human impacts placed on it. Some of these may be positive with local people feeling 'ownership' of the site, and wishing to care for it; some are negative such as the impacts of visitors on vulnerable and sensitive breeding birds.

Work on the reserve over the past 10 years has shown how with relatively few resources, parts of a formerly poorly managed site can be improved as a wildlife conservation resource. This case-study site demonstrates how Biodiversity Action Plan targets may be achieved, and the value of sometimes not doing very much but letting wildlife habitats recover naturally, with a gentle helping hand! The once species-poor areas now have an increasingly rich flora and fauna. The site boasts extensive wetland meadows with numerous locally rare species including several orchids. It is the best site in the region for Odonata and one of the best for butterflies. It is an excellent habitat for mammals such as water vole and harvest mouse, and has breeding kingfishers, sand martins, skylarks and much more. This is all in an area of former gross industrial pollution and with a community that has major social and economic problems.

This site highlights the following:

- the need for partnerships;
- the critical role of individuals in making it happen and keeping it going;
- the vital roles of relict areas;
- the significant (and unknown) potential for recovery;
- the key issues of local ownership, of local empowerment and, hence, local action.

One of the most important lessons to be learned from this case study is the enormous potential for recovery of a site through careful management and targeted restoration. As described by Handley and Rotherham (2000) nothing has been introduced here and all the recovery has been through natural processes.

The Sheffield Area Woodlands

The case studies here represent a series of ancient woodland sites that have been brought back into restoration and conservation management over a period of around 20 years. The process and context were discussed by Rotherham (1996b), and by Rotherham and Jones (2000b). Monitoring and overseeing much of this work has generated a number of important issues and ideas. The current research is based on a regional assessment of woodlands around South Yorkshire and North Derbyshire in England. In particular the following areas have been considered in detail:

- Ecclesall Woods in Sheffield;
- Gleadless Valley Woodlands in Sheffield;
- Grimethorpe Woods in Barnsley;
- the Upper Moss Valley Woodlands in Derbyshire.

Ancient woodlands in the UK are amongst the most valuable of our conservation resources; providing habitat for vulnerable and interesting wildlife species, with many of these being associated exclusively with such wooded environments. There is also an emotional response to wooded landscapes, especially 'ancient' woods, as fragments of a perceived primeval 'wildwood'. This is of course almost totally incorrect; in reality these wooded landscapes are complex palimpsests of human activity and have been shaped over countless centuries. The special importance of ancient woodland is the feeling of walking in the footsteps of the ghosts of people that lived and worked there over thousands of years (Rotherham and Jones, 2000a). Indeed it is these 'ghosts' that have left their mark on the wooded landscape and even on the vegetation itself. It is also important to recognize that many of today's ancient woodlands probably incorporate phases of non-woodland, often agricultural or even settlement use. The soils, the landform, the vegetation, the hydrology and the fauna, all reflect human impact over the millennia.

Ancient woodlands in Britain have been extensively researched and in many cases thoroughly documented. Despite this, there still remains a dearth of collaborative research that considers both the ecology and archaeology of many of these sites. This has major implications for the effective restoration of wooded landscapes. Despite this, much progress has been made in recent years to both manage by reversion ancient deciduous woodland sites that were planted with conifers, and to reinstate conservation management in former decid-

uous coppice woods. Some significant issues and threats still remain (Rotherham and Avison, 1998).

Two major problems are highlighted here in relation to woodland assessment and management. These relate in part to the roles of the professions that oversee and initiate such work. The first relates to woodland surveys and subsequent management being often led by foresters or by ecologists. These two professions often have little experience or training in the recognition or interpretation of landscape or archaeological features. The second problem is concerned with what is recognized as 'archaeology' by professional archaeologists. The key to understanding the nature of particular ancient woodlands is the soil, the ground and surface features, and the trees and other vegetation. These may all hold clues to former management and indeed to former landscapes, and these may be vital to informing current conservation management. However, these aspects of woodland are often ignored by archaeologists, generally either more interested in monuments, earthworks and artefacts than earth and vegetation, or simply untrained to recognize these subtle landscape features. Trained archaeologists tend to recognize archaeology *in* woods, but not the archaeology *of* woods. Both a cause and a consequence of this situation is that there is presently almost no literature to guide the would-be field worker or to inform a site manager in surveying or evaluating the archaeology of their woodland resource. In many cases, the restoration and conservation management of these valuable sites is fundamentally flawed as a consequence (Ardron and Rotherham, 1999).

The presence of woodland on a site may effectively preserve landscape features going back over thousands of years. This is clearly demonstrated by ancient woodlands in England, in some cases in the heart of major urban areas, holding evidence of landscape utilization going back over 3500 years. Only recently has much of this evidence been formally recognized, and there are serious issues for cross-disciplinary collaboration and for effective training and support for field workers. Many of these ancient woodland landscapes are extremely vulnerable to inappropriate management or to intensive recreational disruption. Often unrecognized, they may be lost or degraded very easily and very quickly. This work is based on case studies from South Yorkshire and North Derbyshire in England, but the findings apply widely across Europe and the USA.

The archaeology both *of* woodland and *in* woodland is of huge interest. Much of the vital evidence for the unique site history is in subtle features and these combine human interference, ecological and edaphic characteristics. Since ancient woods often represent landscapes relatively unaffected by gross disturbance, they may hold evidence of cultural and ecological histories spanning many centuries. However, these clues to the past are very vulnerable to damage and destruction through contemporary management. This management is often intended to bring about environmental improvement and loss is usually (though not always) inadvertent. The research has brought into sharp focus the need for:

- more reliable and informative documentation;
- awareness raising – especially for foresters, conservation managers, ecologists and, indeed, archaeologists;
- effective education for the public;
- further research to evaluate and quantify the resource;
- conservation guidelines for site managers.

The woods being restored here are generally ancient woodland used for centuries to provide coppice wood as fuel and as building material to support the industrial revolution in England's heartland. This use was generally abandoned in the 1800s and sites were converted to high forestry often with exotic species and later left as amenity areas for local people. However, the imprint of management is written deep in the landscape and in the ecology of the sites, and understanding this has major implications for restoration and for conservation. Too often this has been misunderstood or

simply overlooked. There is also a very significant issue of the importance of archaeology of the centuries of management now evidenced within the sites in terms of veteran working trees, soil surface structures and features such as charcoal pits and hearths.

Some of the work points to the importance of the trees themselves. Here, significant trees and, in particular, trees of historic interest, especially 'working' trees from the past, may assume importance. This may be within a wood or in wooded landscapes beyond the formal boundaries. The value of such trees may be to do with ecology, but often it is cultural, historical and aesthetic; and all too frequently they are neglected. Wooded heaths and commons, for example, can have significant trees, but are often overlooked.

This leaves some issues and problems and despite regional and local policies and strategies recognizing the importance of old trees, of dead wood, and of habitat for saproxylic fauna and flora, contemporary (50–150 years) economic management of woods has generally left them impoverished in terms of dead and decaying wood. In the study area, our regional woods in particular are depauperate in terms of the dead/decaying wood resource. This is probably reduced to less than 5% of that in a 'natural woodland' and may be less than 15% of that in a traditionally managed woodland. The Sheffield Nature Conservation Strategy (Bownes *et al.*, 1991) noted the rarity of trees over 200 years old in Sheffield, and in developing ideas in the Sheffield Woodlands Policy 1987 it notes that the Authority will continue to implement the policies and proposals set out in this policy. In particular, it notes the EEC Committee of Ministers Recommendation No. R (88)10 'On the Protection of Saproxylic Organisms and their Biotopes' and stresses the importance of dead wood in woodlands. However, Victorian foresters, and then 20th-century amenity woodland managers, liked clean and tidy woods (bad news for deadwood, wildlife and history) and in many cases this trend continues today despite the conservation policies. In

Sheffield we have even had oak trees aged around 220 years (which is very old for the region) felled in order to regenerate oak. Even big trees selected for conservation are still felled in publicly funded management projects.

A further complication is that the our oldest trees may be ones such as holly (*Ilex aquifolium*) clones that are not what people expect or even what most people see as veteran trees. These are relics of former management and are a unique archive of information on woodland and landscape history. These vestiges of the former coppice woods are easily removed by today's management. This is not usually deliberate, but is inadvertent damage through management. Some of these individual trees may not always be of great ecological interest, not always 'veterans' and are not recognized by archaeologists. Indeed many field archaeologists, unless they have a particular interest in woods, cannot actually identify tree species anyway. Furthermore, most field ecologists have not got much idea on these specimens or their interpretation either.

Long-term studies have identified both opportunities and threats to the effective conservation and restoration of ancient, wooded landscapes. The first issue is that of recognition of the resource and then prioritizing zones for management and appropriate tools for conservation. There may be some conflicts between interests and approaches and therefore ways of addressing these within formal management plans need to be established. Current approaches are failing to do this in an effective or systematic way.

In part there may be a need to reinvigorate woods through the intervention of traditional management. In other areas, identification of long-term non-intervention or minimal intervention areas may help to maintain and increase dead-wood content as a natural resource. Sensitive and enlightened approaches need to be applied to re-planting as well as tree or wood removal, since re-planting of a semi-natural woodland can seriously damage its conservation

value. Managed regeneration is a much more satisfactory approach.

There is also a desire to generate new ancient trees – not conserving veterans, but generating new veterans and enhancing dead and dying wood resources. This is inherently long term, but requires action and vision now. If felling existing oaks, for example, then it would be best to take only those up to 85–120 years old, leaving those older than this. For somewhere like our Sheffield study area this means leaving and protecting all trees over 180 years, and if in doubt using the precautionary principle and leaving them.

One fundamental problem is that woodland management is inherently long term, but contemporary short-term grant aid and short rotation employment encourage short bursts of sometimes inappropriate or irredeemable action. There is often a need to be seen to be doing management, even if this is inappropriate in terms of long-term conservation.

Conclusions

This chapter draws on findings from four distinct, but related case studies. These are all part of a coherent programme of research across the South Yorkshire area. Together, they highlight the potential for restoration and recovery of historic landscapes. In some cases the results have been quite remarkable. However, the work also highlights important issues of a lack of holistic thinking and working, and difficult issues of the recovery and restoration of cultural landscapes. These are not natural areas and the impacts of human utilization are etched deeply into their fabric (Rotherham *et al.*, 1997). In such cases, there are legitimate questions about what we are restoring to and why. How do we address conflicts between contemporary sustainable landscapes and their ecology and a need to recognize and conserve the unique historic archives that these areas represent? These studies help inform the debate and they raise issues and questions to be considered further.

References

Anon. (1994a) *An Ecological Survey of the Wharncliffe Area.* Sheffield City Ecology Unit, Sheffield, UK.

Anon. (1994b) *River Rother Wildlife Strategy.* Derbyshire County Council and partners, Matlock, Derbyshire, UK.

Ardron, P.A. and Rotherham, I.D. (1999) Types of charcoal hearth and the impact of charcoal and whitecoal production on woodland vegetation. *Peak District Journal of Natural History and Archaeology* 1, 35–47.

Beynon, H., Cox, A. and Hudson, R. (2000) *Digging Up Trouble – The Environment, Protest and Opencast Coal Mining.* Rivers Oram Press, London.

Bownes, J.S., Riley, T., Rotherham, I.D. and Vincent, S.M. (1991) *Sheffield Nature Conservation Strategy.* Sheffield City Council, Sheffield, UK.

Handley, C. and Rotherham, I.D. (2000) Woodhouse Washlands – a major urban nature reserve. In: *Abstract Proceedings of the South Yorkshire Biodiversity Conference March 2000.* South Yorkshire Biodiversity Research Group, Sheffield, UK.

Harrison, K. and Rotherham, I.D. (2006) A memory re-discovered: map-based reconstruction of the former wetlands of the historic landscape of eastern South Yorkshire. Paper presented at Yorkshire Naturalists' Union Conference, Harrogate, February 2004: The Humberhead levels – their value to biodiversity in wetness. *The Bulletin* (in press).

Jones, M. (1984) Woodland origins in a South Yorkshire parish. *The Local Historian* 16, 78–83.

McCarthy, A.J. (1994) *Woodhouse Washlands Proposed Nature Reserve – Feasibility Study.* Sheffield Centre for Ecology and Environmental Management, Sheffield, UK.

McCarthy, A.J. (1995) *Woodhouse Washlands Management Plan.* Sheffield Centre for Ecology and Environmental Management, Sheffield, UK.

McCarthy, A.J. (1996a) *Wharncliffe Heath Nature Reserve Management Plan 1996–2001.* Sheffield Centre for Ecology and Environmental Management, Sheffield, UK.

McCarthy, A.J. (1996b) *Wharncliffe Chase Conservation Plan 1996–2001.* Sheffield Centre for Ecology and Environmental Management, Sheffield, UK.

McCarthy, A.J. (2000) *Wharncliffe Heath Nature Reserve – A Heathland Management Plan 2000–2003.* Andrew McCarthy Ecology, Sheffield, UK.

McCarthy, A.J., Dulieu, K., Rotherham, I.D. and Milego, C. (1993) The natural history of the Wharncliffe area. *Sorby Record* 30, 7–19.

Rodwell, J.S. (ed.) (1992) *British Plant Communities, Vol. 3: Grasslands and Montane Communities.* Cambridge University Press, Cambridge, UK.

Rotherham, I.D. (1995) Urban heathlands – their conservation, restoration and creation. *Landscape Contamination and Reclamation* 3(2), 99–100.

Rotherham, I.D. (1996a) Habitat Fragmentation and Isolation in Relict Urban Heathlands – the ecological consequences and future potential. In: *Proceedings of the 28th International Geographical Congress: Land, Sea and Human Effort.* August 1996. RGS/IBG, The Hague, The Netherlands, p. 396.

Rotherham, I.D. (1996b) The sustainable management of urban-fringe woodlands for amenity and conservation objectives. Proceedings of the conference on vegetation management in forestry, amenity and conservation areas: Managing for multiple objectives. Association of Applied Biologists Symposium, York, 1996. *Aspects of Applied Biology* 44, 33–38.

Rotherham, I.D. (1999) Urban Environmental History: the importance of relict communities in urban biodiversity conservation. *Practical Ecology and Conservation* 3(1), 3–22.

Rotherham, I.D. (2002) Woodland landscapes in Sheffield, England – reconstructing the evidence of four thousand years of human impact. In: *Proceedings of the International Conference, Florence, September 2002: Analysis and Management of Forest and Rural Landscapes.*

Rotherham, I.D. and Avison, C. (1998) Sustainable woodlands for people and nature? The relevance of landscape history to a vision of forest management. In: Atherden, M.A. and Butlin, R.A. (eds) *Woodland in the Landscape: Past and Future Perspectives.* The proceedings of the one-day conference at the University College of Ripon and York St John, York, UK, pp. 194–199.

Rotherham, I.D. and Cartwright G. (2000) The potential of Urban Wetland Conservation in economic and environmental renewal – a case study approach. *Practical Ecology and Conservation* 4(1), 47–60.

Rotherham, I.D. and Jones, M. (2000a) Seeing the woodman in the trees – some preliminary thoughts on Derbyshire's ancient coppice woods. *Peak District Journal of Natural History and Archaeology* 2, 7–18.

Rotherham, I.D. and Jones, M. (2000b) The impact of economic, social and political factors on the ecology of small English woodlands: a case study of the ancient woods in South Yorkshire, England. In: *Forest*

History: International Studies in Socio-economic and Forest ecosystem change. CAB International, Wallingford, UK, pp. 397–410.

Rotherham, I.D. and Lunn, J. (2000) Positive restoration in a green belt opencast site: the conservation and community benefits of a sympathetic scheme in Barnsley, South Yorkshire. In: *Abstract Proceedings SER 2000.* The Society for Environmental Restoration, Liverpool.

Rotherham, I.D. and Whiteley, D.W. (1995) The importance of relict and created wetlands with reference to invertebrates and vegetation in urban and post-industrial sites. Preliminary findings from the South Yorkshire Biodiversity Research Programme. British Ecological Society Conference: Recent Advances in Urban and Post-industrial Wildlife and Habitat Creation. Leicester, 20–22 March, 1995. Poster presentation.

Rotherham, I.D., Ardron, P.A. and Gilbert, O.L. (1997) Factors determining contemporary upland landscapes – a re-evaluation of the importance of peat-cutting and associated drainage, and the implications for mire restoration and remediation. In: *Blanket Mire Degradation. Causes, Consequences and Challenges.* Proceedings of the British Ecological Society Conference in Manchester, 1997. British Ecological Society and the Macaulay Land Use Research Institute, Aberdeen, UK, pp. 38–41.

Rotherham, I.D., Rose, J.C., Handley, C. and Goodman, K. (2000a) Restoring urban wet meadows: five years of recovery of a major floodplain in urban South Yorkshire, UK. In: *Abstract Proceedings SER 2000.* The Society for Environmental Restoration, Liverpool, UK.

Rotherham, I.D., Rose, J.C. and Percy, C. (2000b) Linking past and future; the dynamic influence of history and ecology on the restoration of a major urban heathland at Wharncliffe, South Yorkshire. UK. In: *Abstract Proceedings SER 2000.* The Society for Environmental Restoration, Liverpool, UK.

Scurfield, G. (1986) Seventeenth century Sheffield and environs. *Yorkshire Archaeological Journal* 58, 147–171.

Scurfield, G. and Medley, I.E. (1952) An historical account of the vegetation in the Sheffield district: the vegetation of the Southall Soake in 1637. *Transactions of the Hunter Archaeological Society* 7, 63–77.

Scurfield, G. and Medley, I.E. (1957) An historical account of the vegetation in the Sheffield district: the Parish of Ecclesfield in 1637. *Transactions of the Hunter Archaeological Society* 7, 180–187.

Vickers, A.D. and Rotherham, I.D. (2000) The response of Bluebell (*Hyacinthoides non-scripta*) to seasonal differences between years and woodland management. *Aspects of Applied Biology* 58, 1—8.

Vickers, A.D., Rotherham, I.D. and Rose, J.C. (2000) Vegetation succession and colonisation rates at the forest edge under different environmental conditions. *Aspects of Applied Biology* 58, 351–356.

16 Comparative International Research on Agricultural Land-use History and Forest Management Practices: the Tuscan Estate of Castello di Spannocchia and Vermont's Marsh–Billings–Rockefeller National Historical Park

G. Latz

Office of International Affairs, Portland State University, Portland, Oregon, USA

Introduction and Purpose

The purpose of this chapter is to report on ongoing research sponsored by the US Fulbright Commission, 'Comparative International Research on Agricultural Land-Use History and Forest Management Practices: The Tuscan Estate of Castello di Spannocchia and Vermont's Marsh–Billings–Rockefeller (MBR) National Historical Park' (Latz, 2001). Other supporters of the research include the US National Science Foundation, the Forest History Society, Portland State University, the University of Firenze, and the Marsh–Billings–Rockefeller National Historical Park (Latz, 2002 and 2004). The project's research objectives are threefold and include comparative study of:

1. Policies and legal instruments in the USA and Italy for stewardship and conservation of sites deemed representative of valuable cultural, environmental and historical landscapes.
2. Evaluation of opportunities for sustainable forestry and agriculture as demonstration sites and for public education.

3. Strategic master planning for sustainable forestry and agriculture, including evaluation of Geographic Information System (GIS) and Global Positioning System (GPS) techniques for construction of a series of multipurpose land-use maps of the Spannocchia estate and their comparison to the MBR National Historical Park (Foulds, 1994; Brown *et al.*, 2000).

Research findings presented here focus on the latter two research objectives as they pertain to the Italian case study, Tenuta di Spannocchia: the role of GIS/GPS land-use maps, compiled for the years 1823, 1954, and 2002, for communicating concepts of sustainable forestry and agriculture in the resource management planning process, and for public education.

Project Background and Characteristics of US and Italian Case Studies

A brief summary of the background to the research project is as follows. During the

2001–2002 academic year, the author conducted research under the auspices of the US Fulbright Commission as a visiting research scholar at the University of Florence. Academic collaboration there was primarily with Dr Mauro Agnoletti, and secondarily with Dr Gherardo Chirici, both of the University of Florence. At the Tenuta di Spannochia estate, proper, the author's primary contact was its owner/manager, Mr Randall Stratton.

The primary objective of the author's sojourn at Spannocchia, 2001–2002, was to conduct a rigorous analysis of the physical and cultural geography of the estate, a step deemed essential prior to proceeding with detailed, comparative research with the US site. The first stage of the research project began with the compilation and analysis of the renowned 1832 Cadastre for Tuscany, the historical foundation for assessment of land-use changes over the past two centuries, focusing on the estate of Spannocchia. Comprehensive, digitized land-use maps have been crafted and interpreted for 1823, 1954 and 2002. These will be discussed later in the chapter.

The justification for and significance of the compilation and analysis of historical and contemporary cartographic data in the cross-cultural comparison of agricultural land use and forest management was endorsed in 2000 with the signing of a 'Memorandum of Understanding' between the US National Park Service and the Italian Nature Conservation Service, Ministry of the Environment. The memorandum pledges, in part:

> to recognize the mutual interest in identifying natural and cultural heritage sites of international significance … and toward that end … [we] support a joint work program in such areas as: information exchange, evaluation of innovative strategies for management of new national parks, preparation and use of geographical information systems, and promotion of environmental education programs.
>
> (US/Italy Memorandum of Understanding, 2000)

This document confirms an unprecedented opportunity to coordinate interested parties and information exchange for the proposed study at both a national and regional level in Italy and the USA.

Comparison of environmental issues in the USA and Italy coincides with the 200th anniversary of the birth of George Perkins Marsh (1801–1882). Abraham Lincoln appointed Marsh, an accomplished lawyer and Vermont legislator, as Ambassador to the newly created Kingdom of Italy in 1861. During Marsh's 21 years as ambassador (1861–1882), a length of tenure yet to be equalled in the American diplomatic service, he was highly regarded in Italy not only as a diplomat but also as a conservationist. Seventeen years of his tenure were spent living in Florence and elsewhere in Tuscany, and his great conservation opus, *Man and Nature: Physical Geography as Modified by Human Action*, authored there in 1864, proved to be as influential in Italy as in the USA in furthering national forestry legislation in both countries (see Marsh, 1965). Today, *Man and Nature* is considered to be the first book to challenge the American myth of the inexhaustibility of the earth; as the fountainhead of the US conservation movement, it has had a profound effect on worldwide perceptions of the human relationship to the natural world and its impact endures in the contemporary debate about sustainable resource management in the 21st century. Parenthetically, it should be underscored that, as the principal namesake of the Marsh–Billings–Rockefeller (MBR) National Historical Park, Marsh and his legacy are an important and understudied connection between environmental movements in Italy and the USA (Hall, 1998a, b).

The Case Studies

The USA

The value attributed to conserving traditional rural landscapes, and explaining their significance through public stewardship and education, is well illustrated by the Marsh–Billings–Rockefeller (MBR) National Historical Park, in Woodstock, Vermont. This protected area is the premier example in the USA of efforts by the National Park Service to focus on the theme of conservation history and the changing nature of land stewardship in America (MBR National Historical Park, Conservation Study Institute, various documents; Sellars, 1997; McClelland, 1998). Bequeathed by the Rockefeller family in 1992 and authorized by Congress as a national park and opened to the public in 1998, the MBR National Historical Park includes 205 ha (550 acres) of forest and, within a protected zone, a privately owned, 40-ha (88-acre) farm, with an active dairy and interpretive museum. The Park also serves as the headquarters for the National Park Service's Conservation Study Institute, an organization that works nationally and internationally to promote education and training, research, and network building for the conservation community (MBR National Historical Park, Conservation Study Institute, various documents). The MBR National Historical Park has been proposed as one of the first pilot projects in the nation for independent, performance-based, third-party certification of forest stewardship on federal land.

In collaboration with the Institute, the MBR National Historical Park is currently initiating a master planning process for development of a sustainable forestry programme that will serve as a demonstration and education site for the public, for K–12 education, and for professional audiences. Indeed, a model for the comprehensive planning to be considered at Spannocchia is suggested by three publications on the MBR National Historical Park: *Land-Use History for Marsh–Billings National Historical Park* (Foulds, 1994); *Report of the Historic Forest Planning Charrette, November 3–5, 2000* (Nadenicek *et al.*, 2000); and the Conservation Study Institute's *Cultural Landscape Report for the Forest at MBR National Historical Park: Site History and Existing Conditions* (Wilcke, 2000). The latter document includes a

definitive example of the value of utilizing GIS techniques for land-use planning in order 'to provide a comprehensive assessment of the cultural landscape of the forest' (Wilcke, 2000, p. 5).

George Perkins Marsh and Frederick Billings, the Park's principal namesakes, were among the most significant 19th-century American conservationists and their legacy is reflected by more than a century of thoughtful land management of the present Park's property. Indeed, the forests contained within the Park are among the oldest continuously managed forest stands in the USA, and these lands, along with the Billings Farm, offer tangible evidence of both the theory and practical applications of land stewardship principles (MBR National Historical Park, various documents). Many researchers have identified Marsh's 1864 book as a prescient example of early global thinking about the potential hazards of ecological destruction in America, and his writing had a profound influence on Billings, who purchased the Marsh estate in 1874, subsequently implementing many of Marsh's ideas in reaction to severe deforestation and overgrazing that characterized the New England region at that time.

In the pursuit of comparative study between the MBR National Historical Park and Spannocchia, one avenue that seems especially promising is exploration of Marsh's early thinking about the significance of creating a museum of rural life. This proposal, which appeared in speeches in 1847, underscores a conception of social history at least half a century ahead of its time, including the observation that,

> history *for* the people must be *about* the people. ... Above all ... history ought to encompass things along with words, not just archives and genealogies but mundane tangible relics ... [which] were more vivid and memorable than written texts.
>
> (Lowenthal, 2000)

Relics of wild nature, as illustrations of human interaction with the environment, are included in Marsh's thinking about the cultural geographical features worthy of stewardship for future generations (Marsh, 1965). Marsh's views about the desirability of a museum of rural life, and the reasons he gave for it, are central to the current agenda of the MRB National Historical Park; in the next section, the chapter turns to review an example of how Marsh's thinking can be applied to the present situation in Tuscany, with particular reference to Spannocchia.

Italy

Conservation is the central objective of all activities on the approximately 450-ha (1000-acre) property of Castello di Spannocchia. Indeed, as described by its owner and managers, the estate is being developed as a living museum of traditional Tuscan rural life, which is fast disappearing after nearly 1000 years of relatively minor changes (Stratton, personal correspondence, 2002). A sketch of Spannocchia's distinguishing characteristics would include: its designation as a historic site; its inclusion within the Riserva Naturale Alto Merse; its status as a wildlife refuge; its certification as an organic farm raising endangered breeds of domestic farm animals and producing wine and olive oil; and its activities as an educational centre, including programmes in archaeology and architectural conservation, sustainable agriculture and landscape stewardship in association with the American not-for-profit Etruscan Foundation and Spannocchia Foundation (the Spannocchia Foundation's annual *Amici di Spannocchia Newsletter;* Stratton and Anderson, personal communication, 1998–2002).

Spannocchia is a valuable example of and connection to pre-modern rural life, particularly given its location within Tuscany, one of the world's great cultural landscapes. As a *tenuta,* or agricultural estate, the property represents in the present day the system by which rural Tuscany was organized and functioned, probably since the 9th century (Cosgrove, 1993; Agnoletti

and Paci, 1998). Until the middle of the 20th century, when there was a mass exodus of rural labourers, the estate proper continued to operate under the *mezzadria* tenant farming system, developed in Tuscany as early as the 1100s (Salbitano, 1988; Sereni, 1997). Changing little over time, the *mezzadria* system represents a form of share-cropping defining the relationship between landowner and peasant that shaped rural life in the region for centuries, encompassing social relationships, cultural practices and agricultural methods (Spender, 1992; Guidi and Piussi, 1993; Nanni, 2000). Throughout this period, forest use was integral to agriculture for the production of wood products complementary to agricultural operations (timber, firewood, charcoal, implements) and food crops, directly and indirectly, wild and cultivated (e.g. berries, other wild fruits, mushrooms, game, nuts) and domestic animals pastured in woodland and nut tree groves (Agnoletti and Anderson, 2000; Stratton and Anderson, personal communication, 1998–2000; Latz, Spannocchia site visits, 1999, 2001–2002).

Interviews with Spannocchia's property owner and managers indicate interest in creating a detailed land-use history, with the aim of developing a professional management plan for Spannocchia's woods, emphasizing its educational value as a historic example of traditional agricultural and forest land uses (Latz, Spannocchia site visits, 1999, 2001–2002). However, despite interest in and commitment to conserving the historical agricultural landscape of this site, these efforts have yet to be organized into a comprehensive land or landscape planning document. In particular, the question of how to identify a sustainable economic base for the property looms large. As one looks to the future of the estate, additional research and policy recommendations must be considered to ensure Spannocchia's survival as a living historical museum for the next generation. These include: cross-cultural comparison of the legal instruments and policies in Italy and the USA for conserving valuable historical and cultural landscapes (e.g. management designation as a public historical park, as a private non-profit organization, or some combination of public and private land stewardship); the need for a comprehensive cartographic inventory of the property, laying the foundation for comprehensive planning; and the specification of an economic and educational strategy that engages the larger community in a dialogue about the conservation of this important historical landscape.

Preliminary analysis of the 1823, 1954 and 2002 land-use maps of Tenuta di Spannocchia

Cartographic analysis of the historical evolution of the landscape of Spannocchia is multi-dimensional and includes archival materials (Tenuta di Spannocchia 1925–1955), historical and contemporary cadastres (*Catasto Leopoldina*, 1832; *Catasto Toscano*, 1998), aerial photographs (1954 and 1996, University of Florence), on-site confirmation of mapped data (2001–2002), interviews (1999–2002) and literature review. The following discussion represents a preliminary summary of the data obtained from these sources (Chirici and Mirra, 2002, unpublished work; Latz, 2002, 2004).

Interpretation of Spannocchia's landscape is based on collection of spatial data describing the physical and human geography of the estate. Cartographic sources are key to such analysis. The basic source of cartographic information for Spannocchia comes from two cadastre (*catasto*) compilations, one from the 1820s and 1830s, and one from the 1990s. Cadastres identify precisely the distribution of specific types of land use on a given parcel of land. Their primary purpose is to organize information about land usage for tax assessment purposes based on the types of improvements to the land, and the value of crops grown or resources harvested. In addition, these land-use data sources are repositories of information about both the social appraisal of resources and the kinds of crops actually produced on a given parcel of land. In the

case of Spannocchia, this distinction is extremely important; the landscape found there represents a culturally distinctive example of human interaction with the environment. Such interaction is multi-dimensional and includes complex strategies for overcoming site-specific limitations as well as adaptation of the local environment to produce commodities which are locally consumed or traded.

The earliest written record of Spannocchia documents the donation of a parcel of land by Zacaria dei Spannocchi in the year 1225 to the monks of the nearby monastery of Santa Lucia for the protection of the soul of his mother, Donna Altigrada. The remains of this monastery as well as the early medieval fortress known as Castiglione che Dio Sol Sa and the Romanesque bridge, Ponte della Pia, are among the elements which still shape the landscape of Spannocchia (Quiviger, 2002). An agricultural estate for at least 800 years, Spannocchia passed in the early part of the 20th century from the Spannocchi family to Delfino Cinelli, a Florentine aristocrat and noted Italian writer of that period. The estate at that time continued to be farmed under the *mezzadria* tenant farming system. It was with the passing of that era in the first decades after World War II that Delfino's son, Count Ferdinando Cinelli, initiated a new course for Spannocchia as an educational and cultural centre through its association with the American non-profit-making Etruscan Foundation (Grosse Pointe, Michigan) and Spannocchia Foundation (Portland, Maine), which include programmes in archaeology, architectural conservation, sustainable agriculture and landscape stewardship.

Spannocchia is located in the Upper Merse watershed approximately 20 km south-west of Siena. The estate comprises a portion of the approximately 2700 ha Riserva Naturale Alto Merse (see Fig. 16.1). The vast majority of Spannocchia is wooded, approximately 83%, consisting predominately of low-density (4%) and high-density (85%) stands of coppiced oak (*Quercus ilex*, etc.), mixed with chestnut (*castagno*) and other species (see Fig. 16.2).

Virtually all of the forests are coppiced, a forest management technique that encourages the sprouting of multiple shoots, and which leaves approximately one mature tree every 7 m to provide shade, encourage regeneration, and make available large-diameter trunks for harvesting on a multi-year cycle (so-called 'coppicing with standards'). A secondary concentration of the forested area is high stand, coppiced chestnut and mixed chestnut with the ingress of associated species (3%). A similarly small percentage of the forested area is low-density, high stands of maritime pine (6%), with associated oak and chestnut species. The remainder (17%) of Spannocchia is pasture for farm animals, or cultivated farm land that raises wheat and other grains, hay, olives and grapes. Wine is produced and consumed on the estate, as are olive oil, honey, faro (spelt) and sausage; with the exception of sausage, each is sold to visitors, but not on the open market. The estate produces the majority of its vegetables in intensively managed gardens adjacent to the main castle. There are seven farmhouses on the property, in addition to the main castle complex; these serve as an important source of rental property revenue, primarily for foreigners (McNamara, 2004; Spannocchia Foundation website, www.spannocchia.org).

This pattern of forest distribution clarifies the two primary historical uses of Spannocchia's forested areas, for building material, and for charcoal production; secondary uses of the forest include wood pasture, wild vegetables and mushroom gathering. Charcoal was especially important to income generated by estate resources until the period immediately after World War II, representing roughly 60% of the gross estate project between 1925 and 1940, with evidence of the significance of charcoal production stretching much further into the estate's past (archives of Spannocchia). This pattern of resource use is well illustrated, for example, by aerial photographs from 1954 (Fig. 16.3).

It is important to note that the smaller quantity and area of the chestnut groves, and their distribution, should not be confused with their huge importance to peasant agriculture. Much like the coppiced stands of oak, chestnut stands also are a primary indicator of human/environment interaction at Spannocchia given the fact that in a natural state, chestnuts do not congregate into groves of trees. The chestnut was and is essential for its fruit (eaten or made into flour, a staple of peasant life into the first half of the 20th century), building construction, agricultural implements and growing poles for vineyards, fence posts and charcoal.

Spannocchia's spatial pattern of land use, vegetative cover and human settlement present the research project with a number of examples of landscape continuity and change over the approximately 175-year

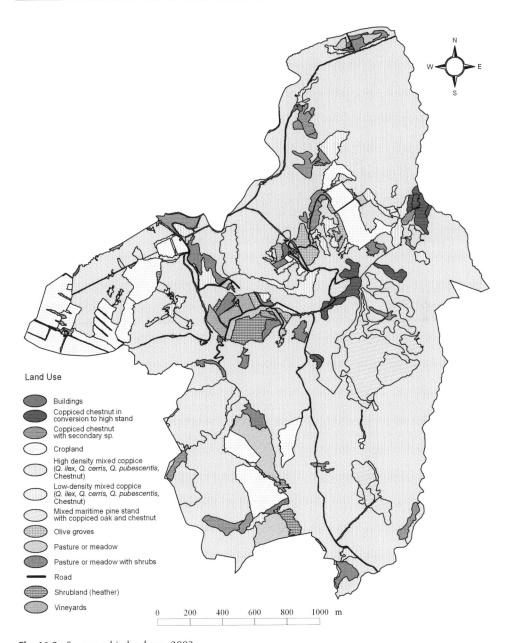

Land Use

Buildings

Coppiced chestnut in
conversion to high stand

Coppiced chestnut
with secondary sp.

Cropland

High density mixed coppice
(*Q. ilex, Q. cerris, Q. pubescentis*,
Chestnut)

Low-density mixed coppice
(*Q. ilex, Q. cerris, Q. pubescentis*,
Chestnut)

Mixed maritime pine stand
with coppiced oak and chestnut

Olive groves

Pasture or meadow

Pasture or meadow with shrubs

Road

Shrubland (heather)

Vineyards

Fig. 16.2. Spannocchia land use, 2002.

time-frame between the early/middle 19th
and early 21st centuries (see Figs 16.2, 16.4
and 16.5). Four findings stand out in particular. The first concerns the relationship
between the maps of 1954 and 2002. The
earlier map date was selected because it

allowed for digitization of the first aerial
photographs of the area after World War II.
This aerial information, in turn, could be
incorporated in and compared to the most
recent aerial coverage of the area, in 1996.
In the comparison of these two 'snapshots'

G. Latz

Fig. 16.3. Patterns of charcoal production, mid-20th century.

of land uses at Spannocchia, there is dramatic evidence of the extent to which the wooded portions of the property supported charcoal production. The circular areas (*piazza*) that characterize the sites where the coppiced wood was carbonized are clearly evident in the 1954 aerial photograph, both in their size and their number (see Fig. 16.3). Indeed, the geometric pattern discernible is so regular that future research is called for (Agnoletti, 2002). This finding represents enduring evidence of the estate's reliance on a 'value-added' resource conversion process evident not only in early 20th-century written archival records, but in Spannocchia's present-day woods.

Comparison of the 20th and 21st century maps also reveals that there has been a slow but steady expansion of forested area, from 77% to 83%, a change in arable land use of about 30 ha. This reflects, but clearly does not come close to mirroring, province-wide trends that indicate Tuscany today is 60% more forested than half a century ago (Agnoletti, 2002; Agnoletti, personal communication, 2003). Indeed, the degree of land-use continuity reflected by the Spannocchia case suggests that the site more clearly reflects the traditional, *mezzadria*-influenced Tuscan landscape. As is true throughout the region, at the same time, Spannocchia-specific findings are consistent with the recent and rapid depopulation of rural areas in Tuscany. In the pre-World

War II period, Spannocchia supported as many as 25–35 tenant families, about 300 people; today, there are no tenant farmers or families on the estate. This past land-use pattern is represented by the more extensive crop land associated with the main castle complex, and especially some of its peripheral properties (i.e. Santa Lucia), areas that by 1996 saw a net increase of arable land that had become reforested (see Figs 16.6 and 16.7).

A third observation concerns land uses at Spannocchia in 1823, and is directly related to the ratio of crop land to forested land discussed above. The research project's earliest, detailed land-use records present a remarkable story in comparison to the 21st century; whereas in 2002 the forested area of the estate stood at 83%, in 1823 the amount of forested land at Spannocchia was 67% of the total area (71%, if one includes 'wood pasture' in the calculation). This finding is significant not only because it is an indication of the degree to which land was used for field crop production at that time. In addition, it allows us to highlight a fourth and final point, that of contrasting the pattern of past forest vegetation with the present. What stands out in the analysis of the earlier, 1823 map is the extent to which the forested areas, though smaller in terms of absolute area compared to the present, are none the less used more intensively and managed more precisely (see Fig. 16.5). Examples of

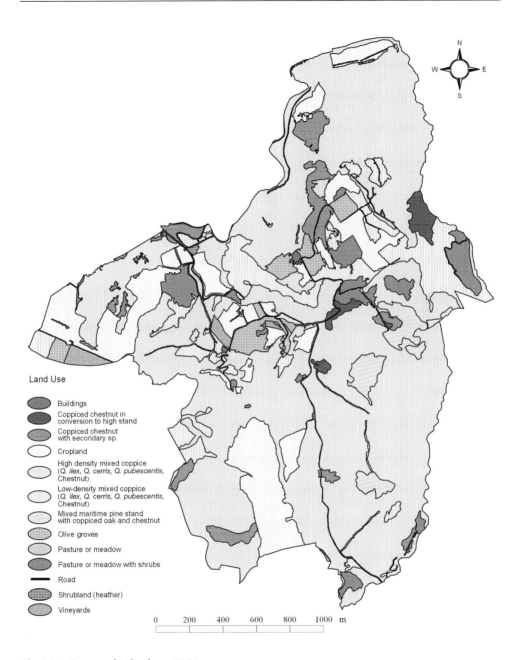

Fig. 16.4. Spannocchia land use, 1954.

the complexity of resource use as a feature of the estate's economy in 1823 can be illustrated by the fact that there were three specific designations for the chestnut tree at this time: *palina* (groves managed for poles); *marroneta* (groves managed for high-quality nuts); and *castagneto* (groves managed for lower-quality nuts). Other examples of continuity or change abound in the close analysis of these maps. These include

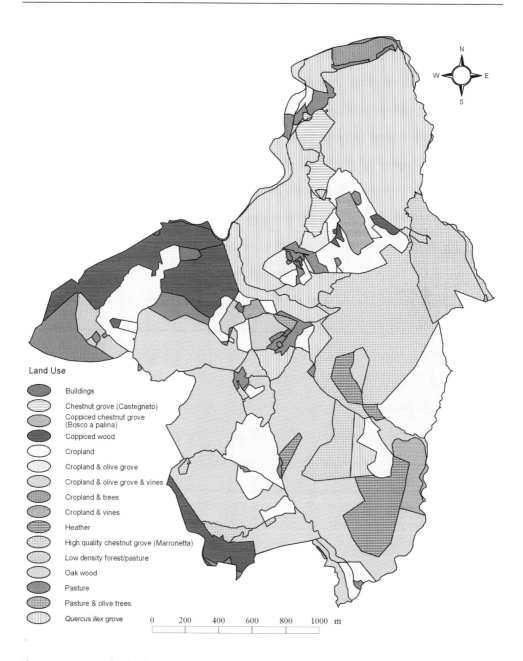

Fig. 16.5. Spannocchia land use, 1823.

changes or continuity in the distribution of olive groves, vineyards, pastureland, the aforementioned wood pasture, along with changes in the very way the estate was knit together by an evolving transportation network linking it internally, as well as to local markets.

Fig. 16.6. Cultivated area surrounding Santa Lucia, 1954.

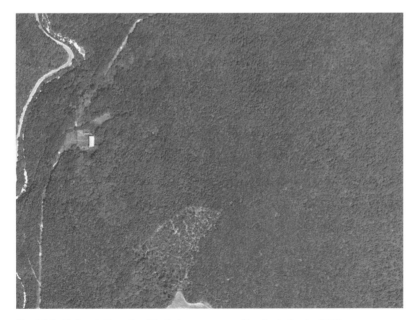

Fig. 16.7. Reforestation of cultivated area surrounding Santa Lucia, 1996.

Broader Impacts and Significance of the Research Agenda: Questions of 'Certification' and Landscape Stewardship

Preliminary thinking regarding the long-range significance of the project centres on the concept of sustainable land management. Here, the thesis is that early modern European agriculture and forestry embodied an understanding of sustained-yield forest management that is analogous to contemporary definitions of sustainable forest management; salient examples are third-party forest certification sponsored by the Forest Stewardship Council (FSC) and Pan European Forest Certification (PEFC) (Sample and Sedjo, 1996; MacArthur Foundation, 1998; Sample, 2002). The methodology for exploring this thesis is examination of presumed linkages between sustained-yield forest management at the community scale for early European forestry and the standards utilized by proponents of sustainable management today (Agnoletti and Anderson, 2000). I would argue that contemporary interest in sustainable land management can be deepened through study of late 18th- and early 19th-century land-use practices of Tuscan estates like Spannocchia. Such territories were in effect 'closed economies' bounded geographically by the needs of the estate and its immediate hinterland; local forests were the primary source of a continuous supply of wood for building purposes and cooking fuel (e.g., firewood, charcoal), as well as a place for livestock grazing and fodder gathering (so-called 'coppicing with standards'; Agnoletti and Paci, 1998; Lowenthal, 2000). These geographically bounded systems of sustained-yield management simultaneously producing fuel wood, timber products for construction and woodland pasture, have been radically altered by modern transportation and market systems. Historical study allows identification of the difference between regionally oriented sustained-yield forestry that aims to meet a variety of localized values and uses, and the contemporary characteristics of large-scale industrial-style forestry management for the global market. Such insight can contribute to a 21st-century model of sustainable land and forest management.

Spannocchia, in and of itself, as well as its detailed comparison to the MBR National Historical Park, offers a unique opportunity to explore these ideas. In the words of one counsellor to the research project,

> what is really intriguing about the MBR/Spannocchia connection is the opportunity to examine the ... resources of two sites whose forests are being managed along the lines of the early European model, one in Europe itself and the other as adapted to an American setting by owners, both Marsh and Billings, who were thoroughly familiar with the European experience. At both sites, the investigation would not be limited just to what is being practiced on the ground at this moment, but would include what each of the sites is trying to create on the basis of historical scholarship. Clearly, the industrial model of intensive timber management [the norm in much of the world today], while useful in some instances, is unlikely to be considered appropriate in many others. If we as a society are more attuned to the multiple-use sustained-yield model, might independent, third-party certification programs such as FSC and PEFC serve as a useful mechanism to get there?
>
> (D. Alaric Sample, President, Pinchot Institute for Conservation, personal communication, 19 December, 2000)

Such interviews with experts in the field, as well as literature review, indicate that there is a pressing need to incorporate site-specific historical data within existing forest certification procedures in order for forest managers, particularly small wood-lot owners, to pursue sustainable forest management in the 21st century. The Tuscan and Vermont sites selected for case-study analysis can serve as prototypes for such investigation of the relationship between historical land-use practices and contemporary forest certification protocols.

Study of certification and land stewardship will focus on understanding a critical omission of contemporary forest certification policy – that is, how site-specific historical study of forest management can shed light on the past methods and decisions of small wood-lot property owners to devise sustainable land management practices (MacArthur Foundation, 1998). Site-specific study of the history of land management in forested areas speaks both to policy and academic research questions now surrounding the debate about certification of forests for sustainable management. On the one hand, scholarship on land-use history – that is, case-study analyses of the traditions, practices, and successes and failures of human/environment interaction – can deeply inform contemporary forest management discussions, with multiple implications for policy-makers. On the other hand, knowledge of land-use history deepens understanding of traditional resource management practices as compared and contrasted to industrial-style forestry management paradigms that shape contemporary timber harvesting practices. Thus, one of the most compelling aspects of the research project, long range, is its potential to contribute both to our understanding of environmental history as well as unfolding resource management policy. In this sense, the project is Janus-faced, looking forward and backward at the same time.

An important aspect of longer-range research is to gauge the impact of historical information on forest certification policy through comparison of two certification protocols, e.g., the Forest Stewardship Council (FSC) and the Pan European Forest Certification (PEFC) programme, with regard to which applies best in geographically, historically and culturally distinct regions in North America and Europe, respectively. In particular, it can be argued that without such information, small land holders cannot be incorporated effectively into present-day certification strategies. Research should proceed by clarifying the essential principles that shaped sustainable land management in the late 18th and early 19th centuries in terms of economic, ecological and social values. Insight thus gained, step by step, clarifies how the estate of Spannocchia managed to meet food,

shelter and fuel needs given its location within the 'reach' or hinterland of the adjacent city of Siena. This question in turn assumes a need to understand clearly what the forest management regime looked like and how it was managed. The lessons resulting from such historical study will identify a set of distinctions between more regionally oriented sustained-yield forestry that aims to meet a variety of localized values and uses, and the contemporary characteristics of large-scale industrial-style forestry management for the global market. Such historical insight can help to build a model of more sustainable land management that further informs the standards guiding the forest certification process in the 21st century.

Conclusions

Comparative research findings will be of value to the World Bank and various NGOs pursuing 21st-century forest certification policies. Rigorous comparative research in the USA and Italy on the certification options for small wood-lot owners creates an opportunity to recommend the establishment of novel demonstration projects incorporating site-level analysis of land-use history, as well as training programmes for widespread dissemination of the lessons learned to relevant researchers and policy-makers in Europe, the USA and other mid-latitude forested areas. Demonstration projects and training sites can further deepen comparative understanding of the complex land-use histories of small-scale wood-lots, thus leading to more effective contemporary management of private lands in diverse environmental and socio-economic settings; they can also grant insight into how historical research can have a constructive impact on certification procedures for small-scale timber marketing by small wood-lot owners.

A final proposed aspect of future research is to evaluate the role of GIS- and GPS-based land-use maps for communicating concepts of sustainable forestry and agriculture for public education. One future

field-work objective is to collect data perti-
nent to the establishment at Spannocchia of
a Tuscan landscape stewardship museum of
traditional and contemporary farm and
forest management practices. This proposal
complements international trends such as
the World Conservation Union's creation of
Category V 'protected landscapes' as well as
provincial efforts in Italy, such as The
Museum of the Woods (Brown *et al.*, 2000;
Molteni, 1993). This expanded research
stage will organize and comment on spatial
data mapped for museum display, to include
an electronic retrieval system granting
researchers, educators, students and the
public the opportunity to study a historical
and contemporary example of sustainable
forestry and agriculture. A distinguishing
characteristic of the proposed museum and
web page is the idea of creating walking
itineraries of Spannocchia that portray the
ground-level resource management practices of Tuscany's traditional socio-economic system.

The proposed museum's displays and
electronic retrieval system will focus on
landscape stewardship in a region of Italy
where traditional resource-use patterns are
fast disappearing after enduring for nearly
1000 years. Field-work objectives in this
regard will include map, text and site-
specific itineraries portraying contemporary and historical resource utilization;
organization and interpretation of the
estate's archival material on the *mezzadria*
tenant farming system (1925–present); designation of extant forest management sites
and arable land parcels; statistics documenting the contribution of these geographically distinct areas to overall gross estate
production; and profiles of tenant families
who managed these resources (Spannocchia archives). The most compelling aspect
of the research project, in my view, will be
utilization and interpretation of spatial data
in this study of environmental land-use history. Map-based analyses of the traditions
and practices, successes and failures of
human/environment interaction and adaptation can deeply inform understanding of
contemporary forest management discussions, with multiple implications for the
public, researchers, educators, students and
policymakers.

References

Agnoletti, M. (ed.) (2002) *Il Paesaggio Agro-forestale Toscano, Strumenti per l'Analisi, la Gestione e la Con-
     servazione* (The Agricultural and Forested Landscape of Tuscany: Instruments for Analysis and Conserva-
     tion Management). Agenzia Regionale per lo Sviluppo e l'Innovazione nel settore Agricolo-forestale,
     Florence, Italy.
Agnoletti, M. and Anderson, S. (eds) (2000) *Methods and Approaches in Forest History*. [Report No. 3 of the
     IUFRO Task Force on Environmental Change]. CAB International, Wallingford, UK.
Agnoletti, M. and Paci, M. (1998) Landscape evolution on a central Tuscany Estate between the 18th and 20th
     centuries. In: Kirby, K.J. and Watkins, C. (eds) *The Ecological History of European Forests*. CAB Interna-
     tional, Wallingford, UK, pp. 117–127.
Brown, J. *et al.* (2000) Landscape stewardship: new directions in conservation of nature and culture. *George
     Wright Forum* 17, 1 (special issue/bibliography).
Catasto Leopoldina (1832) Siena City Archives, Siena, Italy.
Catasto Toscano (1998) University of Florence, Florence, Italy.
Cosgrove, D. (1993) *The Palladian Landscape: Geographical Change and its Cultural Representation in 16th
     Century Italy*. Pennsylvania State University Press, College Park, Pennsylvania.
Foulds, E. (1994) *Land-Use History for Marsh–Billings National Historical Park, Cultural Landscape Publica-
     tion #4*. Olmsted Landscape Preservation Center, National Park Service, Boston, Massachusetts.
Guidi, M. and Piussi, P. (1993) The influence of old rural land-management practices on the natural regen-
     eration of woodland on abandoned farmland in the prealps of Friuli, Italy. In: Watkins, C. (ed.) *Ecological
     Effects of Afforestation: Studies in the History and Ecology of Afforestation in Western Europe*. CAB Inter-
     national, Wallingford, UK, pp. 57–67.
Hall, M. (1998a) Restoring the countryside: George Perkins Marsh and the Italian land ethic (1861–1882).
     Environment and History 4(1), 91–103.

Hall, M. (1998b) Ideas from overseas: American preservation and Italian restoration. *The George Wright Forum: A Journal of Cultural and Natural Parks and Reserves* 15(2), 24–29.

Latz, G. (2001) Comparative international research on agricultural land-use history and forest management practices: the Tuscan estate of Castello di Spannocchia and Vermont's Marsh–Billings–Rockefeller (MBR) National Historical Park. Fulbright Program, Italian-USA Park and Protected Area Twinning Project, #1306, 2001–2002, University of Florence, Italy.

Latz, G. (2002) L'area di studio di Spannocchia (Riserva Naturale Alto Merse). In: Agnoletti, M. (ed.) *Il Paesaggio Agro-forestale Toscano, Strumenti per l'Analisi, la Gestione e la Conservazione* (The Agricultural and Forested Landscape of Tuscany: Instruments for Analysis and Conservation Management). Agenzia Regionale per lo Sviluppo e l'Innovazione nel settore Agricolo-forestale, Florence, Italy, pp. 111–22.

Latz, G. (2004) Comparative international research on agricultural land-use history and forest management practices, Italy and US. INT/West Europe Program, National Science Foundation, Project #0136284, 2002–2004.

Lowenthal, D. (2000) *George Perkins Marsh: Prophet of Conservation*. University of Washington Press, Seattle, Washington.

MacArthur Foundation (1998) *The Business of Sustainable Forestry: Case Studies*. A Project of the Sustainable Forestry Working Group. MacArthur Foundation, Chicago, Illinois.

McClelland, L. (1998) *Building the National Parks: Historic Landscape Design and Construction*. The Johns Hopkins University Press, Baltimore, Maryland.

McNamara, M. (2004) Set for Tuscany. *Los Angeles Times* 16 May, 2004.
    Available at: http://www.latimes.com/travel/la-_tr-italy16may16,0,4827733.story?coll=la-travel-headlines

Marsh, G.P. (1965) *Man and Nature: Physical Geography as Modified by Human Action*. Belknap Press of the Harvard University Press, Cambridge, Massachusetts.

Molteni, G. (ed.) (1993) *Il Museo del Bosco, Orgia* (The Museum of the Woods, Orgia). Protagon Editori Toscani, Siena, Italy.

Nadenicek, D.J. *et al.* (2000) *Marsh Billings Rockefeller National Historical Park: Report of the Historic Planning Charrette, November 3–5, 2000*. Penn State University, Center for Studies in Landscape History, Department of Landscape Architecture, University Park.

Nanni, P. (2000) Forests and forestry culture in Tuscany between the 18th and the 19th centuries. In: Agnoletti, M. and Anderson, S. (eds) *Forest History: International Studies on Socioeconomic and Forest Ecosystem Change*. CAB International, Wallingford, UK, pp. 79–92.

Quiviger, P. (2002) *Spannocchia*. Sator Print, Siena, Italy.

Salbitano, F. (ed.) (1988) *Human Influence on Forest Ecosystem Development in Europe*. Pitagora Editrice Bologna, Bologna, Italy.

Sample, V.A. (2002) Forest management certification: where are we, and how did we get here? Paper presented at the Southern Center for Sustainable Forests workshop on Forest Management and Forest Product Certification, Raleigh, North Carolina.

Sample, V.A. and Sedjo, R.A. (1996) Sustainability in forest management: an evolving concept. *International Advances in Economic Research* 2(2), 165–173.

Sellars, R.W. (1997) *Preserving Nature in the National Parks: A History*. Yale University Press, New Haven, Connecticut.

Sereni, E. (1997) *History of the Italian Landscape*. Princeton University Press, Princeton, New Jersey.

Spender, M. (1992) *Within Tuscany: Reflections on a Time and Place*. Viking Press, New York.

Wilcke, S. (2000) *Cultural Landscape Report for the Forest at Marsh–Billings–Rockefeller National Historical Park: Site History and Existing Conditions*. US National Park Service, Conservation Study Institute, Woodstock, Vermont.

17   Shaping the Landscape: Long-term Effects of the Historical Controversy about the Viennese Forest (Wienerwald)

E. Johann

University of Natural Resources and Applied Life Sciences Vienna, Vienna, Austria

Introduction

The global discussion of the causes of deforestation in the tropics has led to renewed interest in the history of forests in countries which are now considered developed. What factors were the process of deforestation and the subsequent expansion of forest area based on in those countries? What measures, if any, were taken to reverse the direction of change? Can lessons be learned from the experience? The representation of history is more than the sum of information accumulated in the past and presented. By becoming acquainted with the historical (social) manner and behaviour of human beings in relation to forests and analysing it, conclusions can be drawn relating to the diversity of interactions between man and forests. By these means the question can be answered of what kind of relationship between man and forests is able to meet the various human needs concerning the utilization of the forest without destroying its sustainability. Foresters and other scientists often lack the necessary information to integrate the present situation in a long-term historical context with regard to forestry. Therefore, case studies are necessary to improve the knowledge of long-term historical changes in the forest area resource. The scientific value of a sin-

gular case study is less important than the illustration of its linking. In doing so the representation of communicative processes gains high importance. The behaviour of human beings in relation to environment is of a high representative relevance (Küchli, 1998).

The case study presented in this chapter deals with the history of the 'silva viennensis', as it was called in 1332, which nowadays is fulfilling exclusively beneficial functions. The main part of this forest, with an extent of 28,000 ha, is situated close to the western districts of the city of Vienna at an altitude of between 600 m and 160 m above sea level (see Fig. 17.1), with 20% of the forest district situated in the urban agglomeration area.

There are only a very few capital cities that own a landscape of comparable variety to that of Vienna and equivalent extension of landscape potential given by natural factors. With more than 40,000 ha of forest area (situated in the Federal Provinces Vienna, Lower Austria and Styria), the city owns the second-largest area of forest, after the Austrian State, among the forest owners in Austria. The forests have been dedicated totally to public welfare; whereby watershed management and the safeguarding of the forest area for recreation purposes are the main targets (Ballik and Prossinagg, 1993).

Fig. 17.1. View of Vienna from the Viennese Forest (Kahlenberg) in 1888. Source: Kronprinz Erzherzog Rudolf, 1888.

Today the Viennese people take weekend excursions to the recreational forest near the city and clean, fresh water is available at any time. The findings of general inquiries made among the urban population indicate that the forests close to Vienna are the most popular recreation areas of the capital. The historical decisions to establish a water supply system from the alpine region and the resolution to implement a green belt consisting of forests and meadows close to the city mark a milestone in establishing the excellent living conditions in Vienna at present.

The Supply of Wood for Fuel to the Capital

Until the 18th century, the Viennese Forest was owned by the imperial family. Not later than the 15th century, a forest administration office was established and forest utilization was regulated very carefully by several laws and orders. Forest harvesting was carried out for the supply of fuelwood to the rural population close by, the various public administration buildings in the city and not least the Viennese population. In the course of the centuries the demand increased due to the fast-growing population and the increasing fuelwood consumption of the different branches of trade and industry.

Because of its advantageous position close to the river Danube the city was able to cover most of the demand for timber and fuelwood from forests located upstream. However, the inhabitants preferred to purchase the fuelwood they needed from the Viennese forest because the price was controlled and regulated by the local administration and was very low compared to the price of floated fuelwood. With the increasing demand, a lack of local fuelwood came into being mainly caused by lack of transport facilities. The lack of fuelwood and increasing prices of fuelwood coming down the river forced the consumers to switch to new sources, such as coal, from the middle of the 19th century onwards (see Fig. 17.2).

Large-scale use of coal was a decisive development. Coal reduced the urban demand for timber as a source of energy and promoted industrialization. Ancient conflicts over resource utilization disappeared as the pressure on the forests was reduced. The long-heralded vision of forestry as an independent discipline moved closer to becoming reality.

The Protection of Forest Resources: Forest Administration and Forest Law

To ensure forest ownership and to estimate the growing stock, which was considered to

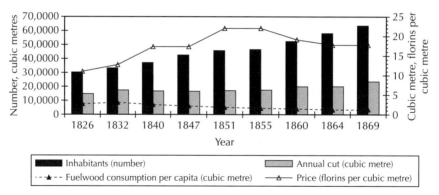

Fig. 17.2. Fuelwood harvesting in the Viennese Forest in relation to fuelwood consumption and price development in Vienna.

be the basis of sustainable management planning, inventories were carried out by the imperial forest officers from the 16th century onwards (Schachinger, 1934).

The first precise inventory dates back to 1720. It was caused by the debts of the Emperor Karl VI who mortgaged the Viennese Forest to the Hofbank, which was taking over his debts. On the occasion of this inventory, in addition to the calculation of the sustainable felling budget the financial yield of the Viennese Forest was estimated for the first time. The inventory covered the total area of the forest including 18 forest districts of a total extent of 46,000 ha.

In 1755, Maria Theresia placed the Viennese Forest into the ownership of the State. Nevertheless, the forest was managed in the same way as before. Forest laws of this time already aimed at silvicultural methods as well as the implementation of special cutting sequences and the sustainability of wood production, but the main target of publishing these laws was to avoid, generally, forest destruction and forest devastation.

From 1811 to 1857 the Viennese Forest was subordinated to an intact administration. Broadleaf stands were regenerated by the coppice system, conifer forests were regenerated after clear-cuttings by remaining seed trees, with pre-regeneration of beech and fir. From 1820 onwards there was a switch to the compartment method imple-

mented by Hartig, accompanied by a regeneration period of about 20 to 30 years.

In 1852 a new forest law, 'Reichsforstgesetz', was implemented in all parts of the country. According to this law, cutting was only allowed with official permission, the forest area was not to be reduced any further, and forest owners were obliged to carry out reforestation (see Fig. 3.3). It also contained further legal restrictions on private ownership taking into account public welfare and public interest in relation to the utilization of forests. Although forest devastation was forbidden by law, forest enterprises were occasionally sold to foreign investors just for cutting, who sold the clear-cuts again in order to save money for reforestation.

The Influence of Sustained-yield Forestry in Relation to the Ground Rent and the Financial Requirements of the State

In 1849, the administration of the Viennese Forest passed over to the Ministry of Landscape Culture and Mountain Affairs (Ministerium für Landeskultur und Bergwesen), which instigated the carrying out of a new estimation of growth and yield of the forest taking into account the Austrian assessment method. When the administration shifted to the Ministry of Finance in 1862, the calculated annual cutting seemed to be estimated too low. Therefore, based on

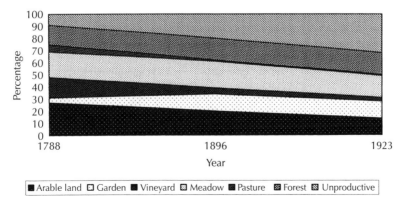

Fig. 17.3. Distribution of cultivated land: development in the western suburbs of Vienna, 1788–1923.

national orders, the rotation period was reduced by 16% (from 120 to 100 years), the felling budget was remarkably increased (up to 140–190% of the former annual cut) and harvesting was delivered up to timber merchants, who were allowed by governmental licences to make large clearings.

Since 1840, an economic system was already practised in some forests considering the operable timber land as capital, which had to bear a fixed rate of interest in forestry. Furthermore, at the same time there was a disadvantageous influence on forestry by the maxims of Adam Smith. The selling of forests administered by the state was caused by Smith's theory, saying that only private timber holders were able to receive a very high benefit through cultivation of land. Therefore, parts of the Viennese Forest should be cleared and used for agricultural purposes (Anon., 1863).

The Protest of the Viennese Population – an Example of Early Participation

As far back as 150 years ago, some foresters began to question the extent to which the desire for profit could be allowed free rein before it posed a threat to the ecological function of forests. They also began to express concerns about protecting the work of previous generations – 19th century

forestry laws (e.g. the Reichsforstgesetz 1852) were partly influenced by these ethical considerations. However, the return of the trees was much more likely a result of the ongoing social and economic changes that took place in the 1850s.

From the beginning of the 19th century, the urban population made use of the Viennese Forest as an excellent recreation area of scenic beauty. Without long travel distances, the region provided sufficient country houses and cottages and offered possibilities to stay for summer vacations as well as for weekend excursions and was open for the rich as well as for the poor (see Fig. 17.4). In 1857, Feistmantel, head of the forest department of the Ministry of Landscape Culture and Mountain Affairs at this time, had already pointed out the importance of forest stands beyond their value for the production of wood. He emphasized the beneficial influence on health, fertility, climate and condition of soil as well as its protective functions. Public welfare was considered to be quite closely connected to the conservation and cultivation of forests (Neumann, 1888).

When the administration of the Viennese Forest shifted to the Ministry of Finance, a place was set up for the sale of fuelwood at the main railway station in Vienna in 1862. The purpose was to promote the selling of wood products from the forest. In 1865, this place was sold to a

Fig. 17.4. Popular recreation area in the Viennese Forest (Baden) near Vienna, 1822. Source: Auracher von Aurach, 1822.

timber merchant and two years later the Minister of Finance also made a 5-year fuelwood and timber sales contract with this merchant concerning the large-scale utilization of the Viennese Forest. In doing so he avoided the hearing and consultation of the foresters who were officially responsible for the management of the forest. These contracts were at this time the most discussed event within Austrian forestry and were analysed with great interest by foresters.

In 1868, the monetary crisis induced the government to undertake the selling of wide areas of state-owned forests to speculators. From 1869 to 1870, 2700 ha of the Viennese Forest, especially those which were provided with transport facilities

(railway and boat), were opened for exploitation purposes.

The transformation of timber from an item of everyday use to a commercial commodity fundamentally started to change the character of the Viennese Forests. This was especially true where chessboard forestry replaced forests which had been used in traditional ways and in which regeneration had occurred naturally. This logging system influenced the natural composition of tree species and increased the proportion of conifers (mainly spruce and pine) on sites originally dominated by broadleaved forests (see Fig. 17.5).

Overexploitation of resources did indeed occur in many parts of central

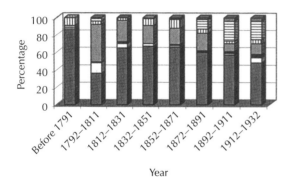

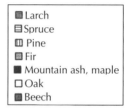

Fig. 17.5. Development of the distribution of tree species in the Viennese Forest, 1791–1932.

Europe in the 19th century, hand in hand with rapid economic development. The demand for timber, accompanied by a rise in timber prices, induced the felling of mature trees with increasing regularity.

The exploitation of the Viennese Forest was promoted by the newly established Minister of Finance in 1868 and in 1870 the project to sell the Viennese Forest passed through Parliament. When the law was published, the Viennese inhabitants started to protest against it for the first time. The current situation of the Viennese Forest was also discussed among the municipal councillors who developed the project to buy the forested area on offer for sale. At that time foresters being employed by the government or local administration did not have the courage to protest against the government's plan to sell the Viennese Forest for the purpose of timber harvesting and clear-cutting, either orally or in writing because of their impending dismissal.

It was a long-lasting, severe and fierce fight between the different interest groups, mainly between the financial interest of the State and the demand of the local population for the multiple use of the forest. The conflict at least was resolved to the benefit of the conservation of the forest, with public participation and the assistance of several newspapers. At that time Joseph Schöffel, an independent scholar and a member of various scientific associations, started a journalistically conducted campaign for the rescue of the Viennese Forest with an article in a Viennese newspaper (*Wiener Tagblatt*). This campaign lasted for more than 2 years and by this Schöffel became the first pioneer of nature and landscape protection in Austria. Parallel to this campaign, the Viennese municipal council handed over a protest note against the destruction of the Viennese Forest to the Austrian Government. The aim of the press campaign was to save and to guarantee for the future the Viennese Forest as a recreation area for public use, taking into account the high demand of the increasing population of a capital of more than 2 million inhabitants (Prossinagg, 1993).

The discussion took place among scien-

tists and foresters engaged in theoretical and practical work. The main important items were theories of sustained-yield forestry in relation to the ground rent and the importance of the forest in relation to its beneficial influence on health, fertility, climate and condition of soil. Because no standardized teaching doctrine in forestry or forest policy existed in those days, the points of view were extremely antagonistic. Already in the first part of the 19th century efforts were undertaken to establish forestry on the basis of science corresponding to other natural sciences and to do research work instead of solely teaching (Killian, 1974). However, not until 1868 were these ideas broadly discussed among German foresters and agriculturists at an international meeting in Vienna. Therefore, science was not able to contribute essentially to the ongoing discussion. In spite of these circumstances, the dean of the forestry college, Josef Wessely, having been nominated to the international committee for planning forest experimental stations, pointed out the importance of forest stands beyond their value for the production of wood. He considered public welfare to be in close connection with the conservation, cultivation and efficient running of operable timber land.

Several representatives of practical work and members of the Austrian Forest Association organized an excursion to the region near Vienna that was already exploited by logging operations. The result of this excursion and the following discussion on the basis of experience after personal review was a petition to the government to stop the devastation of the forest stands (Österreichischer Reichsforstverein, 1870).

Heated debates had already begun to rage among foresters prior to 1870. Those who favoured a natural approach argued that concentrating exclusively on timber production was an unnatural form of forest use for which a price would eventually have to be paid in terms of poor yields. Counterarguments were advanced by advocates of modern forestry techniques, who scoffed at the 'pompous words' associated with natural forest management (Milnik, 1997).

Facing large clear-cut areas around Vienna, the population became aware of the importance of woods with regard to their aesthetic and beneficial values. Thereby, the multiple functions of the forests received a higher rating in public opinion than their economic and commercial value. Taking into account that the prices for fuelwood were relatively high at that time, the appreciation of the Viennese population of the non-economic services of the forests was remarkable (Güde, 1932a).

As a result of the subsequent public protest and the published details of the corruption in which several members of the government had been involved, all agreements with the favoured timber merchant were finally cancelled in 1872. The officers who had been responsible for the felling contracts were forced to retire (including the Minister of Finance), and some of them were imprisoned (for not having proceeded according to the Forest Law). The administration and responsibility for the management of the Viennese Forest shifted to the Ministry of Agriculture. The journalistic fight of Schöffel at least led to the preservation of the Viennese Forest and was honoured by the population of the surrounding communities.

The Creation of a Woodland and Meadow Belt around Vienna

Though the fundamentals of regional planning had not yet been developed in those days, the Viennese municipal council started activities to establish public gardens and to cultivate open areas from around 1858. Forced by the increase of built-up areas, especially along the roads and railway tracks, which had already reached the boundary of the Viennese Forest, a project was developed by the municipal council to create a belt of gardens and parks around the densely populated city area (see Figs 17.3 and 17.6).

This project sponsored the idea of designing a woodland and meadow belt around the capital. The plan was supported by the Lord Mayor Dr Karl Lueger and legally dedicated by the Viennese municipal council in 1905, causing a sensation in international public opinion in those days. In protecting the Viennese Forest against destruction, its beneficial functions could be preserved for the urban population. Thereby clean air and recreation areas of sufficient extension should be offered to the inhabitants. The area originally enclosed 4000 ha, but some adjustments had to be made in the course of time.

After World War I, the Viennese Forest suffered from so-called 'wild' timber and fuelwood logging, which took place for the supply of the population everywhere. Large clear-cuts came into being again and former well-forested areas turned into waste land (see Fig. 17.7). People also settled and built houses within the protected area. The housing development caused severe damage to the forest stands especially until 1930 when a law came into being generally prohibiting

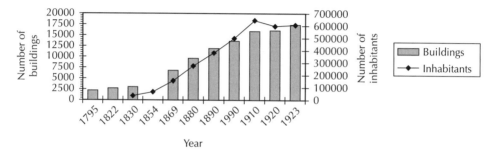

Fig. 17.6. Development of buildings and inhabitants in suburbs neighbouring the Viennese Forest, 1795–1923.

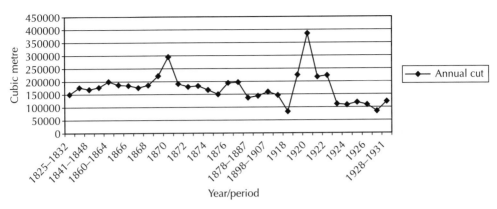

Fig. 17.7. **Fig. 17.7.** Fuelwood harvesting in the Viennese Forest, 1825–1932.

the construction of new buildings except those for recreational purposes (Güde, 1932b).

Because of the lack of available building areas to satisfy the demand for housing and industry, the protection of a large part of the forest owned by the capital was considered to be a 'luxury' in the 1930s. Pressure was again put on the Municipal Council to sell these areas to the population and to several other institutions and to industry; but the Municipal Council of Vienna still had in mind the former idea of installing a green belt around Vienna and therefore not only refused the sale, but also stimulated the creation of additional gardens in densely populated parts of the city.

The Implementation of Scientific Findings and Knowledge in Forest Policy and Forestry Practice

No other landscape in Austria has been of higher interest for forest science and forest policy and has been more discussed among experts than this forest area around Vienna. From the very beginning the administration of the Viennese Forest has always tried to improve forestry by implementing scientific knowledge and the results of scientific research. At present the most important target of forest policy is the provision of the beneficial functions of the Viennese Forests. By this the implementation of alternative silvicultural methods, developed

with the assistance of scientific institutions, is of high importance.

Nature reserves – natural forest reservations – national park

Owing to the activities mentioned previously, the Viennese Forest today is one of the richest forests in Austria in relation to its biodiversity. This statement is proven by an investigation into the botanical biodiversity in Austria, which has been published on behalf of the WWF, discerning more than 20 existing vegetation types still characterized as close to nature (Flesch and Fraissl, 1994).

Taking into account the increasing instability of the forest caused by external stress factors (e.g. air pollution, global warming, diseases), comprehensive protection of the Viennese Forest appeared to be of great importance at the beginning of the 1990s, just as 100 years earlier a widespread discussion took place concerning the establishment of wilderness areas inside the Viennese Forest. Having signed the convention for biological biodiversity (UNCED/Rio de Janeiro 1993) and the resolutions of the Ministerial Conference for Protection of European Forests (H2/Helsinki 1993), the Austrian government is obliged to establish a network of climax forests, natural forests and special forest types. Natural forest reserves are considered as a suitable tool to assume these liabilities. For the district of the Viennese Forest a working party was established composed of representatives from the regional government, forest owners, NGOs, scientific institutes from the University and the Forest Research Station. The scientific results of the brainwork of this study group were the formulation of recommendations and criteria to establish natural forest reserves. Today, these recommendations and criteria are not only implemented in the Viennese region, but also serve as a model for the whole of Austria in relation to the programme of creating natural forest reserves.

In the Viennese Forest about 160 ha have been put out of utilization voluntarily and are managed under scientific consideration. In this area the forest stands remain untouched in accordance with their natural dynamics. At present, scientific revision is evident for more than 30 years regarding the development of these natural forest reserves. The continuation of this documentation will show the way to interesting and new acknowledgements.

In the total area of the Viennese Forests great efforts are made to improve the biological structure by way of adequate silvicultural methods and scientific considerations. The scientific basis (e.g., careful site mapping of open and forested areas or the composition of wildlife management plans) is provided by projects supervised by Austrian universities. A successful example of these efforts is the coordinated management of the Lainzer Tiergarten as a unique reservation situated within the city boundary (Janda, 1993).

Forestation in relation to the beneficial functions of forests – the enlargement of the urban recreation area

Since the 1950s, the afforestation policy of the Viennese Forest Administration has focused on the beneficial functions of the forested area by enlarging the green belt of forest and meadows around the capital. This afforestation happened partly under extreme conditions, for example, on former industrial sites (Laaer Berg). Further suc-

cessful afforestation plans were realized recently on the islands in the Danube River. The new cultivated forests on these islands and in the lowland forest area are good examples of the productive teamwork of theoretical and practical knowledge. The afforestation of a former brickyard and garbage deposition area (Wienerberg) together with the shaping of a natural recreation area in the southern district carried out by the Viennese Forest Administration concerning 85 ha has been the last forestation for beneficial purpose within the city boundary to date. In this way, new forested areas of more than 350 ha have been created in Vienna within the last few decades (Haubenberger, 1993).

Today, the main emphasis of the city's forest policy is the maintenance of forest stands close to nature as well as the enlargement of the urban recreation area. One hundred years after the creation of a woodland and meadow belt around Vienna and the protection of the Viennese Forest by local people, the main target of the urban forest administration of today is the establishment of additional recreation areas and gardens of an extent of 1000 ha as of 2005. Thereby the Forest Administration aims to improve public participation and public relations with regard to silvicultural methods practised in the urban forests and other environmental activities.

Conclusion

The case study of the Viennese Forest gives evidence of the high importance of public participation for the survival of forests being under pressure from different interest groups. The sale of the Viennese Forest and the planned logging operations would not have led to the complete destruction of the woodland (because of the liability to reforestation), but would have increased the percentage of coniferous trees alien to the site. It is quite clear that the built-up area of the growing city would have diminished the forest land to a large extent. The green lungs of Vienna and the recreation area would have been lost.

The historically experienced awareness and responsibility for the shaping of the surrounding landscape is still in the minds of the Viennese population. Therefore, the sensitivity of the population with regard to changes to the environment might possibly have a higher value than in other European cities. This has been proven recently when, 100 years after the controversy about the Viennese Forest, the lowland forest down the River Danube was dedicated as a storage lake for the water supply of a planned power station. Once again, public opinion, a press campaign and the engagement of the younger generation forced the government to cancel the planned destruction of the forest, making possible the creation of the Donauauen national park (Schreckeneder, 1993).

The case study of the Viennese Forest is only one example among others. History shows that individual communities can succeed in opposing intervention by higher authorities, and even in resisting powerful market forces. It is notable that near-natural, high-quality forests often survive in such communities. Local resistance is on the rise in many marginal areas of the world, where the effects of centralization and global economic penetration are now being felt. Groups in civil society and non-governmental organizations often play an important role in advocating the rights to which local people are entitled. These groups are frequently rooted in the urban world.

It has been proven that scientific findings have always played an important role in forest policy decisions and also in the history of the Viennese Forest; but it has also to be pointed out that long-lasting scientific discussions about a certain problem have often been overrun by current affairs, if scientists have not been able or did not propose to implement the results of their research in ongoing political decisions.

References

Anon. (1863) Der kaiserliche Wienerwald. *Österr. Vierteljahresschrift für Forstwesen* 13, 58–61.

Auracher v. Aurach, J. (1822) *Perspektivische Ansichten der landesfürstlichen Stadt Baden und derselben Umgebungen*. Carl Gerold, Wien, p. 22.

Ballik, K. and Prossinagg, H. (1993) Wald und Forstwirtschaft. In: *Wo Wälder sein müssen. Magistratsabteilung 49*. Forstamt und Landwirtschaftsbetrieb der Stadt Wien, Wien, pp. 127–138.

Flesch, P. and Fraissl, C. (1994) *Naturwaldreservate im Wienerwald*. Vorstudie im Auftrag der MA 49. Forstamt und Landwirtschaftsbetrieb der Stadt Wien, Wien.

Güde, J. (1932a) Die forstliche Bewirtschaftung des Wienerwaldes in ihrer geschichtlichen Entwicklung unter besonderer Berücksichtigung der Tannenfrage. *Centralblatt für das gesamte Forstwesen* 67(7/8), 156–158; 9, 170–172.

Güde, J. (1932b) Aufgaben und Ziele der Forstwirtschaft und des Forstbetriebes im Wienerwald. Referat, erstattet anlässlich der Tagung des Österr. Reichsforstvereins und des Niederösterr. Forstvereins in Wien am 3. Juli 1932. Österreichischer Reichsforstverein, pp. 40–63.

Haubenberger, G. (1993) Der Wald als Erholungsraum. In: *Wo Wälder sein müssen. Magistratsabteilung 49*. Forstamt und Landwirtschaftsbetrieb der Stadt Wien, Wien, pp. 139–157.

Janda, G. (1993) Die Forstverwaltung Lainz. In: *Wo Wälder sein müssen. Magistratsabteilung 49*. Forstamt und Landwirtschaftsbetrieb der Stadt Wien, Wien, pp. 101–105.

Killian, H. (1974) Die Gründung der k.k. Forstlichen Versuchsleitung in Wien. *Centralblatt für das gesamte Forstwesen* 91(3), 129–152.

Kronprinz Erzherzog Rudolf (ed.) (1888) *Die Österr. Monarchie in Wort und Bild. Wien und Niederösterreich, 2. Abt. Niederösterreich*. k.k. Hof- und Staatsdruckerei, Wien, p. 9.

Küchli, C. (1998) *Forests of Hope. Stories of Regeneration*. New Society Publishers Ltd, London, pp. 226–231.

Milnik, A. (1997) *Hugo Conwentz 'Naturschutz, Wald und Forstwirtschaft'*. Brandenburgischer Forstverein, Berlin, pp. 9–19.

Neumann, F.X., v. (1888) Volkswirtschaftliches Leben in Niederösterreich. In: *Die Österreichisch-ungarische Monarchie in Wort und Bild. Wien und Niederösterreich. 2. Abt. Niederösterreich*. k.k. Hof- und Staatsdruckerei, Wien, pp. 317–360.

18 Recovery and Valorization of a Historical Fruit Orchard: the Kolymbetra in the Temple Valley, Sicily

G. Barbera, M. Ala, D.S. La Mela Veca and T. La Mantia

Dipartimento di Colture Arboree, Università di Palermo, Palermo, Italy

Introduction

In southern Italy, particularly in Sicily, citrus orchards are traditionally called 'gardens'. This has been the case ever since the introduction of some citrus species in the Islamic gardens (there is evidence of the presence of bitter orange and lemon at the end of the 11th century), even though citrus monoculture has been established since the mid-19th century in coastal areas, in order to supply the European market (Barbera, 2000).

The word 'garden', used both for the mixed orchards and for the specialized ones, shows that the citrus species had been appreciated either for their productive or cultural functions based on aesthetic and sensory pleasures, such as the shape of the trees, the colour, shape and taste of the fruits, the showy and scented flowers, and the shade of the crown. In 1929, Ernst Jünger, walking 'in a garden of thick lemon grove' during a Sicilian trip, experienced 'such feeling of observing an exotic fruit maturing that we have known ever since we were children. There is a presentiment of the heaven garden' (Jünger, 1993). In eastern Sicily, according to Trischitta (1983), today the citrus orchards are still called *paradisi* and in Pantelleria, in order to celebrate both utility and beauty, they even

refer to imposing, dry stone buildings that contain a single orange or lemon tree as 'gardens' (Barbera and Brignone, 2002). The title of garden when applied to a citrus orchard is probably linked to the phenomenon of reflowering that characterizes the trees (lemon above all). Probably referring to this, Assunto in 1973 wrote: 'the perfection of the landscape is simultaneity of flower and fruit, the flower for which every landscape appears a garden but also the fruit for which appears useful land'.

The systems and the landscapes of the traditional Italian citrus fruit area (suburban lowlands and Sicilian terraces, Sorrento coast, Gargano peninsula, Ligurian coast, *Limonaie* of Garda) have been left to progressive decay for years. In areas suitable for intensive systems, which are not damaged by urbanization, we can see the diffusion of monoculture with banal and homologous landscapes. On the contrary, in unsuitable areas such as terraces, is verified a decay which involves crop, environment and landscape. Accordingly, the systems and the landscapes of citrus tradition disappear, in their ecological, agronomic and historical complexity, even if depositories of biodiversity, ancient knowledge, productive, environmental and cultural values still remain today (Barbera, 2003).

On the other hand, awareness of the

importance of protecting the landscapes of traditional agriculture is increased, and the multi-purpose character, such as environmental, cultural, ethical and aesthetic functions, is recognized for possible protection. The protection and the exploitation of cultural landscapes will be possible if the yield is matched by quality and typicality and supported by linked activities to environmental and cultural services.

Here, we present the case of a citrus orchard set in an area of archaeological interest, the probable site of the Greek Kolymbetra in the Temple Valley in Agrigento (Sicily), which is today a regional park that preserves the imposing remains of the Greek civilization and a traditional landscape, in a country that has been the destination, for centuries, of trips and meditations that have contributed to forming the European landscape culture (Cometa, 1999; Barbera and Di Rosa, 2000). Since 1997 the Temple Valley has been on the UNESCO list of World Heritage Sites.

In 1999, the Garden of the Kolymbetra, in a deep valley which divides the Temple of the Dioscuri from the Temple of Vulcan, was granted in trust by the Region of Sicily to the FAI (Fund for the Italian Environment). The following year, after an intervention for recovery, it opened to the public. The interventions of recovery and multi-purpose exploitation are related to other recent experiences of the systems and landscapes of traditional Sicilian citrus fruit-growing (UE Life Project for the suburban agriculture of Palermo, 1997; Master Plan for the Favorita Park, Municipality of Palermo, 2002).

The Project

The awareness of the high cultural value of the area, its characteristics of agricultural landscape and historical garden in the same place, the preservation of the traditional productive function and the improvement of the cultural function for visitors were the basis of the project. It was initially founded on research about the history of the place (soil utilization, historical and literary doc-

uments, iconography, maps, oral sources), the environmental character (hydrogeology, climate, soils, flora, fauna, vegetation), cultural techniques, evidence of archaeological ruins and rural buildings. In order to have a computerized cartography that picks up useful elements for the visitors and enables a maintenance programme to be created, a GIS has been prepared.

The Temple Valley of Agrigento is an ample tableland, consisting of yellowish calcareous of inferior quaternary interposed from sandy clays, next to the southern coast of Sicily. Towards the south it is bounded by a hilly zone (the so-called 'hill of the Temples', due to the presence of numerous Greek ruins dating back to the 5th century BC), and to the north by the high ground of Rupe Atenea and by the hill occupied by the contemporary city. In the east and west there are two rivers, called the Akragas and the Hypsas, that define a fascinating cultural landscape in which the ruins of the Greek civilization cohabit with the mixed traditional field of almond and olive trees. Pindar, in the 5th century BC sang of Akragas, the ancient city, 'the most beautiful city which belongs to the mortals'.

A stream called Baida Bassa, which forms part of the hydrographical grid of the valley, flows into the Hypsas. It flows partly in a deep valley that was once occupied by the Greek Kolymbetra, which gets wider toward the confluence with the Hypsas and becomes cultivable due to the presence of alluvial soils and terraces; the total area is 6.29 ha (Fig. 18.1).

At the base of the calcareous walls that bound it on the northern side are outlets of numerous hypogeals (today there are 12, although there were 18 in the past), draining tunnels dug in the rock that allow, today as in the past, the irrigation of crops (Fig. 18.2).

Thanks to a microclimate that mitigates the dry Mediterranean climate, vegetables and orchards (citrus fruit, above all) can be cultivated, and because of the high calcareous walls, there is shadow enough to protect the crops and the evapotranspiration level is low. It is protected from cold winds in winter too. The opening of the valley

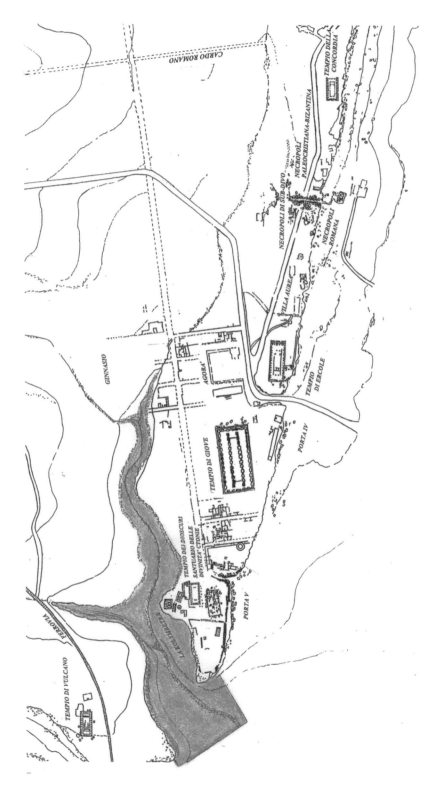

Fig. 18.1. The Kolymbetra in the Temple Valley.

Fig. 18.2. Carrying water from the Kolymbetra (circa 1920).

towards the south-west guarantees the entry of moderate winds of libeccio and sea breezes. The mild climate has been changed by intense rainy phenomena, such as that in 1971 which has provoked intense erosion.

Climatic conditions show that the original climax vegetation consisted of referable shrub communities referred to order *Pistacio-Rhamnetalia Alaterni* and *Oleo-Ceratonion* alliance. It deals with thermophile and basophile association characterized by *Olea europaea* var. *sylvestris* and *Euphorbia dendroides*, still today widely diffused on the scarps and in the semi-rocky habitats, which give place to discontinuous formation.

Generally, the site is identified, despite some contrary opinions, with the 'sumptuous basin' of which Diodorus Siculus writes when describing the works 'that embellished the city and the territory' completed by the slaves taken after the battle of Himera (480 BC) against the Carthaginians:

> these cut stones with which not only the greatest temples of gods were built, but also were built the aqueducts for leading of

water in the city. These aqueducts were called as their builder Feace. People from Agrigento also built a sumptuous basin that had the circumference of seven stadiums and the depth of twenty cubits in which were conducted water of the lakes and sources, becoming so a hatchery, that furnished many fish for feeding and for taste; and because of many swans flew down toward it, its sight was delightful. But subsequently neglected, it was obstructed, and finally, destroyed and the inhabitants transformed the whole region, that was fertile, into planted lands of grapevines, and other kind of trees, in order to draw incomes of it.

In the 1st century, when Diodorus writes, the site of the basin already had the agricultural utilization that would be maintained in the future and the presence of a reed thicket – mentioned in 1225 in a parchment which writes of some concession to the Bishop about a 'land in which there was a reed thicket near to the caves of the giants' (the old name of the Temple of Jupiter), which confirms the presence of a landscape not dissimilar to the actual one.

The ancient possession of the area by the church is the origin of the denomination 'Badia Bassa' or 'horti Abbatie', as the 16th-century historian, Fazello, writes. We have more recent information from travellers on the Grand Tour who had chosen Agrigento for discovering classical antiquities (De Miro, 1994). Swinburne, in 1777, (cited in De Miro, 1994), observed that the basin 'now is dry and used as a garden'; Saint-Non, in 1785, (cited in De Miro, 1994), observed that

> water still flows in this canal and irrigates some luxurian gardens that occupy the bottom of the basin ... runs in a small valley today that for its amazing fertility, it resembles to the valley of the Eden, or to a part of the Promised Land.

In 1821, de Foresta (cited in De Miro, 1994), found it 'destined to the vegetable cultivation'. In 1896, Vuillier observed that

> the sight from the edges of the basin is superb. The ancient temples show their columns through the orange trees and beyond there is the endless sea. I have remained there for a long time, weak for hot weather, with the lost look among the trembling leaves that sparkle in the irregular puffs of the sea breeze and my wandering thought went back to past.

The historical iconography has also been important to the discovery of the history of the place (Figs 18.3 and 18.4).

Even though the orchard has not been cultivated for about 20 years, the Kolymbetra shows the evident characteristics of a suitable area for fruit cultivation conducted by dry farming where water is not available, and citrus orchards and other kinds of fruits and vegetables where irrigation is possible. Concerning fruit trees, the most represented group is citrus (492 sweet orange, 47 bitter orange trees, 62 lemons, 59 mandarins and tangerines). Among the other tree species, in the dry area there are olive (86 trees) and almond trees (144). Numerous other species of fruit trees are present and testify to an elevated specific biodiversity: azarole, banana, locust, quince, fig, prickly pear, white mulberry, black mulberry, kaki, apple, pomegranate, Japanese medlar, winter medlar, pear, peach, pistachio and sorb. In general, every kind is represented by ancient varieties that are no longer cultivated in modern fruit orchards. Besides the proper values of a cultural landscape, the presence of archaeological features increases the interest of the place: along the walls, in fact, there are different caves (one is hypothesized to be a rural church), edges of a prison and places carved into the rock that have shown the presence of Greek-age materials.

The basic idea of the project has been to consider the Kolymbetra as part of a historical rural landscape: not only a citrus orchard, but also a garden. The intervention

Fig. 18.3. The temple of the Dioscuri from the Kolymbetra, today.

Fig. 18.4. The temple of the Dioscuri from the Kolymbetra in the past (end of 19th century).

had the aim of preserving, conserving traditional species and cultivars and cultural techniques, the landscape of traditional agriculture, and supported by small interventions (pathways, rest places) enhancing the visitor's experience and acquaintance with the place.

In summer 2000, the first intervention of recovery consisted of the elimination of invading flora (Fig. 18.5). Only then was it possible to study the garden in all of its constitutive elements: the original order of citrus fruit, the traces of the traditional irrigation system and the plan of dry-stone walls. Interventions concerned the restoration of the citrus and fruit orchard, the care of the natural spaces, as well as the restoration of dry-stone walls and of the archaeological ruins, the recovery of the traditional irrigation system, the cleaning and the

Fig. 18.5. The valley before recovery.

retraining of the stream, the recovery of paths with the restoration of an old staircase carved in the tuff-made wall and the creation of a double crossing of the river to facilitate visitors.

For the restoration of the citrus orchard, traditional rootstocks were planted and grafted using cultivars of the traditional Sicilian citrus industry. Cultural techniques were those of the traditional Sicilian citrus industry (La Mantia, 1997a, b). The recovery of the citrus orchard had made an extraordinary pruning necessary: for this operation specialist pruners were needed who were able to recover the old plants safe-guarding as much as possible their original shapes. Extraordinary prunings have also been carried out on other trees using tree climbing techniques.

The recovery has also involved the terraces built along slopes, partly occupied by citrus, olive and almond trees, many of which had collapsed. The irrigation system has also been mostly recovered and today the traditional basin irrigation is possible using the water derived from hypogeals and preserved in old ponds (Fig. 18.6).

Particular attention has also been paid to recovering the original paths. In order to aid enjoyment, some new paths to bring the visitors to panoramic or culturally meaningful places (hypogeals, monumental plants, archaeological ruins) have been constructed: for example, the Kolymbetra can be crossed via a path that brings the visitor from the Temple of the Dioscuri to the Temple of Vulcan. Along the paths, simple benches made of natural materials (blocks of tuff and boards) have been built in panoramic or shaded places. There are two bridges made of wood and iron crossing over the stream (Figs 18.7 and 18.8).

The definitive recovery of the garden foresees some interventions on the river to reduce the risk of overflowing in case of severe rain, the creation of aids for cultural tourism and ecotourism, and the restoration of an old rural building that will become a centre of services for the enjoyment of the garden (reception, restoration and cultural activities).

Fig. 18.6. Traditional horticulture.

Fig. 18.7. The Kolymbetra, today.

Fig. 18.8. The garden.

References

Assunto, R. (1973) *Il Paesaggio e l'Estetica.* Giannini, Napoli, Italy.

Barbera, G. (2000) *L'Orto di Pomona. Sistemi tradizionali dell'arboricoltura da frutto in Sicilia.* L'Epos, Palermo, Italy.

Barbera, G. (2003) I sistemi frutticoli tradizionali nella valorizzazione del paesaggio. *Italus Hortus* 10(5), 40–45. (Special issue on the 50th anniversary of SOI.)

Barbera, G. and Brignone, F. (2002) Il giardino di agrumi di Pantelleria. *Frutticoltura* 1, 39–44.

Barbera, G. and Di Rosa, M. (2000) Il paesaggio agrario della Valle dei Templi. *Meridiana* 37, 83–98.

Cometa, M. (1999) *Il Romanzo dell'Architettura. La Sicilia e il Grand Tour nell'età di Goethe*. Editori Laterza, Roma-Bari.

De Miro, E. (1994) *La Valle dei Templi*. Sellerio, Palermo, Italy.

Jünger, E. (1993) *Viaggi in Sicilia*. Sellerio, Palermo, Italy.

La Mantia, T. (1997a) L'evoluzione delle tecniche nell'agrumicoltura. In: *Il Progetto LIFE per il Parco Agricolo di Palermo*, Unione Europea DG XII, Città di Palermo, Confederazione Italiana Agricoltori, Palermo, Italy. pp: 53–58.

La Mantia, T. (1997b) Tecniche colturali nella frutticoltura periurbana della Conca d'Oro di Palermo. *Atti delle III giornate Tecniche S.O.I.*, Cesena 13–14 novembre 1997, pp. 47–53.

Trischitta, D. (1983) *Toponimi e Paesaggio nella Sicilia Orientale*. Edizioni Scientifiche Italiane, Napoli, Italy.

Vuillier, G. (1897) La Sicilia, impressioni del presente e del passato. Fratelli Treves (ed.) Milan.

Index
